The Epistemological Quest

The Fate of the Fallible Cousin of the Chimpanzee

A Call to Sexual Selection, Morality, and Racism

Aaron Adekoya

Copyright © 2023 **Aaron Adekoya**

ISBN: 978-1-916820-20-3

All rights reserved, including the right to reproduce this book, or portions thereof in any form. No part of this text may be reproduced, transmitted, downloaded, decompiled, reverse engineered, or stored, in any form or introduced into any information storage and retrieval system, in any form or by any means, whether electronic or mechanical without the express written permission of the author.

Illustrations by Chris DiBiase

Contents

PREFACE 3

THE NUMBER 1.6 3

ON AN INTRODUCTION TO KNOWLEDGE 7

CHAPTER 1 8

CHAPTER 2 31

PART I 43

OF SEXUAL SELECTION 43

CHAPTER 3 45

PART II 77

OF GOOD AND EVIL 77

CHAPTER 4 80

CHAPTER 5 103

PART III 144

A BRIEF DISCOURSE ON RACISM 144

CHAPTER 6 147

CHAPTER 7 169

ON A CONCLUSION TO KNOWLEDGE 193

CHAPTER 8 195

BIBLIOGRAPHY 212

NOTES 223

The Epistemological Quest

Preface

The Number 1.6

So I am what philosophy might call a philosophical naturalist, meaning I believe the natural world to be all that exists. For philosophy, it may be a strong stance to take, but it's simply a relationship I have maintained with knowledge. For instance, I have no idea if it will rain tomorrow, and there are other things I have no idea of tomorrow, such as having a fierce argument with my boss that will lead me to get fired from work or being hit by a bus on crossing the road. Given all unbeknownst, I somehow manage to conjure up things I know about this incredible world.

Three useful types of knowledge I have used in my personal ratiocinations are *doxa*, *episteme*, and *gnosis*. To put them into context, and on analogising these concepts, I will tell you of a summer vacation I had last year. For up to four years, I lived in many different cities in the remarkable Spanish Iberian peninsula. During my time in Spain, and like many of its expatriates, I availed of the opportunity to travel the country and its neighbours. But yet strangely enough, and until visiting this mysterious country, I had never met a person from Andorra. The seemingly landlocked nation sits between France and Spain while cloven on the Pyrenean mountain range. So without evidence of its people before arriving to this mystery, how could I say I knew Andorra to exist? Let's take at look at my *ratiocinations*:

Doxa: While working for a company from 2015-16, I taught in an IT company on the Mediterranean coast. Techs purportedly skied on the Pyrenees and **told me** they often went to Andorra and vehemently suggested I should go. Fast-forward to the few days before vacationing five years later, and I have an amicable student going out of her way as she lists famous spas, restaurants, and landmarks I should visit, and so Andorra must exist and is real.

Episteme: I set forth to Andorra, go over demarcations, timely **measure** topographies, altitudes, landscapes, count the population as they sing their national anthem 'El Gran Carlemany' and ultimately declare Andorra to exist. I surely don't have the time or resources to achieve this.

Gnosis: After personally indulging in gastronomy, low-tax shopping, wonderful hikes, and an **experience** I encounter swimmingly, I can now profess Andorra to exist.

Now look at how these three types of knowledge could have entered both our societal and biological genome. Through the amalgam of society and two memory modules, namely the two hippocampi in the temporal lobe, we could imagine how we *know* something like Andorra to exist.
If through lore or somebody telling you, *doxa* suggested that if you were bitten by a viper damage was fatal, would *episteme* truly help in such matters? In the days of hunting and gathering *episteme* would have been both a timely and costly enterprise.*[1] Thus *doxa* predominated, *gnosis* was secondary, and when the importune moment arrived, *episteme* tertiary.
As civilisations emerged and the domestication of our plant and animal species gifted others with time to spare, they would write, hone, carve, smoulder, solder,

---

[1]* It may have been wiser to spend time actually... hunting and gathering.

The Epistemological Quest

sketch, weigh, time, and use countless systems of measurability. Little did those *Homo sapiens* know that in a small lapse of the Holocene epoch, we'd be compelled to transpose *episteme* as primary and *gnosis* as tertiary for civilisations to flourish, a time best known as the *scientific revolution*. But before all that is spoken of, it suffices to address the subtitle of this work, what I wish you to use as the token that allows you to continue reading, or simply close this book if deemed too unpalatable. That is the epistemically sound dictum we are one of four great ape genera, othered by three extant species of orangutans, two of gorillas, and two fascinating species of chimpanzees.

We purportedly branched off from chimps around six million years ago, gorillas ten, and orangutans sixteen. Albeit gibbons are our closest relatives outside the great-ape-family, (from around twenty million years ago), they boast twenty species of their own fascinating idiosyncrasies that shed light on apish behaviour. But the point at hand is the number 1.6. That is we share up to 95% of our DNA with gibbons, more with orangutans, and as much as 2.3 separates gorillas from both chimps and humans. Between the two chimp species, the pygmy and the common (or bonobos and troglodytes), a meagre 0.7 divides them while 1.6 of the chimp and human genera. This tells us two significant things, that a) we are genetically closer to chimps than they are to gorillas, orangutans, and gibbons, and b), everything that astoundingly distinguishes us from chimps is locked in the number 1.6.

As per the story of sexual selection, there are three mammalian anomalies for our females worthy of pondering on, that a) the vast majority of mammalian species are fertile until decease, except ours, b) our females uncannily conceal ovulation, and c), seem to copulate in private. Yet of the great ape testes, chimps boast an average of four ounces to our one and a half, while of the phallus, we're longer, and sport bigger breasts of the other three genera. Of our distinguished longevity, menopause, art, language, and territorial practices, it's all in that number, 1.6 (Diamond, 1993).

~

On drafting the proposal for this book, a close friend of mine (on offering well-appreciated criticism), rightly challenged me on asking, 'Why are the three parts separated into sexual selection, morality, and racism particularly? How do they pertain to an 'epistemological quest?' In the hope of an answer to my friend and those who may share the same question, is the notion of ultimate and proximate causes. I profess as both a causal determinist and philosophical naturalist, the three parts of this book address *ultimate* causes that stand for and appertain to our most hostile divisions among the *proximate* and more than necessary political sciences. As figure 1.0 will illustrate:

The Epistemological Quest

Sexual Selection	Morality	Racism
*Abortion	*Free will	*Asylum
*The gender pay-gap (finance or the accumulation of resources)	*God/Atheism	*Politics
*Domestic violence	*Logic	*History
*Sexual harassment	*Ethics	*Franchise
*Rape	*Worship	*Slavery
*Hypergamy	*Religion/Blasphemy	*Segregation
*Contraception	*Scripture	*Marriage
*Prostitution/Soliciting	*Segregation	*Religion
*Sex	*"Terrorism"	*Conspiracy
*Gender	*State	*Free will
*Homosexuality	*Social Constructionism	*"Responsibility"
*Homophobia	*History	*Murder
*Marriage (and interracial marriage)	*Politics	*Affirmative action
*Divorce	*Euthanasia	*Housing
*Virginity/Purity	*War	*Education
*Single parenthood	*Murder/Manslaughter	*Professionalism
*Polygamy		*Homophobia (specifically endogenous to certain races or religions)
*Hierarchies and Competence	*Capital Punishment	*Animal Cruelty
*Sports and Sex	*Incarceration	*Biology
*Family	*Drug Use	*Hierarchies
*Sexual Paraphilias (specifically paedophilia who tend to be predominantly male)	*Discrimination (of any type)	*Cultural Appropriation
*Bestiality	*Disability	
*"Love"	*Evolution	
*Cohabitation	*The Domestication of Plants and Animals (veganism and vegetarianism)	
*Technology (dating apps)		

Figure 1.0

So where would all causes lead back to for a determinist? Precisely, a Big Bang singularity. Plato rightly concluded no person can claim to indubitably know as absolute certainty, quite simply, cannot exist, and as we approach the halfway mark of a third decade of the twenty-first century, it seems as though philosophical games need concessional rules.

Richard Dawkins, in his 10th year anniversary introduction of *The God Delusion*, says 'The temptation here is to abdicate responsibility to think clearly and consistently about morality, and instead take the lazy route of slavishly following an ancient book of rules, rules invented by fallible men (and they *were* men) and tailored to very different times and conditions' (2016, p. 17). I believe there to be a red pill of natural selection that explains for actions as vicious as sadistic torture or genocide, to reasons we find toddlers, kittens, and puppies as cute as they are. This book concerns a quest of epistemological knowledge, being *doxastic* epistemology or *gnostic* epistemology; knowledge founded on measurability, i.e. scientific experimentation. I thus divide the book into three parts; sexual selection, morality, and racism, fairly including epistemology as knowledge as a *whole* which I hope does not confuse the reader (as I believe one's best left to their own conclusion). As a determinist, naturalist, and humanist, I aim to pull things back a little closer to the drawing board and contribute to the 1.6 and beyond. I truly hope you enjoy this book.

Aaron Adekoya

The Epistemological Quest

On an Introduction to Knowledge

'The Rebuke of Adam and Eve' (Natoire, 1740)

The Epistemological Quest

Chapter 1
ἐπιστήμη

On Rene Descartes, First Meditations on First Philosophy: *'How do you know that you know that you know you are knowing?'*

'Je pense donc je suis'

'I have such good reason for offering this work to you, and I trust that you will have such good reason for taking it under your protection, once you understand my intention in writing it, that I could recommend it here in no better way than by saying briefly what my aim was.

I have always thought that the two issues of God and the soul were the most important of those that should be resolved by philosophical rather than theological means. For although it is sufficient for us Christians to believe by faith that the human soul does not perish with the body and that God exists, yet it seems certain that unbelievers cannot be convinced of the truth of religion, and scarcely even of any moral values, unless these first two truths are proved to them by natural reason. And since often in this life there are greater rewards for the vices than for the virtues, few will prefer what is right to what is useful, if they neither fear God nor expect an afterlife. And although it is completely true that we should believe in the existence of God because it is taught in the holy scriptures, and by the same token that we should believe the holy scriptures because we have them from God – since, faith being a gift of God, he who gives us the grace to believe the rest of religion can also give us the grace to believe he exists – there is no point in asserting this to unbelievers, because they would call it arguing in a circle. And indeed I have observed that not only do you and all other theologians affirm that God's existence can be proved by natural reason, but that also the holy scriptures imply that the knowledge of him is much easier to attain than that of many created things: so easy, in fact, that those who lack it do so by their own fault. This is clear from the passage of Wisdom 13: 'They have no excuse. For if they are capable of acquiring enough knowledge to be able to investigate the world, how have they been so slow to find its Master?'* And in Romans 1: [20] they are said to 'have no excuse'. In the same chapter [1: 19], the words 'What can be known about God is perfectly plain in them,'* seem to be pointing out that all that can be known of God can be shown by reasons derived from no other source than our own mind. How this comes to be true, and by what means God may be known more easily and with more certainty than the things of this world, I thought it would be appropriate to investigate.

And as regards the soul, even though many authors have judged that it is very difficult to discover its nature, and some have even dared to say that human reasoning convinces us that it perishes along with the body, and that we believe the contrary by faith alone, nonetheless because the Council of the Lateran* held in the reign of Leo X condemns these people (session 8), and explicitly enjoins Christian philosophers to refute their arguments, and to make every effort to prove the truth, I did not hesitate to tackle this issue as well. Besides, I know that most of the impious refuse to believe that God exists and that the human mind is distinct from the body, for no other reason than that they say that these two points have never been proved to anybody up to now. I do not agree with them at all in this: on the contrary, I think that nearly all the reasons adduced by great thinkers in this debate, when they are sufficiently grasped, have the status of demonstrations;* and I can scarcely persuade myself that any proofs might be found that have not been already discovered by someone else. Nonetheless I think that I could achieve nothing more useful in philosophy than to perform a careful search, once and for all, for the best arguments put forward by anyone, and to arrange

The Epistemological Quest

them in so clear and precise an order* that from now on everyone will accept them as having the status of demonstrations. And finally, since there are several people who know that I have developed a particular method for resolving all difficulties in the sciences – not indeed a new one, for nothing is older than truth, but one they have seen me use with some success in other areas – they have insistently urged me to do this: and therefore I decided it was my duty to make an effort in this area as well.

Whatever I have been able to achieve is all in this treatise. It is not that I have sought in it to bring together all the different arguments that can be adduced to prove these two points, since this does not seem worth while, except where there is no argument considered sufficiently certain. But I have gone into the primary and most important arguments in such a way that I now dare to offer them as demonstrations that are as certain and evident as possible. I will add that they are such, that I do not think there is any path open to human intelligence along which better ones can ever be found: for the importance of the issues and the glory of God, for the sake of which this whole book was written, compel me here to speak a little more freely about my own work than is my custom. Yet, however certain and evident I think them, I do not for that reason convince myself that they are capable of being grasped by all. In geometry there are many arguments by Archimedes, Apollonius, Pappus, and others that are regarded as evident and also as certain by everyone, because everything they contain, considered separately, is very easy to know, and the later sections are fully coherent with the earlier ones; yet because they are on the long side and demand a very attentive reader, they are grasped by very few people indeed. In the same way, although I think the arguments I use here are no less certain and evident than the geometrical ones, indeed more so, I am afraid that many people will not be able to grasp them sufficiently clearly, both because they too are on the long side, and one part depends on another, and above all because they require a mind completely free of prejudices, and which can readily withdraw itself from the company of the senses. And it is certain that the capacity for metaphysics is not more widespread than that for geometry. And there is another difference between the two. In geometry everyone is convinced that nothing is written down, as a rule, without a rigorous demonstration, and so the unskilful more often err in approving what is false, since they want to be thought to understand it, than in challenging what is true. On the other hand, in philosophy, since it is believed that one can argue on both sides of any question,* few search for the truth, and many more seek a reputation for intelligence on account of their daring to challenge the soundest views.

And therefore, whatever my reasons are worth, because they deal with philosophical issues, I do not think they will have a great impact, unless you help me with your patronage. But since everyone holds your faculty in such high and deep-rooted esteem, and since the name of the Sorbonne has such authority that not only in matters of faith there is no group of men, after the holy councils, that has greater influence than yours, but also in human philosophy no one can think of anywhere where there is greater perspicacity and seriousness, and a greater integrity and wisdom in passing judgement, than among yourselves, I do not doubt that if you deign to take an interest in this work – first, by correcting it (for mindful of my own humanity and above all of my own ignorance, I do not claim that it contains no errors); secondly, where there are gaps, or imperfections or parts that need further explanation, by adding to it, improving it, and clarifying it, or at least, pointing out these defect to me so that I can undertake the task; and finally, once the arguments contained in it, by which the existence of God and the distinction between mind and body are proved, have been brought to the degree of clarity to which I trust they can be brought, so that they can be considered as absolutely rigorous demonstrations, by agreeing to declare this and bear public witness to it – I do not doubt, I say, that if this takes place, all the erroneous views that have ever been held on these questions will swiftly be erased from people's *minds. For the truth itself will readily bring other intelligent and learned men to subscribe to your judgement; and your authority will bring atheists, who are generally pretenders to knowledge rather than genuinely intelligent and learned, to lay aside their urge to contradict, and perhaps even to give their support to reasons that they know are regarded as demonstrative by all people of intelligence, in case they might seem incapable of understanding them. And finally everyone else will readily believe so many testimonies, and there will be no one else in the world who dares to question the existence of God or the real distinction between the human soul and the body. How useful this would be, you yourselves, with your outstanding wisdom, will judge better than anybody; nor would it be seemly for me to recommend the cause of God and religion any further to you, who have always been the staunchest support of the Catholic Church.'* (Descartes, 2008, pp. 3-6)

As Descartes would eventually seal and sign this draft after careful, timely, assiduous, and repetitious redaction, the letter was delivered to the Dean and Doctors of the holy Faculty of Theology at the University of Paris in 1641. This was prior to a set of metaphysical underpinnings akin to the intrepid collection of scientific theories about what he believed to be the true nature of what contemporary scientists refer as *episteme*... knowledge... science... ἐπιστήμη. Those theories were propositioned painstakingly by René Descartes, a French geometer whose background in mathematics and physics laudably rendered him a controversial philosopher of the 17th century, and his meditations have been posthumously recycled as the foundation for epistemological philosophy hitherto. Descartes' theories, along with a plethora of other scholars, were closely surveyed by the Church as questions on the cosmos interspersed amid the scientific community germane to early 17th century scientific revelation by European astronomers such as the illustrious Galileo Galilei and German counterpart, Johannes Kepler. Today, it seems philosophers and linguists ponder on axioms of knowledge as is often discursively argued, that, and most especially as a species, we hold a keen predisposition to believe, know, and simply be sure of those things thus *believed* or *known*. We roll dice, flip coins, wager bets, prognosticate, and then bout as we infer our existence; embellished with doctorates, fellowships, and prizes then claimed to be the lot in our civil contribution.

Meditations on First Philosophy came with a series of objections and replies from competent readers. Solicited by Marin Mersenne, a French theologian, philosopher, polymath and friend of René Descartes, a network of intellects poised to challenge his propositions of the sense-based dictum that aimed to prove the existence of God. *'Je pense, donc je suis'*, 'I think, therefore I am', was his axiomatic claim we refer as Cogito (which I may also refer to throughout this chapter for brevity).

His empirical claim entailed the oscillatory polarisation of a monotheistic entity; 'Supreme being', perfection of God, and a nihilist nonentity of principally regarded nothingness. He thus believed one's imperfection, errors, or impurity as fundamentally privation of the necessary sort,[1] a supernatural entity, *being* metaphysically supreme. It's the Old Testament narrative of Adam and Eve, tempted by the forbidden fruit as pursued *episteme,* knowledge, and wisdom.

As Genesis so gleefully tells, and, fundamental to the Tanakh, Adam and Eve are granted paradise. Upon receiving paradise they are granted the luxury to eat from wherever they wished and drink from wherever they liked as paradise was enslaved to their whims. Paradise notwithstanding, the serpent's irresistible offer of *episteme*, knowledge, and wisdom induces them to eat from the tree. Through a sudden and unprecedented cognizance, they now *knew* themselves to be naked.

Descartes' postulation was of a supernatural substrate in idea, thought, and concept. Those postulates were distinguished amid three categories of meditated ideas. The first were simply *innate* and derive within the *soma*, one's body. So according to Descartes, concepts; things, truths, or thoughts must only be *true*. The second were *adventitious*, deriving phenomena extraneous to the body and may only be experienced sensually, an *a posteriori* phenomenology, i.e. phenomena that may frankly not exist. The third *factitious*, in which ideas may be invented or fabricated. His example of *factitious* was of the legendary creature, 'Hippogriff', a concoction of the mythological creature, the Griffin, and vertebrate animals such as the eagle and mare. Adduced to the factitious is the 'Siren', the Greek mythological creature who sung ships to its wreck across perilous and rocky coasts. Ideas are self-evidently *factitious* in the eyes of Descartes. In what one defines in Descartes categories, namely, Cartesian philosophy, are the vastly unventured metaphysics purported as God. Jean-Baptiste Morin, then a French Mathematician, exclaimed, 'You have in fact no idea of God, any more than you have that of an infinite number or an infinite line; and if you had such an idea, this would not imply that an infinite number exists, which is impossible' (Descartes, 2008, p. 85).

One may simply venture spaces of Descartes' meditations through experience of the linguistic sciences. Ambiguities of synonymous verbs *know*, *discover*, *find out*, *reveal*, *realise*, *get to know*, etc. (in so far as one can evade the illogical nature of ambiguity) may help us define what's knowledge. Given the treatise was originally written in Latin translators are tasked with understanding Descartes. Michael Moriarty distinguishes *cognoscere/cognitio*; *nosse. notitia*; *scire/scientia*,[2] to aid in those ambiguities. Through Descartes' meditations, we are tasked to alleviate the mind of prejudice, *belief*, and *knowledge*. Ontic conditions of subjects, predicates, and properties of his warranted trajectory will purportedly lead to *Cogito*. The sad reality is, given capital pressure from the Church (seen through the obsequious letter), we can never rely on Descartes' obscure meditations and its nuances. But what we can do is take a position, both apply and test it to the core of knowledge and see how it fends for itself. This book's position is *episteme*, which is solely concerned with measurement. As experience and knowledge cannot be displaced, we're always effaced with the problems of knowing. So, the best *metric* to test if such knowledge is truly epistemological throughout, is with the simple question, 'Would I get the same result if I repeated the experiment?'

It was the English philosopher Thomas Hobbes, summoned by Mersenne as an offer to procure his expertise, that replied as such: [3]

'For I am thinking' is not inferred from another thought: someone may think that he has thought (which is the same as remembering he has thought), but it is altogether impossible to think that one is thinking, just as one cannot know one is knowing. For there would then be an infinite string of questions: how do you know that you know that you know that you are knowing' (Descartes, 2008, p. 108)?

Thus granted the opportunity of inquiring into Descartes' presupposition he refers as Cogito, I may ask Descartes... 'How do you know that you know that you know *that you know* you are knowing? We are then effaced with the issue of Jean-Baptiste Morin with a burden of proof for infinitely knowing a thing that leads to infinite regress, as it then becomes victim to the burden of proof for knowing that thing, then that thing, then that thing, *ad infinitum*. Thus the logical question but still remains, 'how does one know that one is conclusively knowing?'

The Epistemological Quest

A Biological Flagship

What we can gladly attribute to such perspicuity by philosophers alike, is that knowledge of natural philosophy and sciences enlightened imminent and subsequent thinkers, as *episteme* headed unabatedly into the latter stages of the nineteenth century. A heavy toll on foundations theological institutions hitherto commandeering the west was taken,[4] as *episteme* was becoming ecclesiastically intractable. Revolutionary work from the Swedish taxonomist, Carolus Linnaeus, imputed to an alacritous spell in both zoological and botanical *Scientific Classification* in mid-eighteenth century, set forth a new trajectory for the biological flagship. Then work from Jean Baptiste-Lamarck on suspicions of inheritable characteristics and Gregory Mendel's famous pea-alleles, led biologists to understand both the dominant and recessive. For former stages of nineteenth-century naturalism this took another spin at the helm. But it was arguably the most prominent naturalist of our time, Charles Darwin, from early writings of evolved traits in the Columbidae family that led to evolutionary claims collated in the polemic work, *On the Origin of Species*. The origin went on to be published in 1859 and was heavily influenced by his correspondence with his contemporary, Alfred Wallace.

Gallant and supporting Darwin after *the origin* was published was the British biologist and anthropologist, Thomas Huxley. Going on to debate Bishop Samuel Wilberforce at the 1860 meeting of the British association for the Advancement of Science, Oxford University, the Bishop satirically asked Thomas Huxley whether on his father or mother's side he claimed descent from a monkey. Huxley famously retorted, 'between a man that abused his intellectual capabilities and an ape, he would rather claim relation to the ape' (Hauserman, 2013).

The philosophical vessel may have anchored when neuroscientists began to hypothesise on those manic and elongated cells that exponentially placed *Homo sapiens* on top of the food chain in the Quaternary period, albeit significant questions would superfluously remain.

The governing idea for most of the nineteenth century, was that, all neurons were connected as one super single neuron, principally known as 'synctitium'. Neurons encompassed axons, arboreal dendrites, and terminals, until the Italian neuroscientist, Camillo Golgi, invented a way of staining brain tissue in 1873.

It was using Golgi's technique that the Spanish neuroanatomist, Santiago Ramón y Cajal, ingeniously discovered a microscopic gap between the terminals of one and spines of another; the *synapse*, which thus concluded the *neuron doctrine* and the *synctitium* war.

Shedding new rays of light onto epistemological dicta did work of life science achieve from an early enlightenment. But the biological flagship may have got lost at sea until electron microscopy in the 1950s unveiled electrical excitation in the brain. It would be Francis Crick, James Watson, and Rosalind Franklin's discovery of the double-helix in deoxyribonucleic acid molecules in 1953 that would further empower *episteme*...

So, how will epistemology appertain to the purpose of this book? Or how would the anachronistic nature of seventeenth-century knowledge entail obtruding questions of the 21st century? The letter sent to the Faculty of Theology by René Descartes alludes to much more than mere supplication to the ordained authorities as a strategic move for power in the philosopher's game of chess.

The letter is one of the many examples of pre-enlightenment religious trepidation indicative of a threshold of metaphysics scholars happily dispel of today. Thus further questions must be asked, 'where is knowledge best reliable?', 'who regulates knowledge?', 'who would be certified to wield the authority to regulate knowledge in the first place?', 'how can we ensure knowledge is free from corruption or conflicts of interest?', 'who can we trust and thus who can we rely on?[5] Such questions appertain to *doxastic* episteme, i.e. scientific knowledge you've been *told*. But before we venture a vast scope of sexual selection, morality, and racism that compose the three parts of this book, we must first ask ourselves the necessary questions that may free ourselves from the biases we so often fall guilty of, well at least in so far as we possibly can.

A First Question of Free Will

Throughout a relatively slow geological timescale, an apparition of metaphysical questions began to concern those dextrous, clever, and linguistically loquacious primates that slowly worked earth to their behest in the Holocene. Hitherto a Greek golden age, the ethereal question had slowly gained purchase and manifested itself into the scientific and computational revolution, spanning from the English pioneer and polymath, Charles Babbage in the early nineteenth century, to a large network of technological engineers and neuroscientists witnessed today. The question being: Does free will really exist? Or is causal determinism *compatible* or *incompatible* with our free will? To a cornucopia of scientists, quantum mechanics, other engineers and philosophers today, our actions, events, and behaviour may be relative to the positioning of loci in the universe, i.e. in one event (giving that time and space is one entity in a four-dimensional universe with the possibility of many more), thus significantly limiting free will.[6] Some important questions are invoked when conceptualising the scope of free will. For instance, is it mere coincidence that we reside on one planet as a species 4.6 billion years into the life of the earth out of nine possible planets in our Solar System? Or that we don't find holiday or residence brochures on immigrating satellites? Well, let's say no. We could suggest that the atmosphere of other planets thwarts our opportunity to do so akin to a scarcity of oxygen in extraneous atmospheres for the ability to survive. Planets would frankly need trees. Good, okay. The next inquiry could be, is it a coincidence that our species are highly endogenous to land than sea, even though water makes up seventy one percent of the earth's surface that at least 2.2 million species capitalise on (Williams, 2014)?. Well, again, no. Natural selection has bequeathed *Homo sapiens* bipedal qualities with a 'freedom' to terrestrially forage for pockets of energy, food, and find mates on land in a war against entropy and sexual selection. If this were not the case, we simply would not exist! Is it then a coincidence to find *Homo sapiens* concentrated in some areas more than others on terrestrial land, which we intersubjectively define as cities? Again, no. What we define as cities allow our species to access social labour in a capitalistic *benefit-cost* spinoff and contractual agreement that offsets distances we would otherwise have to travel and animals and plants we would have to domesticate. It suffices for an urbanised trading power in a shape of intrinsically valueless entities that allow us to sustain and access even more benefits germane to such social cooperation.

Okay, we get the gist so let's get back to knowledge. We must now flirt with the more complex task of free will concerning thought and speech. In order to achieve such an inquisitive task, as a form of a quasi-philosophic meditation, we are now compelled to inquire into how thought and speech could come to limit itself. So let's do the same and think of what those important questions would be:

'Do we really have the agency to utter anything we wish which would *free* itself of any teleological endeavour?' 'Or is such teleology that which would destabilise ethological *evolutionary games* of surviving and reproducing, inhibiting the proliferation of our genes given speech and thought derive from organs, both the vocal tract and the brain? No, certain things socially uttered could significantly increase probabilities of social rejection, injury, humiliation, excommunication, sexual rejection, and even death. In wrong *spaces* and *time*, rejection can decrease probabilities of survivorship. Social bonding releases necessary neuropeptides that preclude us from isolation, as *Homo sapiens* successfully evolved with a compensatory set of peptides rendering us the cooperative type we appear today.[7] Hence we punish *Homo sapiens* with isolation. Try giving a platypus or a polar bear time-out and see if they actually learn from their mistakes.

A progressively longer question is then asked. Is there an *evolutionarily stable strategy* amid epistemically unconscious *thought* that controls the will of what is said or not said, commensurate with intellectual faculties? Let's think of an analogy, and in this analogy you are a twenty-year-old British female undergrad student.

Your friend, a classmate of your social studies class, has kindly invited you over to her house for dinner with her family. You were brought up attending Church as a child and slowly became agnostically sceptical in your latter denarian years through theology classes in sixth form and conversations with friends. As amicable as your friend is, you have never seen her dressed without a hijab and haven't conferred the topic of Islamic female headgear for more than five minutes, and so you reasonably suspect her parents are Muslim as you know she is. You also know that your friend's parents are Omani and the farthest you have travelled east of the United Kingdom is Vienna, Austria, on a school trip to practice GCSE German and explore the Austrian capital. You have never read the Quran and confess to being ill-informed on the Sunna, caliphates, the Islamic golden age, the southern Middle East, or the country Oman in and of itself. Assuming you have a proclivity for agreeableness, you want to appear polite, educated, and inoffensive.

In this simple scenario, would the free will of speech and thought be completely unscathed? If you were to have dinner with this particular friend's family every evening for the rest of your life, for instance, how would the will of your thought and speech be influenced? Now, if you were born in Oman and almost everybody around you consecrated their lives to the traditions of the Hadith, how would this influence what you *know* within a certain Cartesian framework of epistemology? So, information may play more into the driving force of *doxa* that may manifest thought and speech into ethological *evolutionarily stable strategies* that would prove congenial in the right environment, thus the right *spaces* and *time*.[8]

Reason

Since the distinguishing industries of stone technology and the yielding efficiencies of copper and bronze, information, for the most part, was rather bottlenecked. If you were to have sought information 6,000 years ago, albeit illiterate, the likelihood information pertained to a meagre agriculture, metallurgy, communal or parochial politics, architecture, virtue, folklore, deity, or gossip, would be highly probable. The work of a librarian would stupefy a Chalcolithic Sumerian farmer into utter disbelief with a simple tour of the Library of Congress with its more than 173 million items![*2]

Conducive to the age of enlightenment, information began to proliferate in par with a scientific revolution and enlightenment thought, which was, for those contenders, fundamentally harnessed by reason. Reason encompassed philosophers contending with incontrovertible dicta of metaphysics and the status quo of social laws and

---

[2*] https://www.loc.gov/about/general-information/#year-at-a-glance

politics.[9] At such a time, in order to be successfully disassembled, they needed a tool that would plausibly concern itself with a social philosophy that could duel with issues such as women's rights, slavery, classism, science, and politics. The tool that began to pervade was reason.[10]
Steven Pinker, the Canadian American psychologist and cognitive scientist, in what some scholars brand a contentious claim, argues there is a truth we must ultimately come to terms with in the twenty first century, that quite frankly, we are undergoing an enlightenment of our own.
In *Enlightenment Now*, Pinker ascribes reason, science, humanism, and progress to the enlightenment era. For Pinker, it was reason that took precedence when issues were conferred, discerned, or discussed, weaning thinkers from falling back on delusion. As he wrote precisely:

'Foremost is reason. Reason is nonnegotiable. As soon as you show up to discuss the question of what we should live for (or any other question), as long as you insist that your answers, whatever they are, are reasonable or justified or true and that therefore other people ought to believe them too, then you have committed yourself to reason, and to holding your beliefs accountable to objective standards. If there's anything the Enlightenment thinkers had in common, it was an insistence that we energetically apply the standard of reason to understanding our world, and not fall back on generators of delusion like faith, dogma, revelation, authority, charisma, mysticism, divination, visions, gut feelings, or the hermeneutic parsing of sacred texts' (Pinker, 2019, p. 8).

Science was a distillation of reason as a reliable entity that could stand on its own.[11] Science was seen as a methodical approach via methods such as open debate and empirical testing, with the tacit understanding that ideas had the potential to be confounded. Human nature saw us endowed with an innate sense of *sympathy* and capacity to commiserate. With the capacity to seek out the well-being of others and in turn summon the cruelty of violence, slavery, and sadistic punishment of brethren. Progress became the fundamentally tacit understanding of a social contract between institutions such as governments, laws, schools, markets, and larger spaces of international bodies. The parturition of enlightenment reason rendered the will of children limited to compulsory education, amendments amendable, policies revokable, risks assessed, seatbelts manufactured, death penalty decline, gay marriage commence, war conscription decrease, public health intervene, and the generally vast improvement of human welfare. Such changes are attributed to the ideals of reason, science, humanism, and progress, rendering an *enlightenment now* many unwittingly subscribe to.

It's the Context That Counts

Akin to the amalgam of progress, classical liberalism, and a human proclivity for knowledge, the torrential deluge of information since the preliminary stages of enlightenment thought engendered inevitable ramifications for classical liberalism to run awry, coming to fruition in the realm of what the political sciences call conspiracy. But what is ever conclusively conspiratorial when concerning knowledge?
Popular Hollywood films such as the 1998 *The Truman Show*, laudably plays on the alternative worlds of knowledge when successfully misplaced. They breech intraspecific conditions of a lengthy social contract and thus doctor with the ethics

of just how far *doxastic* knowledge can run awry. The film rewires phenomenological experience in the main character as Truman Burbank, the first child legally adopted by a corporation, is compelled to live a life completely unaware that his days are broadcast to millions of people around the world. He marries, lands a job in insurance, and through pure scepticism, intuits a dearth of phenomena when hit with a sudden wanderlust, *coming to know* that everyone around him knew something he didn't, namely, that he was the protagonist of 'The Truman Show!'[12] Another Hollywood example of misplaced *doxa*, in the historical sense, breeches conditions that had but gradually evolved into the nineteenth-century social contract. Directed by Quentin Tarantino, *The Hateful Eight* depicts racial tensions amidst the American revolutionary war. Eight travellers seek refuge in a Wyoming haberdashery, in which each member compels the other to show evidence that they *are who they say they are*, as a bounty hunter endures a perilous mission to yield his bounty to the sheriff in Red Rock, the next town along. The bounty hunter, compelled to spend a night in the haberdashery due to the unforeseen snowstorm, is the first to demand identities of who he'll be spending the night with as he wishes to retain his bounty. It turns out to be a ruse by renegades to retrieve the bounty, ending in the gruesome-bloody-Tarantino style with riveting plot twists.

As far as fictional extremes may go, thus cleverly conveying the problems of *doxastic* knowledge, we must return to *epistemic* misuse of knowledge. The type of doxastic epistemology we trust when we step on a plane. The type of doxastic epistemology we trusted in those tedious geoscience and physics lessons we may have slept through during secondary school. That pertaining to the curve of the earth. Look at what the flat earther, Samuel Shenton, has to say about the curvature of the planet we reside:

'It isn't surprising, then, that people believe so strongly that the Earth is a sphere. We are bombarded every day of our lives with information. Television, radio, books and the Internet all compete to tell us things. Society agrees that some ideas are worth debating and that others are not. The idea of a spherical Earth falls into that second category. At some point, our society decided with great certainty that the Earth is a sphere and, consequently, that further consideration is unnecessary and anyone holding an opposing viewpoint is unworthy of debate. That the Earth is spherical is a 'fact' and we are, from an early age, told to accept it without question and in the face of our own first-hand experience' (Shenton, 2009).

This excerpt I extracted from an essay compiled on the Flat Earth Society's library; a resource as young as 2009. The library contains a quasi-archival compartmentalisation of writings that hark back to 1865, principally canonising the lives and work of the English inventor, Samuel Rowbotham, and the founder of the International Flat Earth Research Society, Samuel Shenton.

The pseudoscientific community yet champions extant societies of believers wielding very clever and articulate ideas, and through charismatic leaders obliterate frustrated apologists of the common-sense *epistemically* held view that the earth is spherical, using esoterically crafted pseudoscience and a semblance of artificial *a priori* and *a posteriori* presuppositions:

'The sun and moon, in the Johnson version, are only about 32 miles in diameter. They circle above the earth in the vicinity of the equator, and their apparent rising and setting are tricks of perspective, like railroad tracks that appear to meet in the distance. The moon shines by its own light and is not eclipsed by the earth. Rather, lunar eclipses are caused by an unseen dark body occasionally passing in front of the moon' (Schadewald, 1980).

One should ask here, what philosophical rules are broken when concerning knowledge? Well, as we'll come to better understand in part II when we cover logic, these flat earthers play on what we call the *burden of proof* and what I mentioned in the preface as *gnostic* knowledge. But before we throw the burden back in their direction, let's first look at how and why they've become so numerous. Maybe it's the less science part, and a little more philosophy.

In April 2018, Senior Lecturer in Philosophy at the University of Birmingham, Nick Effingham, wrote an article online he named, *'To argue with flat earthers, use philosophy not science'* (Effingham, 2018). He argues predispositions in duelling with flat earthers, is, to summon the common-sense science that you *know* to exist. To Effingham, one's common-sense science is the very red herring that sordidly plays you into the hands of the philosophic provocateur. Effingham gives an example of common-sense science rebuttals using the 1994 television sitcom *Friends*. Ross, a palaeontologist, endures the relentless pursuit of convincing Phoebe that evolution is not really an opinion but fact. Meanwhile, Phoebe argues she simply does not buy into the idea of evolution as it seems to be, 'too easy'. Eventually irritated by Ross' persistence, Phoebe presents her case by asking Ross:

'wasn't there a time when the brightest minds in the world believed that the earth was flat? And up until, like fifty years ago, you all thought the atom was the smallest thing, until you split it open, and like this whole mess of crap came out! Now, are you telling me that you are so unbelievably arrogant, that you can't admit there is a teeny, tiny possibility that you could be wrong about this?'[13]

Ross is compelled to concede admitting there is a small possibility he might be wrong. His concession surprises the others and almost absolves Ross of his arrogant contention that evolution is fact![14] Nick Effingham called this *'Epistemic Contextualism'*. Contingent upon the context you may be in, knowledge is manifold from context to context as flat earthers are only right in a context where testimonies of hundreds are disregarded, widely accepted facts among the scientific community don't count, and photographic evidence inadmissible.

Through an evolutionary lens, the *great leap forward* rendered the becoming of our modern cognitive faculties competent, coupled with a speed-train of contexts in which consensus arrived at platforms of reason. Through an increase of average volume in the brain and old ideas too aberrant to refute, our modern species came to master contexts of knowledge in infancy, long before reaching adulthood.[15]

When infants imitate superheroes, play with toys, shoot toy guns, inhabit toy houses, and build snowmen, they do not need a constant reminder superheroes don't exist, toys are simply plastic objects, guns aren't real, toy houses are completely inhabitable, and snowmen are simply a mound of snow. Notwithstanding the socially subversive, we would treat this as an 'epistemically contextualised' truth that those given realities are true in each scenario. We wouldn't suspect anything defective of the child's development when disillusioned with such realities, as it is a fundamental period of cognitive operational development in infancy.[16]

Throughout the modern species' experience, spanning from those early operational stages through to maturity, the unencumbered train of epistemic contexts continue manifold, to the extent that novels, games, politics, virtual reality, religion, and reality in and of itself suffer over-conflation and intersect.[17]

So let's give a simple example that were the corollary of the caveats, problems, and dangers of epistemic contextualism plaguing progress since the fall of theocracies, monarchies and autocracies.

When a lady walks into a church, the context is that Jesus was the son of God, in so far as denominations divide. When she walks into a synagogue, the context is that Jesus was not the son God. When she walks into a Trump rally, the context is that Donald Trump is the optimal candidate for the next election, and maybe that the Republican parties' policies are maximally better than the Democrat's. Examples of these types of contexts vary. Thus people villainously utilise contexts in benefit-cost *games* when concerned with serious matters such as employment and criminal law, capitalising on age, nationality, sex, disability, and SES (Social Economic Status). They use malicious acts such as underpay employees, manipulate tax laws and severances, stop and search practices, and general misinformation amid both employment and criminal law. Imagine being burdened with a legal war on foreign soil. Irrespective of language problems, the vast contexts your adversary is cognisant of also fall to your disadvantage.[18] Apropos of the varying contiguous contexts, all of the caveats, problems, and dangers quickly touch the surface when varying contexts become intertwined, inter-crossed, and interwoven. Thus impositions of *apodictic truths* manifold, a recipe for the 'spoken' chaos of blasphemy, reductionism, pseudoscience, sexism, racism, homophobia, transphobia, segregation, and sensitivity.[19] It's that which I will attempt to carefully unravel throughout the parts that follow in this book. But before we venture those parts, we must first peruse the origin of what caused information and such proliferating contexts to overflow. That which should be historically understood as both the independent and dependent domestication of plants and animals, food production in its geographical position across temperate, tropical, and subtropical climates.

If you were to cherry pick from the numerous bands of hunter-gathers to be inducted around 13,000 years ago, the most expensive liability on your free will balance sheet,[20] like all other animals, would have been reduced to the production or seeking of food. Even if taking a seasonal hiatus due to: a) well-produced harvests from regrowth of crops, b) altitudes that allowed for farming in summers, c) burned landscapes that propel edible seed plants to sprout from fires, or d) the replacing of stems and tops of tubers for regrowth in wild yams, human experience would still be founded on the production or seeking of food.

As it stood, contingent upon many still-contentious circumstances (likely predicated on either culture or efficiency, or both), an agricultural revolution ensued around 11,000 to 12,000 years ago. In temperate regions of Southwest Asia with its dry and hot summers we found the very first domestication of plants and animals. Those plants were wheat, peas, and olives, while animals sheep and goats. In temperate China with its hot summers and plenty of rainfall, by 9,000 to 10,000 years ago, agriculture saw the domestication of plants; rice and millet, and animals; pigs and silkworms. On arid steppes and deserts of hot and cold Mesoamerica, we saw corn, beans, squash, and animal; turkeys, domesticated by 5,000 to 6,000 years ago. By the same time, savannahs, rainforests, tundra, and deserts of both the Andes and Amazonia (some regions with seasonal monsoons) domesticated plants; potatoes and manioc, and animals; llamas and guinea pigs. A temperate eastern U.S by 4,000 to 5,000 years ago, domesticated plants; sunflowers and goosefoots, but not yet animals. The southern steppes and savannahs of the vast Sahel (with a comprehensive area significantly larger than the U.S.) by 7,000 years ago, domesticated plants; sorghum and African rice, and the animal; guinea fowls. West African savannahs by 5,000 years ago saw the domestication of plants; African yams and oil palm, but not yet animals. Ethiopia (with exceptionally temperate regions) first saw the domestication of plants; coffee and teff, but no animals (with an also disputable timespan). New Guinean rainforests by maybe 9,000 years ago domesticated plants; sugar cane and bananas, but not yet animals. These areas are where animals and plants became independently domesticated.

The Epistemological Quest

Local domestication subsequent to the arrival of *founder* crops, i.e. from elsewhere, entered the temperate regions of western Europe between 8,000 to 5,500 years ago. They domesticated plants; poppies and oats, but not yet animals. Arid deserts of the Indus Valley domesticated plants; sesame and eggplants, and animal; humped cattle 9,000 years ago. Saharan Egypt saw the domestication of plants; sycamore figs and chufa, and animals; donkeys and cats 7,000 years ago (Diamond, 2017).

Why those earlier modern species adopted agriculture is still a question we ask today. Once opting for agriculture, why would the work of tilling a land prevail? To the extent that most human societies would adopt the conversion?[21] After careful consideration I thought of the possibility of *opportunity cost*. Hitherto the agricultural revolution, in so far as some like to flirt with optimism in believing hunter gathering societies to be egalitarian in practice (which may have been partly true), for what it's worth, ancestors preceding those earlier modern species were profoundly structured in hierarchies. Such hierarchies hark back to a post-Cambrian-Palaeozoic era that moved into early mammalian evolution throughout the warmer Mesozoic era 200 million years ago, where temperatures from the equator were more indifferent. Thus a hierarchal structure was most likely embedded in that pre-agriculturalist-hunter-gathering thinking and may have been determined by competence, sex, age, physical prowess, captured slaves, debtors or religious and spiritual dogma. Important in encompassing those determinants would be language. Simply, where the opportunity cost of a hunter-gathering life and tilling your own land is greater for an individual higher in the dominance hierarchy (than, say an individual of lower status), the opportunity to engage in an activity more stimulating duly subsides. We may have found some of the incipient stages of slavery in the exchange for agriculturalist life on the free will time sheet, turning a timely liability of food gathering into an asset on someone else's expense, but unfortunately for that minority, via husbandry.

~

Granted the myriad contexts that constitute a philosophy of knowledge, thus engendering epistemological debate, the question must then be asked, 'how and when in time did we organise a system of reliable information (whatever that could possibly mean)? One that could save time wasted, protect trustworthy information, and one in which one specialised human could continue where the other left off, a system that could be free from adulteration and charlatanry?

Over a scientific revolution in the west, concessions seemed to concern a filtering of a dangerously vast overflow of information. Information that would undermine the work of science in the event that one single context ensues, thus creating a platform for reliable information in the scientific community under a process familiar to us today, *peer review*.

Institutions sought boundaries as information exclusive to universities, court culture, private patronage, and informal societies were pre-empted by what became a secure correspondence system. During the seventeenth century, letters were the most common form of writing as they were 'swift, certain and cheap'. Not only were letters convenient in Europe, they could be sent and received within weeks, until combining with diplomatic couriers that went as far as the Levant towards latter stages of the seventeenth century, establishing an even wider network of correspondence.[22]

In specifically scientific correspondence, rapid evolution ensued, exclusive to a system of 'intelligencers' emerging as 'classified' science. Among the community, letters were copied, forwarded, circulated, and read aloud in meetings that concerned revolutionary work by major figures such as Copernicus, Tycho Brahe,

Kepler, Galileo, Descartes, Pascal, Hobbes, Flamsteed, Huygens, and Newton. Like Descartes' letter that opens this book, scientific freedom of thought was limited. It was respective of the underlying threat of apostasy and treason that polluted the network of correspondence, as malpractice was risky in any conflicting state of affair. To the American historian Elizabeth Einstein, whose work concerned the transition of letters to printing, she points out 'Intelligencers' as founding fathers of the periodical press. Those *intelligencers*, in principle, were the pioneers of scientific peer review. They endeavoured to create one singular epistemological context free from adulteration, as opposed to the mishandling of scientific information and the inevitable conflicts of interest (Hatch, 1998).

As correspondence evolved, it wasn't until the later years of the eighteenth century exclusivity would start to dissipate through enlightenment thought. With monarchies challenged and an industrial revolution imminent, it's when a populace would start to think about learning how to read and write. So let's summon the data that altered the precedent of human history, played out as military shells stacked with information in mortars, as the projection onto human society would besiege later years of the nineteenth century.[23]

Max Roser, founder and director of Our World in Data, updated an article in 2019 originally published as *The Short History of Global Living Conditions and Why it Matters that We Know*.[24] He gives a quantitative analysis of six areas; poverty, literacy, health, freedom, population, and education, collating resources from the World Bank, OECD, UNESCO, and the research institute, IIASA. When thinking about information in the last two hundred years, it's necessary to first think about a system endogenous to the possible gentries as the world's population in 1820 had 9 of every 10th person older than 15 illiterate. By 1930, literacy increased to every third person, and as of 2015 soared to 86% (Roser, 2020). What was it that inversed an illiterate population since 1820?

Principally in north-western Europe, and evocative of causes such as temperate acres of land to harvest, opportunities to culturally diffuse and resist disease was eventually granted. Once the Vikings, Angles, Saxons, Jutes, and Franks would settle their differences and Byzantium would fall in the east, time was rewarded for political chambers and institutions to flourish. As a scientific revolution ensued and Edward Jenner would learn to dispense small doses of cowpox in 1796 to his patients, a better understanding of the behaviour of antigens would allow us to think about a more prosperous living. Human progress was simply concomitant with a scientific and technological revolution. From little over a billion people, a juxtaposition of population and education on the literary trajectory was galvanised as the world population increased seven-fold, peopling the planet to the 7.8 billion today. As far as education, enlightenment thought for the species started to necessitate education for minors. Those succeeding secondary education increased and over half of the population would reach elementary schooling. As for the UN's endeavour, education remains the fourth goal of *'The 2030 Agenda for Sustainable Development'*, to ensure 'inclusive and equitable quality education and promote lifelong learning opportunities for all' (United Nations, 2015; Roser, 2020).

Forces such as the Zetetic community shouldn't be misconstrued, as they simply opened the doors to the freedom of scepticism. That's to say the Zetetic community ought to exercise every right to contend the earth is erroneously flat. Scientific peer review was inculcated in a capitalist market, harnessed by publication fees in an appellate system implored to university boards for the possibility that scientists research.[25] As a result of the revolutionary practices of both literacy and education, such causes facilitated an interspersing of information that both academic and research institutions must sift and collate. It's why knowledge has called for a philosophic restructuring against the undercurrent of peer-reviewed science today.[26] But how would this be reconciled in politics?

~

Politicians and scientists certainly aren't immune to the cherry-picking of sciences confirming their biases. So, who is it that can arrogate the authority to be conclusive on matters where an event purportedly concerns their own expertise?[27] As long as we exist we fall victim to physiology, the type that comprise cognitive biases and emotions. Take this example:

A brutal murder takes place as a black woman is murdered by her white boyfriend. The crime happened to have taken place in the state of Missouri, U.S.A. As the law of the Missourian land holds, the defendant is charged with first degree murder and is immediately considered for the death penalty.[28]

When assessing whether his sentence is commutable and relative to the nature of the crime committed, conclusively, who's knowledge takes precedence? Criminologists because of the crime? Feminists because of the sex? Political scientists because of the 'race'? Lawyers because of the law? Neurologists because of the brain? Or philosophers because of the 'will'? We may soon be compelled to summon historians, geneticists, quantum mechanics to this list, and so on.

They may all disagree with each other on the corollaries of the crime, and within these distinct backgrounds on the nature of those, and within each body on the nature of those, *ad infinitum*. The role of science and politics may be useful in so far as the next event differs. When philosophising on an inextricable network of epistemological dicta, it may be useful to think of knowledge in terms of *epistemological points* or *purchase*.

Let's say you begin to notice you are developing a rash on your chest, you notify your spouse. Your spouse tells you that she saw the same type of rash on her older brother who complained of wearing a certain type of shirt in the hot weather, so she suggests you should use a certain cream her brother used to dispel of the rash... the cream does not work. Let's give her 1 epistemological point. You now appeal to your doctor, who, other than studying many years for her doctorship, is vastly experienced. Your doctor takes a good look at your rash and tells you that it could be allergens, but she is not too sure, let's give her 7 epistemological points. The doctor refers you to another type of doctor, a dermatologist. Your dermatologist diagnoses you with *pityriasis rosea* (Christmas tree rash) and prescribes you a lotion to stop the itchiness, which gladly expedites its disappearance. Let's give the dermatologist 9 epistemological points as the rash may reappear and present itself as something else!

Incumbent upon the individual may be the onus of edifying oneself on the testing, training, and severity of science that ramify into its humbly distinguishing areas, an ology in and of itself. The most preeminent physicists don't tend to be sitting around rolling spherical objects with heads in equations. Most likely, they'll be under pressure by a plethora of other competent physicists in the same specialty. They'll be compelled to elaborate loose hypotheses about the interaction of matter and energy in the universe, conduct costly scientific experiments, or help design electron microscopes, particle accelerators, and lasers to conduct ever more costly scientific experiments. Endemic to the field of physics, distinguished physicists specialise in many of their own areas to an esoterically surprising degree, whether concerning general relativity, dimensional physics, or the speculatively studied string theory. Biology will encompass of microbiologists, ethologists, primatologists, entomologists, and ornithologists who publish critically reviewed papers for the sole cause of scientific progress. It may be the case that one ornithologist holds a grudge against another as they quarrel about binomials of taxonomic differences in species of birds as there are arguably up to 18,000 species of birds to come to reasonable concessions with (Barrowclough et al., 2016). Psychology will entail of experienced clinicians who spend decades with thousands of clients. Such psychologists may in turn understand the science of the individual to laudably surprising degrees. Other than seminars and lectures to loan-stricken students, historians, archaeologists, and anthropologists get admissions to classified archives handling significant relics and artefacts.

So, does this suggest specialists are conclusively infallible or *de facto* correct before a proposition has even been tested? I am sure we can all testify to when professionals and experts can be egregiously wrong.[29] It may be sensible to think of right or wrong in terms of epistemological points, or whether something has '*epistemological purchase*' the more severe an event.*[3] There is a calling for a better foundation of intraspecific consensus when conferring the most important questions on progress. These grounds point to the direction of more than the Occam's razor in doubting a close-one's diagnosis, but point to evolution, sex differences, sexuality, religion, ethics, morality, and an anthology of other grounds of the cornucopian information institutions today. In Steven Pinker's recent work, *Rationality, What It Is, Why It Seems Scarce, Why It Matters*, Pinker appeals to Bayesian reasoning and the lack of a base rate neglect quintessential for upholding rationality where probability theory and statistics are concerned (Pinker, 2021).[30] So is *the known* fundamentally a tug of war between science and religion? An evolutionary battle between sexes? A tempestuous duel between the political right and left? I argue that to think about knowledge, we mustn't look over the importance of time. A vast period of time to consider that which has been scarcely considered. A brief narrative on time in order to take the first step forward in the same direction...

Time and the Universe: I Guess It's Relative

For time immemorial to human history, thinkers have quarrelled on the question of time. Today, you can read distinguished beliefs of both fiction and non-fiction, venturing the surrealism of singularities, afterlife, creationism, an alternative universe or multiverse. Amazing about the nature of time is how possible *a priori* judgement of time ensued. A type of *a priori* in constituents of what natural philosophers define, in simpler terms, *self-explanatory*.[31]

One of the many egregious assumptions was that space and time were mutually exclusive.[32] But once flirtation between space and time succeeded its dalliance, it only paved the way for advances in science and a landslide of newer questions to proliferate. By the latter stages of the seventeenth century (at a time when a divorce between space and time was common consensus), the English physicist, Isaac Newton, postulated his law of universal gravitation. It stipulated all bodies in the universe were attracted to others by a force more influential the more mass it contained, paving a perilous road for scientific reason as the *special theory of relativity* was eventually articulated by a twenty-six-year-old physicist, Albert Einstein.

---

[3]* What mathematics simply calls probability.

Amassing space and time into a single entity (published in a paper on September 26th, 1905), he stated that energy was the same as its mass once multiplied by the speed of light to the power of two ($e=mc^2$), creating a cul-de-sac for alternative philosophies on time.

As for the inseparability of time and space, our progenitors bequeathed *Homo sapiens* the profitable acquisition of a capacious memory, thus allowing us to dance with time better than any other species to date. Sumer studied the heavenly bodies and found stars to return to their original position after 60 days. Babylon later found circles divisible by the six identical triangles evenly divisible by 60 degrees. Notwithstanding secretory sleep-cycles of melatonin (within a reasonably prompt circadian rhythm), we accept our fate and work with time. We count the three hundred and sixty-five spins around our closest star, shelving the additional hours to a four year cycle while found a civilisation on the hands of time...

The measurability of time is partial to the domestication of crops which allowed for civilisations to first emerge, so whether canonising an Ancient Egypt, Mesopotamia, Indus Valley, Sumer, Akkad, or others, we know history is partial to agricultural histories. But as of the twentieth century, and with the newfound tool in 1913 (episteme), the question re-emerged when the British geologist, Arthur Holmes, along with the discovery of isotopes,[33] showed the earth to have an age of at least four billion (Lewis, 2001).[34]

As for the geological time scale and scientific consensus, the beginning of time dates to 13.8 billion years. Be that as it may, many reconcile faiths of both science and religion, achieved through what they claim to be interpretation and faith.[35] Albeit 'reconcilable', we find many academics to practice their professions with utmost competence.

The theoretical physicist, Stephen Hawking, in his popular bestseller, *A Brief History of Time*, outlined a simple cosmological chronology of time. As for the chronicles of time, Hawking dates the timeline of cosmological knowledge back to 340 B.C. [36] In a Greek golden age, in which Aristotle unwittingly became one of the first philosophers to establish the prescient idea that the earth was round, he observed the spherical shadows of the earth on the moon, and the North Star appearing lower in the sky when viewed from the south.*[4] Ptolemy, in the second century AD and augmenting to the idea, saw a cosmological model of the earth being central to the universe. With eight spheres surrounding the earth as its centre, it included the five known planets at the time (Mercury, Venus, Mars, Jupiter, and Saturn, along with the moon, the sun, and stars). Copernicus, replaced the model in 1514, claiming that planets moved in circular orbits around the sun as the centre, not the earth. Almost a century later, the Italian astronomer, Galileo Galilei, found small satellites (moons) orbiting Jupiter, hence adopting the Copernican model of the sun as the centre, whilst Kepler contended orbits to be *elliptical* in nature. But Johannes Kepler failed to understand the real *forces* of influence, until Isaac Newton saw the gravitational relationship between mass and its body, galvanising Einstein to finalise the relationship between space and time. Ultimately, through the *theory of relativity* that was continuously studied, additional curvature came a decade later in 1915, known as *the general theory of relativity* (Hawking, 2016).

So what does relativity really mean about knowledge, or in this context, epistemology? Let's get the first thing straight, we can't separate knowledge from experience, and likewise, experience from time. We can't separate knowledge from experience, and henceforth, experience from space. To get a true grasp of how this question is ultimately answered, let's briefly peruse the distinguished relativities

---

4* Albeit Aristotle erroneously believed the earth to be the centre of the universe

before we can start to understand how life could have emerged.[37] So we know Newton's laws work as astronomical forces throughout the universe, leading to Einstein's observation of space and time as the single entity. But what does this effectively mean?

The first prerequisite is classical relativity, which means, there is no such thing as absolute motion or rest throughout the universe, as everything is moving or resting relative to one thing or another. Let's take an example. You sit quietly on a train whilst drinking a cup of coffee and read about an analogy of relativity in a book. A lady on the other side of the aisle is people watching, and glances over at you reading this page. Relative to her, you're stationary, in fact, according to her, you're sat quietly on the other side of the aisle. As the train continues on, you peek outside of the window and see an elderly couple sat calmly on a bench watching trains go by. You catch the elderly lady take a nosy glance as you both lock eyes. Relative to her, you're moving as quickly as the train, and relative to you, the couple are rested. But are the elderly couple really rested? As classical relativity has it, the answer is no. If an astronaut observed, she would see the earth to be moving roughly a thousand miles per hour, so the elderly couple are not as stationary as you suspect, just as the earth, sun, solar system, and Milky Way cannot be said to be in absolute motion nor rest. It is a question of relativity, in so far as the perspective changes from you to the people watcher, elderly couple, astronaut, whoever; the observer. Hence, the theory on time becomes classical relativity.

The theory of special relativity is adduced with two core scientific proponents, *time dilation* and *length contraction*. Time only becomes interesting in special relativity when we start to think of objects moving close to the speed of light![5] Nuancing a popular example of special relativity, imagine you have two sets of two mirrors facing each other, one set is stationarily placed on the ground on a field and the other set travelling, close to the speed of light in space. Let's call the set of mirrors on the ground F (for field) and the second set of mirrors S (for space). In both sets of mirrors; F and S, there is a beam of light that reflects the opposite mirror, creating a back-and-forth effect in the light beam. Now the beam of light reflecting S in motion must become diagonal, engendering farther distance, as greater **distance** is covered compared to the light beam in F. It also creates greater **time** the light beam reflects the opposite mirror, *respectively*. It could be logically induced that the time the light beam is moving in S is greater than the light beam in F as the distance is greater. However, this is not the case. The time for the light beam to reflect the opposite mirror in both S and F are the same. This is achieved by what Einstein had called *time dilation*, in which the time for the light beam to reflect in S had slowed down to compensate for the distance the light beam had travelled, allowing the light beam to travel the greater distance, in what was articulated as **speed** being equal to its **distance** divided by its **time**. Meanwhile, time does not sufficiently dilate for the speed to be the same for both S and F when considering that S is travelling close to the speed of light. So, what compensates for the rest? It is what is then called *length contraction*. As objects approach the speed of light, objects contract. Length contraction allows for speed to be equal to its distance divided by its time. This is what Einstein dubbed as the theory of special relativity.

With principles of classical and special relativity agglomerated, we then look at general relativity and continually understand time and space to be essentially singular in entity. We think of time not belonging to any part of the universe, but only that *massive* objects in the universe, like the earth, significantly warp what is best described as 'spacetime'. Two proponents of general relativity help

---

[5]* Why? Because the speed of light is constant for all observers in the universe.

The Epistemological Quest

conceptualise physical principles of time dilation, that being *gravitational* and *relative*. The concept of gravitational time dilation entails of time being slower the higher the gravitational potential there is. The usual confounded notion of when and where on earth and in space one could expect time to be slower or faster, conducive to great sci-fi movies such as Christopher Nolan's 2014 *Interstellar*, lucidly demonstrates the gravitational time dilation phenomenon. You could expect the closer an object is to a body that contains enough mass to warp spacetime, i.e. gravity, the slower the time will become for that object, relative to an object that may be orbiting a massive object, like an astronaut around the vicinity of the earth as opposed to someone standing on the surface of the earth.

In terms of relative velocity time dilation, again, a significant amount of velocity is needed, precisely, nearing the speed of light for time dilation to truly take place. This is when the physics acquires our attention. To think about this, it's useful to think of a simple question, namely, 'what is speed or velocity?' It is simply expressed as speed being the amount of time an object can get from A to B within a limited space. The subsequent question, how does that something get from A to B faster than another something throughout space? The answer, simply, is energy. Energy is what accelerates, and additional energy is required for motion maintain for a limited amount of *time* throughout space. However, are not space and time indivisible?

Unlike Newton's belief that time moved forward, the special theory of relativity proposes time to both stretch and contract throughout the universe (with respect to its velocity). To achieve a more accurate representation of velocity throughout the universe, an energy-filled package of planets may be useful in perceiving how general relativity may work, being that each planet would share two qualities, one, they are spherical, and two, they are massive! Let's imagine you were to intersperse this package of different-sized planets throughout a relatively empty space within the universe, and let's add a huge star in the package. The star being the object of most mass, all interspersed planets would find itself orbiting the star comfortably, while some planets would lose its status as a planet, as it finds itself orbiting another; thus becoming a satellite. As all planets and the star are spherical in the interspersal, dimensions coming away from the surface of each planet and star variegate into where you would position point A to B. The planet with more mass slows time down on its pull on gravity, while the smallest planet has the smallest of gravitational influences *with respect* to where you plot point A and B. It's that velocity is the relative phenomenon, as different bodies contain different masses and thus encompass different gravitational strengths, meaning that objects will be accelerated at different rates throughout the universe. Apropos of space and time's monopoly over velocity, in so far as velocity becomes each of them, objects approaching the speed of light become victim to a *relative velocity* time dilation phenomenon, due to all masses in spacetime.

The problem re-emerges when we think of what may occur when an object achieves a velocity greater than the speed of light. This results into what the common physicist purports to be time travel, albeit the amount of mass and energy to achieve a greater velocity becomes inconceivable when approaching the speed of light. Once achieving such speed, objects increase in mass which give way to velocity, requiring an amount of energy to achieve the very unattainable velocity in order to be ever greater than the speed of light. The essence of time dilation is, whether *gravitational* or *relative velocity*, that the tool of light is most significant in experimentation. As light remains constant, space (speed or distance) and time only become that to be offset, thus leading to *appearances* of the phenomenal world.

What is then deduced as relativity is *a priori* of spacetime. It's a fabric of the three dimensions experienced as rudimentary principles of classical relativity play out into a fourth dimension. The proclivity to augment time to the three-dimensional understanding of space may trouble the mind, as divisible entities happen to not be so mutually exclusive. Experience of space and time is a profitable phenomenology, that being merely through the evolutionary prism of an organism.[38] Modules of the brain that thus create phenomena perceive it as limitless,[39] until scientists hypostasised limits into what was eventually proclaimed to be an infinite universe expanding. But before such hypostases succeeded, there had to be the obvious question asked, 'how could one ever come to know of a limited or limitless world?' The answer looks something like this... it would be akin to a simple breakdown of what we lionise as humanists; reason.

Tis' Absurd! A Critique

Immanuel Kant, unarguably one of the most distinguished philosophers of modernity, approached these philosophical contradictions, publishing, with a set of brave metaphysical propositions, his most famous and obscure *Critique of Pure Reason* (Kant, 2007). As we saw in the preface of this book, knowledge with experience we define *gnosis*, but how is it one is able to truly separate knowledge from experience? The British philosopher John Locke, suggested before Kant, that 'We can have knowledge no further than we have ideas' (Locke, 2014, p, 531). Thus Kant went on to organise such contradictions into what he described as antinomies, accounting for two *worlds*[40] that may quite reasonably exist, a *phenomenal*, i.e. that experienced by our senses, and a *noumenal*, that which isn't. That which is sensually experienced (both *a posteriori* and *a priori*) in the phenomenal world represents itself as *appearances* through the senses (irrespective of consultation), while the noumenal world as one of *transcendental idealism*, quite frankly, that which transcends the phenomenological world.

Kant believed antinomies could be reasonably argued with both a thesis and antithesis and yet show itself to be true on either side, in which a thesis would corroborate that of a phenomenal world, and its antithesis, that of a noumenal. We may find such a strategy when one is compelled to argue the opposing argument and thus reason its way to absurdity. Writing four antinomies, Kant proves prove its converse to be true by both its thesis and antithesis. I thus pose those antinomies and argue each case in its most brief and succinct respect (with respect to the complexity of Kant's writing).

1. *Space and time are finite, the world has a beginning in time.*

Thesis: The world must surely have boundaries within space and limits within time, which kindly agree with a phenomenology of all matter experienced. If space and time were infinite, a *limited* amount of time or a *limited* amount of space would have to elapse in any two given points for a certain event to occur, otherwise one could not possibly perceive any phenomenal event as succeeding any other, thus an infinite space and time would be simply absurd!

Antithesis: If the world had a limit within space and time then the world would have simply come from nothing, and so if nothing was purely nothing, something could not have possibly appeared, which proves that a something cannot be purely nothing, thus proving an infinitely successive synthesis.[41] The world as finite is quite frankly, absurd!

2. *Everything in the world is made of a composite.*

Thesis: It's impossible to sensually experience anything as constituently unestablished, as every composite will have a composite, as one cannot eventually break a thing down to such a thing that would find no composite. Nothing cannot be a composite of something else, otherwise its composition would also be nothing. That which has no composite is simply absurd!

Antithesis: A thing within space cannot be ultimately composed of another, and thus composites can be divided until no such composites exist. Either simple parts do not necessarily comprise of all things, or composites are finite, otherwise one would find composites of the minutest of things. That which must entail of a composite leads to absurdity!

3. *Causality is supplied by the laws of nature and for its own being. A transcendent world allows for such causality to emerge.*

Thesis: Spontaneity can be found in the world, substantiating a possible domain transcendent to the causal laws of nature, thus reconciling the idea of a successive synthesis. If I simply get out of this chair and walk three meters in front of me, it would be an act of spontaneity. A world reduced to causality is quite simply, absurd!

Antithesis: Causality can only be a representation of *appearances* in a phenomenal world, and only the laws of nature allow for such causality to exist, representing itself as proximal causes, and such proximal causes are found to be analogously ultimate. Anything transcending such natural laws are quite frankly, absurd!.

4. *Is there a necessary being, and if so, is this being a representation of the world?*

Thesis: Quite frankly, there is a necessary being attributable to the spontaneity we experience in the world. It only transcends what we perceive as causality. That being, actions and or any event. That what cannot be defined as part of a necessary world is absurd!

Antithesis: There is simply no necessary being, as we know something cannot emerge from nothing, and if there was a necessary being, then it would be manifest in space and time and thus show itself to reject all laws of nature. A truly reasonable existence of a necessary being within all universal laws of nature is, fundamentally absurd!

Before Kant, it was believed by philosophers, particularly the Scottish philosopher David Hume, there were two distinguished forms of knowledge; *a priori* and *a posteriori*. The latter concerned with sensical experience, *a priori* without the need for consultation of our senses (or previous experience). For instance, two plus two is equal to four. One wouldn't be compelled to consult one's senses to justify such knowledge nor consult the experience of yesterday's equation, and so such knowledge remains *a priori*. Two fundamental characteristics are held when concerning *a priori* knowledge, one, is that such *a priori* knowledge is **necessary**, meaning true in all worlds or circumstances, as opposed to contingent, which would be true in some worlds or some circumstances. Two, all *a priori* knowledge is **universal**, meaning such knowledge is true irrespective of time or place. There would be no exceptional circumstance, i.e. time, or place in which two plus two would not be equivalent to four. It is believed the discipline of mathematics is a significant instance of *a priori* knowledge.

Conversely, *a posteriori* knowledge would be knowledge compelled to consult the senses and previous experiences in order to be ultimately justified. For instance, you would have to look out of your window to see if it was raining, open your drawer to see if your house keys were left there, taste a mushroom to know whether it's a poisonous species or not, or turn on the television to know if the electricity is functionally working. Natural sciences such as those that constitute the humanities rely on *a posteriori* knowledge as knowledge is obtained through the consultation of one's senses, empiricism, or at least someone else's empirical experience.

As two forms of knowledge pertain to such epistemology, two forms of judgement must now be instantiated also, that being *analytic* and *synthetic* judgements. In order to understand such judgements, we must understand judgments through both subjects and predicate concepts. The former concerning itself with who or what is doing the action, the latter with what is being done and thus its relationship with the subject. In any analytic judgment, the concept of the predicate can be found in the subject. For instance, *a man is a human*. The concept within the predicate, *a human*, can be found in the concept of a subject, *a man*. Such instances become true by definition, i.e. that all men are human. Now let's look at its counterpart, synthetic judgements.

Synthetic judgements introduce new concepts to the predicate which cannot be originally found in the subject. For instance, *a man is a fool*. The predicate concept, *that the man is a fool*, cannot be found in the concept of the subject, as not all men are fools. Thus philosophers believed synthetic judgements achieve the fundamental characteristic of being **ampliative**, meaning such judgements provide new concepts within predicate concepts not originally introduced in the subject. Thus synthetic judgements find themselves to be untrue by definition.

So by David Hume and others, it was therefore logically presupposed for all analytic judgements to constitute *a priori* knowledge and thus all synthetic judgements to likewise constitute *a posteriori* knowledge, conducive to a necessary consultation of one's own senses and experiences in order to justify such pending knowledge. That notwithstanding, Kant did not believe it so conclusive. He believed there was a possibility for independent synthetic *a priori* knowledge to exist, founded on what one may term the feasibility of such predicate concepts to be **ampliative** in nature, thus providing new information, and yet both be **necessary** and **universal** in origin. A most useful instance is the claim that all angles of a triangle add up to 180 degrees. In such an instance, the concept of the subject, *all angles of a triangle*, is not as analytically *a priori*[42] in origin. The predicate's concept, *add up to 180 degrees*, is neither implicitly analogous nor obvious to its subject, meanwhile both concepts are known to be necessarily and universally true in all possible circumstances or worlds.[43] Thus such claims do give new information in so far as they are ampliative, and yet entail the same characteristics as *a priori* knowledge in being both universally and necessarily true. Synthetic *a priori* knowledge is what Kant believed to compose the discipline of metaphysics as well as mathematics, thus we move knowledge back to the drawing board yet again on how one can truly come to know a thing, or of one, in which acolytes of a transcendent world rejoice in re-entering the ever irreconcilable debate (Kant, 2007).

As we've acquainted ourselves with space and time as pure *a priori* intuitions, which may lead one to discover necessary synthetic *a priori* truths, metaphysics pertaining to the limits of space and time were peevish in troubling eighteenth-century naturalism until its final encapsulation in antimonies. After Einstein concluded space and time to be its singular entity within the *special theory of relativity* in 1905, a further conclusion was imminent, and as has been thus far shown, the sciences now had the necessary wherewithal to progress.

Thus now succeeding a general theory of relativity in 1915, over a decade later a worthy astronomer with a keen eye for interstellar activity lured, observing the universe to be expanding and presenting his observations in a paper published in 1929. Like the geologist Arthur Holmes, and other scientists endemic to such fields, the American astronomer endemic to galactic, extragalactic, and observational cosmology, Edwin Hubble, was able to contribute to the speculative question of whether space and time was truly infinite. Using what is known as the Doppler effect, Hubble made use of prisms to observe the spectra of light passing through it, as different wavelengths of light can be observed by what appears to be different colours in the electromagnetic field. The distance between what is known as one wave crest to the next, the wavelength, is at a constant distance from the observer, so, wavelengths shifting as far as the red end of the spectrum result in being longer than those appearing at the blue end of the spectrum. Hubble used the Doppler effect as a useful tool in observing interstellar activity against Einstein's proposition that antigravity worked concertedly with spacetime's inbuilt tendency to expand, creating what Einstein knew to be a 'cosmological constant' (thus the universe erroneously never expanding nor contracting). Meanwhile, with Hubble having effectively used the Doppler effect, he showed most galaxies to be moving away from us as most galaxies appeared to be red shifted, opposed to what could have been expected to be more kaleidoscopic spectra.

The Russian physicist, Alexander Friedman, elucidated on Hubble's premise with a series of models to test his theory, along with American physicists Arno Penzias and Robert Wilson, who inadvertently discovered radiation coming in from distant galaxies due to a rather sensitive microwave detector when testing in 1965 (Hawking, 2016; Bahcall, 2015).

With ever more ingenuity pervading the former stages of the twentieth century, Hubble's discovery only fitted itself quite nicely within such a pervasion, as the Hubble Law, the Hubble Constant, the Hubble Time, and the Hubble Space Telescope became the tributes of his prescient discovery. And so it was the Hubble Constant, of which the universe expands at the same rate at every location (in so far as expansion rates change through time), that aided scientists in completing an accurate age of the universe resulting to the specific number 13.8 billion, coinciding with the age of some of the oldest stars in the universe (Bahcall, 2015). But what exactly happened on this... day, 13.8 billion years ago for a whole universe to start? If you've ever wondered why scientists were obsessed with black holes and what can be found within them, you may now begin to understand why, singularities.

One could flip the script and say scientists may not be necessarily interested in black holes per say, but are interested in the behaviour of infinite density, gravity, and gas clouds, as singularities appear to be that in which physical laws of the universe become indistinguishable and space and time fail to act. What one observes in space and beyond is rather a narrative of our universe and the behaviour of black holes seem to be illegible paragraphs within the story of the universe. Despite this, Stephen Hawking was one who possibly rendered entire chapters legible, as his famous discovery in 1974 on his observation of black holes proliferated among the scientific community and beyond.

From rather careful observations of black holes, it was heretofore known nothing could escape from the gravity of a black hole once it passed the *point of no return*, also known as the *event horizon*. Hawking, having acquainted himself quite competently with principles of quantum mechanics, predicted particles to slowly leak from black holes, notwithstanding a point of no return, becoming the 'Hawking Radiation' the scientific community lauded him for and what he is most commendably canonised for today (Grant, 2020).

Be that as it may, the ingenuity of Hawking's work continues on as researchers at Technion- Israel Institute of Technology, testing Hawking's theory, were able to recreate an artificial black hole in a laboratory setting to test his prediction of what was emitted, for the very first time. According to Hawking, the emission from black holes is spontaneous, meaning such emission arises from nothing while being stationary, meaning such intensity does not change much over time. Using rubidium atoms, researchers used a small analogue black hole to observe its emissions. As rubidium atoms travel faster than the speed of sound, sound waves should be quickly swallowed up, which would be similar to light waves swallowed up by a black hole in space as the current would be too strong for the waves to retreat. As Hawking ingeniously predicted, unarmed with the advanced tools acquired at the institute, radiation occurred, corroborating its tendency to act spontaneously. Like regular stars that radiate a constant type of radiation, such as black bodies that radiate a constant amount of infrared radiation, the emission of light particles were predicted to travel inside of the black hole while at the same time travel out, creating the stationary effect of not changing much in intensity over time. This is exactly what those researchers observed. Interestingly enough, contrary to the event horizon existing as the outer layer of the black hole, there also exists a smaller sphere on the inside, the 'inner horizon'. Once falling through an inner horizon, one would be believed to be trapped inside of a black hole as predicted, meanwhile the intensity of gravity would then significantly subside creating much more of the 'normal' environment experienced on earth (Fadelli, 2021).

Hawking's theorem also professed a rather important claim central to the laws of black holes, predicting the area of an event horizon to never shrink. Researchers at the Laser Interferometer Gravitational-wave Observatory (LIGO) in 2015, observed gravitational waves to see if such laws are held to be true. They observed two inspiralling black holes rippling across spacetime that both entailed of a large amount of energy. As per Hawking's claim, the event horizon of the new black hole should not be smaller than the total horizon area of the parent black holes. The researchers discovered that after a cosmic collision the total area of the event horizon did not decrease as predicted. Hawking's theorem was merely mathematical until observation of such gravitational waves was confirmed, revolutionising a new approach to gravitational-wave data in black hole mechanics. As it seems, with the behaviour of black holes still observed and time and space yet to make its necessary concessions, the debate can be quite easily extended to far-sighted extragalactic questions. However, in so far as such topics may be fascinating, they do not fall into this book's endeavour, the case in point is, as for an epistemological purpose, that our history stretches back far more than a mere six thousand years, and so a return to a history on earth may thus prove more useful for our continued epistemological quest, that which must question how life emerged.

The Epistemological Quest

Chapter 2

Tierra! Tierra! Tierra![44]

If you've ever moved house, or at least sought property to reside long-term, you've probably endured that not so easy process of distilling criteria down to the type of place you wish to reside. Maybe safety is an essential criterion if you have children, access to public transport if you dislike the bother of driving to work, an unfurnished house if you managed to amass furniture over years, a close-by super grocery store if you tend to necessitate the convenience of good food, and so on. Those criteria notwithstanding, what you find along the journey, given you come to adjust to the competition along with a commensurate financial budget in the property market, is that your criteria inevitably change. Well, what if it didn't? What if, along with your initial budget (which was most probably parsimonious), you insisted upon all initial criteria being met? It may be sufficed to say you would probably never find property! Well, it seems that for life to appear on a given planet in the universe, and extraterrestrial life to be then audaciously considered by evolved intelligent life on that planet, criteria is just as inflexible.
Like with the evolution of our domesticable plants and animals across the Fertile Crescent, paving its way to a pervasive agricultural revolution, certain criteria were necessary for the switch from hunter gathering to successfully take place. That would have been a congenial distance from the equator to cultivate land, a worthwhile collection of mammalian candidates to elect, enough rain to resist droughts, sun to resist floods, and an assortment of wild cereals and other crops to feasibly domesticate. Remember Australian Aborigines never switched to agricultural life until Europeans had first found their subcontinent as little as a few centuries ago!
Like agriculture, yet differing to a larger extent, a *Goldilocks* condition for life to evolve on earth was met with criteria, but like the initial criteria for the property you wished for, is both inflexibly and obstinately demanded.[45]
After 9.2 billion years of relative inactivity from a big bang event, our planetary solar system formed from various dusts and grains to what became small rock-like figures, namely chondritic meteorites, around 4.6 billion years ago. It thus led to an aeon of mostly intense volcanic activity, tumultuous bombardment of asteroids, comets, and meteorites, with extremely high temperatures in an atmosphere that rendered the earth both uncongenial and inhabitable for life to emerge. From theories of how planets formed, thought to be planetesimals (material pulled together by gravity to form small chunk-like planets), such planetesimals carried the materials that would also go on to bring its stable elements of hydrogen and oxygen as well as chondrites, thus forming earth's earliest waters.
Within vast stretches of geological time, namely from the inception of the Hadean aeon to as far as the latter stages of the Proterozoic 541 million years ago, scientists have disputed the origin of life, endogenous to disciplines of stratigraphy, fossilisation, biology, palaeontology, chemistry, quantum mechanics, geology, and many more.
To give an example of the nature of such theories, one theory concomitant with the origin of life harks back to as long ago as the Hadean aeon (4.6 to 4 billion years ago), principally observing the behaviour of cosmic rays. Cosmic rays may have played a role in the emergence of life due to a fascinating process of occurrences that would allow for life to come about in a rather peculiar manner, but how?
As subatomic particles are high-energy cosmic rays that travel from the sun and distant galaxies, they may have produced supplementary particles in a shower of

'secondary information', thus traversing the atmosphere and reaching the earth's surface comfortably. Ionisation, the process occurring between the atmosphere and cosmic rays, increase activities such as lightening. This then vastly changes chemistries throughout the atmosphere thus becoming a catalyst for some of life's possibly earliest forms.

A process of either too high mutation rates or too high radiation may have made the environment a little too complicated and hot for life to emerge from the surface of the earth (Laine, 2016). However, ionised radiation, in which particles or electromagnetic waves carry enough energy to detach electrons from atoms or molecules (chemical reactions), would create better conditions below the surface for the building blocks of life to mutate and thus self-replicate at a comfortable rate, free from extreme chemical hindrances. Be that as it may, considering the timely 600 million years of extreme high temperatures hot enough to melt the earth's rocks across the entire span of the Hadean aeon, such a theory proves to be extraordinary. Meanwhile, such theories help in giving a taste of the caveats each theory will invariably efface.

In the 1950s, after almost a century passed from Charles Darwin's *On the Origin of Species* of 1859, research akin to the building blocks of life was preserved, as it came to be communally understood for carbon-based chemicals to be the likely catalyst of life to dissolve in oceans. Such chemicals had to be sufficiently thick and concentrated for them to be a part of a 'primeval soup'. So amino acids, found to be chains of proteins in 1902, would mimic a primordial ocean and atmosphere that should allow for a process of self-replication in gaining momentum.[46] However, the jump from carbon-based chemicals or amino acids to structures of a ribonucleic acid (RNA) system of self-replicating entities, further rendered ideas of the formation of life tenuous, as a necessary link between amino acids and nucleobases,[47] being nucleic acids pairing to become RNA, was still elided in theories germane to this gap. It was yet Charles Darwin in a letter dated 1st February 1871 to the naturalist Joseph Dalton Hooker, who proposed the idea that life may not have started in the open ocean, but rather, in a pool. A pool would have allowed for dissolved chemicals to be concentrated after water evaporates, and chemicals would also assimilate a healthy combination of light, heat, and chemical energy to form. Ultraviolet radiation, a key ingredient for chemicals to form in wet-dry cycles, encompass enough light to form on a mineral surface as chemicals are more likely to react and come into contact with one another.

As a result, the wet-dry cycle thus proposed was laboriously experimented by John Sutherland, who, in the MRC Laboratory of Molecular Biology in Cambridge, UK, surprisingly found carbon-based chemicals to precisely form two of the building blocks of RNA in 2009, and with slightly different treatments made the fatty lipids that form the outer membranes of cells. After Sutherland found all four building blocks of RNA to emerge, follow-up research undergone in Germany simply placed carbon-based chemicals in hot water on a mineral surface, subjecting chemicals to repeated wet-dry cycles. Then research at the university of California, Santa Cruz, found wet-dry cycles to form protocells and amino acids to link up into chains of proteins on a mineral surface, while all shedding light to Darwin's prescience in 'the need for a range of chemicals to become concentrated in a small space, and the need for an energy source that could drive chemical reactions'. Lena Vincent, of the university of Wisconsin-Madison, claims the essence of his theory is that those very same unseen wet-dry cycles of proteins or similar forming could be happening as we speak, nevertheless, these processes would evidently go unseen as hungry bacteria would quite happily gobble them up (Marshall, 2020).

So it may be owed to these very wet-dry cycle-geneticists that the discussion of an emerging life obtrudes for discussion, and as we became parasitic to earth's raw materials and vast-flowing waters, it seems such amenities weren't given free of charge as life was dealt the heavy hand of natural selection, all on its plea for permanent residency. We know the permanence sought for by those primordial carbohydrates, lipids, proteins, and fats became the gene, but the nature of it was never so clear, until the ethologist, Richard Dawkins, wrote *The Selfish Gene* in 1976. So then how has this gene thus far been defined? As Dawkins believed: 'A gene is defined as any portion of chromosomal material that potentially lasts enough for generations to serve as a unit of natural selection' (Dawkins, 2016, p. 36). The foundation of Dawkins' definition still holds, while the gene is also seen as a replicator with 'high-copying fidelity'. Nevertheless, such understandings of a gene perplex the discipline as not a single geneticist community is conclusive in defining a specific portion of chromosomal material,[48] effectively the composition of the molecular nucleotides A, G, C, and T. Different portions of chromosomal material work with other portions of said material, which then result in expressions of observable traits such as eye colour or height. Expression of such genotypes become its phenotype as the contribution of genes, the genotype, also lead to the expression of personality,[49] in so far as the environment then plays its part.[50]

But before we continue to see how complex biological processes occur throughout animal behaviour that thus entail the gene, we must first get two almost incorrigible misunderstandings that particularly concern the fields of genetics and ethology straight, and then biology as a whole.

The first unforgivable misunderstanding is that the selfish gene renders all organisms selfish. That what Dawkins described as the 'Survival Machine', for humans, a person, is what becomes the central body of selfishness. Instead, the idea of a selfish gene can be safely reduced to a gene that uses the survival machine to ensure its immortality within its vast gene pool. Starting from a point of evident bias and selfishness in earlier stages of evolution, evolving to an organism incognizant of its altruism to other members of its species within *evolutionary games* after billions of years elapsing, is what may render a gene *selfish* instead of an organism. For many social species, one must also consider the evolutionary constituent of time within such selfishness. If a social organism in a given environment were to be evidently selfish for a short period of time within its life expectancy, the organism would reduce its chances of surviving and reproducing over its entire lifetime within the species, thus belying the nature of a gene that aims to self-replicate.

The second misunderstanding is that of Charles Darwin. Excoriated by some in distinguished academic fields outside the life sciences and unhelpfully branded as racist, it has thwarted many from accessing his erudition with such a reputation besmirched. So two questions could be asked in plausibly contextualising the work of Darwin:

1. Was Darwin really an anthropologist?

No, and in fact, he never did claim to be. At a time when slavery had not been thus far abolished and disparate groups had coalesced as a result of highly oppressive political regimes affecting most members of our species, it must be tacitly remembered Darwin was contemporaneous with sub-Saharan African racial types such as Frederick Douglass, who, despite being a quadragenerian when slavery was abolished in 1865 in the US, is canonised as a nineteenth century American intellect.[51] If Darwin was in fact a vehement figure of scientific racism, maybe he would have had more of an interest in negros such as Douglass across the Atlantic, instead of venturing as far as the Galapagos islands in his earlier studies. It was his collection of writings published as *The Descent of Man: Selection in Relation to Sex* in 1871 that dabbled into the characteristics of mankind and gave a poor anthropological account of speculative racial differences throughout brief sections of his book, which should be admitted, were merely immature attempts at adducing the theory of sexual selection. French naturalists such as Georges Cuvier and American physicians such as Samuel George Morton, along with others in western academia, may be better targets at expending such disquietude for the reprehensibly tenuous work undergone regarding what they referred to as polygenism,[52] as creationist ideas thus professed effectively polluted philosophies of logic and reason across enlightenment thinking, and thus progression as a whole.

2. What did Charles Darwin really have to say about race?

You may find yourself rather disappointed when reaching the latter stages of his famous work, *On the Origin of Species,* when you discover Darwin to say nothing about not only race, but humans altogether. Whilst making a case for gradual changes, adaptation, and a struggle for existence as he crossed the botanical sciences, ornithology, entomology, and the geological sciences to adduce the audacious and gallant claim that all life fell victim to a process of natural selection and thus evolved, you will find Darwin's pre-eminence was ascribed to his work as a naturalist. This is contrary to the more spurious misunderstanding that his initial endeavour was to propagate the idea that it was a case of a *survival of the fittest,* and that he must have *de facto* been a member of such an understanding.

Such misunderstandings have kept many unfairly evasive of biological sciences that do concern essential theories of human behaviour, more than likely appropriated in the incognisant guise that thus adduces a political agenda. In the case of Darwin, it might be helpful to believe Darwin would be rather surprised with the intellectual varieties we see within our vast spectrum of the race today. Be that as it may, what Charles Darwin would have thought about race today should not keep one up at night and given the torrent of twentieth-century science and knowledge that separates us from Darwin, frankly, it should not matter at all.

A War on Entropy

As our quest on epistemology continues, it will be useful to home in on an essential approach concerning knowledge, absolute certainty, and belief. I therefore put propositions into a *categorical syllogism* that entails the major and minor premise, and conclusion; as well as the middle, minor and major term:

All belief is uncertain,
All knowledge is belief,
Therefore all knowledge is uncertain.[53]

All knowledge is of belief, and such *absolute certainty* can therefore not exist, and so the only type of knowledge that can exist in the world is *justified true belief,* as no knowledge can be unbelieved, untrue, and therefore unjustified.[54] If you were to argue with your friend about whether the wall was blue or purple, you would first have to believe the wall to exist, its existence to be true, and that such an existence can be justified *a posteriori* before any argument on the colour of the wall could commence. But what if your friend decided to deny the existence of the wall?

Like the wall, such presuppositions in a history of science ensued, and in layman's terms may look something like common sense, thus believed, true, but justified through the means of reason in place of faith, which had fuelled epistemic face-offs between science and religion.

One of those face-offs came from a Dutch scientist's observations on rocks in the latter stages of the seventeenth century, Nicholas Steno. He saw that wavy lines in sedimentary rocks form over time, but not simply any time, a time that would significantly predate Genesis, which had troubled Steno being that Steno was a deeply religious man. With Steno now burdened with observations that belied religious thought, a set of principles and presuppositions were necessary in order to carry out his observations assiduously, a set of principles and presuppositions we even take for granted today. The wavy lines in rocks were the first documented observations of the earth's strata that bore the field of stratigraphy, and so what could be those principles and so-called presuppositions? Let's take a look.

Deep Time – Deep time performs as a time that not only predates humanity, but inconceivably predates humanity. I'm sure you have read this following number before: 1,000,000,000; which reads as one billion, right? But I'll be quite sure you have never counted to one billion before, and if, in the unlikely event you have achieved this task, I'm sure you won't be willing to do it again. It won't be the number causing concern, but the evident time it would take to reach the number. Now let's look at a larger time lapse than the time it would take to count to one billion, ten years. The perception of ten years is a significantly large amount of time for our *a priori* experience, one that would be akin to *acceptance* over patience if on misbehaving you were sentenced by a judge in a court of law. Times that ten years by another ten and we reach a centenarian lifetime in which rare experiencers of such a large amount of time achieve such human perception. So when human perception endeavours one thousand, one million, tens of million, or one billion, one must logically concede to its infeasibility.[55]

Time arrow – the assumption time travels from what we call the past to the future, and quite importantly, that it has not travelled in any other direction hitherto, and therefore will not travel in any other direction.

Uniformity principle – the assumption that laws of nature have always been the same and apply to every part of the universe, and therefore has never been any other way.

Uniformity of rate – the assumption geological laws have operated under the same forces, and therefore has never been any other way.

Steady-state principle – the assumption the state of the earth does not change over time, and therefore has never been any other way.

Germane to such arbitrary principles and presuppositions, Steno reasonably assumed that if one stratum was below another, it must be older than the stratum above, therefore adhering to the very principles and presuppositions noted. But as the science philosopher Gadi Kravitz conveyed, the *uniformity of rate* and *steady-state principle* egregiously erred, as what was discovered as entropy in early nineteenth century science would yet redisclose the blueprint of thermodynamics and thus how energy truly comported.
To Kravitz, the two latter principles of uniformity violated what was the second law of thermodynamics, as it follows that entropy is always increasing through a statistical set of laws, as opposed to mandatory laws. Hitherto the very first singularity we call the Big Bang, the universe was in a state of low entropy before vastly expanding, meaning the universe would have also been highly concentrated, while particles vastly disperse throughout the universe and entropy would so-call increase through a *time arrow*. But what does this mean? As Steno's strata was of

observable macrostates, what he failed to see was how energy comported in a world of manoeuvring microstates.

Confined to a world of macrostates, Isaac Newton's laws of motion would speculate a world of atoms in sharing features that were likewise symmetrical, thus maintaining the principles of both a *uniform rate* and *state*. But could a state bely a *time arrow* principle?[56] Ludwig Boltzmann and James Clerk Maxwell would solve this in the early stages of nineteenth century science, as Boltzmann saw macroscopic states as large configurations of several possible microscopic states in a thermodynamic system (Kravitz, 2014). In atomic bonds that compose phenomena such as solids, energy is stored in indivisible units we call quanta. The more of these units this solid has, the higher its temperature. The quanta between objects of this kind can be configured into so many different states, that which we call microstates. But each microstate has its distinguished probability of transpiring in a system, and most observable phenomena have numerous atomic bonds and thus numerous quanta. As there are more ways for energy to disperse in its myriad microstates, and thus achieve higher entropy, energy finds itself predisposed to spread out, and so the chances of energy not spreading out is strictly improbable given the numerous bonds and quanta between objects.

A biological war on entropy welcomes both flora and fauna to compete for pockets of energy in a generational plight to be naturally selected, as profitable mutations graduate and lead species in becoming ever more competent appropriators of energy in its gene pool. A tern will nocturnally scout the surface of the ocean as fish rise to the surface and forgivingly yield energy to the tern. Marine plankters yield energy to varied planktivorous fish they fall prey to, in what entropy appears as life's extraordinary cycle in the distinct biota of planet earth. Nevertheless, one may expect a relatively low state of entropy to result into a featureless gruel, but to the human eye, 'is enlivened with galaxies, planets, mountains, clouds, snowflakes, and an efflorescence of flora and fauna, including us', as Steven Pinker likes to describe it (Pinker, 2019, p. 17).

How Are We Going to Get Rid of the Body?

Geologists and those of concomitant disciplines may enjoy a blueprint of such principles and presuppositions that now include entropy and its manoeuvring microstates today, which contribute to knowledge and thus that of our epistemological quest. But why was Steno's initiation on the earth's strata so pivotal in the scientific revolution? Well, we know carbon-based chemicals appeared to form as the building blocks of the very first stages of life in what may have begun in Darwin's pool-like wet-dry cycles. It's *known* carbon-based food molecules such as glucose, or carbon and oxygen; carbon dioxide, successively built proteins, fats, sugars, and nucleic acids that form the RNA and DNA molecules necessary for life to evolve into both the flora and fauna we observe today.[57] With the demise of such flora and fauna buried in the earth's crust, they thus serve as pieces to a larger puzzle hitherto unknown, which then serve an even larger puzzle, and then an even larger puzzle, and so on, as out of the three forms of rocks, that being sedimentary, igneous, and metamorphic, sedimentary rocks have given the multifarious earth scientists a deluge of pieces to life's varying puzzles, and thus Steno's observations of strata contribute to the story of time in which it might serve as useful in seeing how science came to extract such knowledge. Such extraction refers to a radiometric, radioactive, or radioisotope dating. So how does it work?

The radioactive isotope, carbon 14, simply observes the level of decay since an organism's demise. Decay results as cosmic ray protons fire nuclei into the atmosphere, producing neutrons that in turn react with the voluminous nonradioactive nitrogen 14 in earth's atmosphere, thus forming the radioactive isotope, carbon 14, which likewise holds the same ratio as carbon 12 in living organisms and the atmosphere (Diamond, 2017, p. 91; Lewis, 2001). When an animal dies, carbon, consumed by animals from plants and CO2 in the atmosphere, is not renewed. What takes place is a physical process referred to as *beta emission* (or beta decay), that conversely reverses carbon back into nitrogen 14. As is held in the strong nuclear force, atoms are at war with electromagnetic forces in an effort to pull the atom apart, and the nucleus can only win with an optimal number of protons and neutrons composing its mass. As per beta emission, what occurs is a more than suboptimal number of protons and neutrons that render atoms far too large, and thus neutrons must find a way to be alleviated of its mass and can only achieve this by offloading itself in a turn to lighter protons, the weak nuclear force.[58] It then manifests itself into what is termed a 'half-life', conducive to half the amount of carbon 14 already decayed that calculates for such organisms. Carbon 14 has a half-life of 5,730 years, and so when more than ten half-lives have elapsed, i.e. approximately 57,300 years, reliability in such a method subsequently subsides. Fortunately for scientists, carbon 14 isn't the only radioactive nuclei with a half-life, other nuclei, such as potassium, argon, lead, and uranium, consist of isotopes that date back significantly longer than carbon 14.[59] So while science is compelled to ensure such radioactive nuclei agree, in that which increases its reliability, the geological time machine is able to travel back billions of years in time.

I Came into This World Alone and I Will Die Alone; Trust Nobody.

'Between ourselves, it is not at all necessary to get rid of "the soul" thereby, and thus renounce one of the oldest and venerated hypotheses – as happens frequently to the clumsiness of naturalists, who can hardly touch on the soul without losing it'
(Nietzsche, 2018 pp. 21-22).[60]

What could possibly diminish an idea of the soul? What could one quite possibly *know*? Such questions yet trifle the mind in which an accurate measure of the working units of natural selection are best professed within scientific literature. Populations may be known to be groups of organisms endemic to particular environments of the earth. Organisms, known to be a make-up of an identifiable set of distinguished organs, are made up of tissues such as fat, or skeletal muscles that perform functions serving cells. Cells, like both uni-and-multicellular eukaryotes, thus self-reproduce living matter. Such matter entail organelles within cytoplasm, such as chloroplasts and mitochondria performing functions within cells made up of molecules. Chlorophyll, a good example of a molecular structured organelle, create the 'green' photosynthetically pigmented chloroplast in plants, and we know the DNA molecule discovered in the 1950's presents itself in its double-helix form, sitting inside of the eukaryotic nucleus. On what part natural selection plays on life forms and in what exact way is widely debated among scientists, while from extraneous fields is often tendentiously accused of reductionism on rather complex biological matters, as Richard Dawkins wrote quite succinctly:
'Living matter introduces a whole new set of rungs to the ladder of complexity: macromolecules folding themselves into their tertiary forms, intracellular membranes and organelles, cells, tissues, organs, organisms, populations, communities and ecosystems. A similar hierarchy of units embedded in larger units epitomizes the complex artificial products of living things- semiconductor crystals, transistors, integrated circuits, computers, and embedded units that can only be

understood in terms of 'software'. At every level the units interact with each other following laws appropriate to that level, laws which are not conveniently reducible to laws at lower levels' (Dawkins, 2016, p. 171).
Thus the ways in which units of natural selection work on life, often debated from as large as group selection down to as small as alleles on a gene, have increased the distal valley that lies between biology and the laity in all its esoterism.[61] Such a valley is an epistemological valley of trust, that, in so far as it may not prove to be perniciously deleterious, does cause discrepancy, which circles reason back to where in fact the locus could be in which one can claim a *soul* exists. I therefore deem it useful to ponder on an analogy of how trust imprints itself on knowledge, whilst hopefully killing the second bird with the one stone akin to an implausibility of the reductionistic attack. So looking at three problems, while agreeing many complicated problems can be best understood as relatively *de facto* presuppositions that should be teleologically and thus relatively solved, we shall see how trust imprinted on epistemology is unfurled.[62]

On what was a fine spring day, eight friends and I tirelessly played tag in a neighbourhood consisting of nine blocks in a three-by-three area, with all blocks separated by small alleyways. Many hours later, precisely at 6:20pm, all having left the area, I realise I have lost a gold ring that my friends understood to be very precious to me, as it was a ring given to me by my late grandfather. I was told by my mother I could only wear this ring on special occasions, but in meretricious style, I decided to foolishly flaunt the ring, leading me to foolishly losing it too! Now, my curfew was set to end at 7pm and it takes me exactly ten minutes to travel to and from the neighbourhood. If I lost this ring, my mother made her unequivocal promise that my summer holiday would be cancelled, and I would be grounded for two weeks. Likewise, if I exceeded my curfew, I would endure the same punishment, as my mother was rather indignant about my recently poor performance in school. So, on incisively deciding to look for my ring in the hope that it hadn't yet been stolen or misplaced, my friends and I returned to the neighbourhood.

On arriving at the neighbourhood at 6:30pm, I assign every friend *one* of the other eight blocks to encircle, as I assign myself the first. It takes 10 minutes to circle the block once, and so, I wouldn't have been able to check the blocks alone due to a time limit of 6 minutes and 36 seconds for each block, meaning I would inevitably infringe upon the time limits of my curfew.[63] As it turns out, all alleyways will be logically searched twice by two of my friends as each alleyway evidently meets two of the nine blocks. The outer part of every block meets the road, save the middle block, hence will only be searched once by each friend.

Result: All of my friends return and tell me they could not find my ring. A bad result! However, given the severity of my punishment, and knowing that the cancellation of my summer vacation was at stake, I still have 10 minutes left of enduring uncertainty with the thought that my ring could quite *possibly* be there, if looked for properly! I find myself not so obliged to trust every friend's endeavour as my trust can be calculated. Every alleyway has been searched twice, so my trust can be multiplied by two,[64] being that I am compelled to induce there being a lower probability of my ring being in an alleyway. Be that as it may, an alternative solution could be to walk around the perimeter of all 9 blocks. Being that it takes 6 minutes and 36 seconds to walk around one block, dividing that time by 4 (1 minute and 39 seconds) and multiplying the time by the neighbourhood's perimeter, means that it will take me slightly over 19 minutes and 48 seconds to circle the perimeter of the neighbourhood.[65] That's too long. I will have to resort to trust. Who do I least trust? Let's look at three problems. I ask for a report from each friend akin to my utter desperation and disquietude, and three contrasting reports touch the surface. The first problem I call 'Olivia', the second I call 'Sarah', and the third I call 'Chelsea'.

Olivia: What we should first know about Olivia is that, while she lived with her mother in our neighbourhood, her father lived in one of the houses in this very neighbourhood with a reputation for collecting an eclectic number of tools, and so to my great advantage, Olivia was successfully able to acquire a metal detector her father had always kept in his garage. However, Olivia tells me she didn't find anything even though she'd claimed to meticulously search the block. I have always known her to be the assiduous type of friend who would avail herself of the opportunity to excel in any type of mission. Her father also has a shed full of flashy tools that we often see her help him clean across the spring and summer holidays. So, as a collector, I find he most probably holds the necessary knowledge about a practical metal detector to buy, and that Olivia subsequently inhered his interests in such antiques.

Olivia Reviewed: Now, should I trust Olivia had looked properly? Yes, I do find her well trusted. In fact, I trust myself that I trust she looked properly, given the assimilated experience of her and the *foundations* regarding the metal detector and her search, the father. But do I truly trust the metal detector she used? The foundation of the metal detector seems to serve as an *institution* that would go through its own rigorous process of filtering for the tool to be bought by her father in the first place. There are people who hold expert knowledge about the way in which metal detectors work, in which such knowledge becomes commodified, and the very experts may then seek to benefit from that knowledge with little cost, in the shape of investment. They then allow their ideas to come to fruition by cooperating with others who can help replicate these objects to be sold onto the market, contingent upon the quality of the metal detector, and so forth, until someone such as Olivia's father sees the benefit of having such a metal detector outweighing its financial cost. The foundation of the metal detector thus earns its necessary *epistemological purchase*, which helps me trust my ring isn't sitting somewhere on the very block that Olivia claimed to 'meticulously search'. I have also translated my assimilated experience of Olivia into algorithms, calculated from what was incognizant trust, to what is now purely cognizant.

Sarah: Now, what should be known about Sarah is that she is a little older than the rest of us, so I guess she has that little more deference you would naturally give the older kid, which I have considered in my reasoning. After what she claimed was a 'proper' search of her assigned block, Sarah had a little more information to impart. Sarah tells me the block she encircled has a family, the Jacksons, known to 'run the whole neighbourhood'. She tells me that they're a family of boys, and apparently, the oldest boys are in and out of prison for crimes incessantly committed. The teenage boys in the family apparently 'rob people for fun' and are known to steal people's mobile phones and jewellery right off of them, even in 'broad daylight'. The teenagers are also apparently big for their age, in fact, so big that they've been heard to fight grown men as opposed to teenagers from other neighbourhoods. Sarah tells me that even some police officers evade the vicinity of the Jacksons. Worst of all, was a story she told about the 15-year-old in the family, Lee. He was subjected to an organised fight with a 30-year-old for being accused of stealing his younger brother's bike. It turned out that Lee did actually steal his bike, and, as even as the fight had apparently been, many say Lee won the fight. Sarah's boyfriend is also quite close with the Jacksons, and so I guess Sarah is one of the first to know about the nature of Jackson history and current affairs that surround them. Sarah also then told me that the boys usually come out at around 4/5pm and if my ring had fallen out on that specific block, I should consider my ring as long gone, as I shouldn't run the risk of accusing one of them of stealing my ring, that's if I happen to run into any of them!

Sarah Reviewed: Should I trust what Sarah said? This time I'm a bit more confused about what I should *believe*. First of all, Sarah is older, and so naturally she has a later curfew than the rest of us, and so she has also had the experience of becoming much more acquainted and knowledgeable about what transpires around the more expanded vicinity of the neighbourhoods in the city than the rest of us. I guess that gives her a fair number of epistemological points. Her age also gives her an almost institutional authority as well, as she's in the same school year as the 15-year-old aggressor, Lee Jackson, and so is naturally engrossed in the network of gossip surrounding what happens around our neighbourhood, being that her boyfriend can be considered to be an almost primary source of information. I guess she has a little more *epistemological purchase* as a result of this. However, there is something important I ought to mention. I had been told by other friends in the past not to believe a word she says. They tell me Sarah often likes to hyperbolise anything she can and is a bit of a gossip, most especially within the realm of storytelling, which has apparently rendered her suspicious in previous affairs. And so I hope this time I am not falling victim to a concatenation of her exaggerated stories, in order to fulfil an unconscious social acceptance game she may be incognisant of. So, I guess she loses epistemological points in my judgement because of this. This has not helped me with my mission of retrieving my ring, and not only do I feel the evident trepidation of the Jacksons, quite frankly, I feel confused.

Chelsea: Surrounding the neighbourhood, Chelsea tells me there are magpies that fly around the area and reach this very specific block at exactly 3:20pm every day in the spring, as the local migration time of magpies is apparently 'flawless'. The magpies have been trained by nearby marauders who call themselves the 'birdy bandits' and use the magpies to steal as many useful items as they can possibly garner, especially gold. The gold is then kept in a secret chest in a cave that the leader holds the key to, and the leader, Pablo, is the only person who holds the key to that chest. Chelsea tells me that if I had any hope of retrieving my ring, it would probably be to find whoever could convince Pablo to open that chest.

Chelsea Reviewed: Irrespective of the fact that magpies come in numbers in the area, and that I may have seen a magpie steal a thing or two in my lifetime, I don't believe a word she says. It's also notwithstanding the fact that I am no birdwatcher or ornithologist who would have the wherewithal to refute her claim that the 'local migration time of magpies is flawless', whatever that means. In fact, I'm quite livid she would waste my time on such a ridiculous story. We know Chelsea to be quite superstitious in her ways as it is. She's a big fan of star signs, and, in my time I've known her, I've often had to endure her both painfully and garrulously waffle on about horoscopes to the rest of us. Would it be fair to brand me a reductionist should I reduce her claim to a completely false one, in a city life free of treasure chests, caves, and bandits? Of course not. To put it in simple terms, she hasn't cooperated or helped me at all, and so it's probably best I conjure up some extenuating circumstances in the hope that I have my punishment reduced by my mother.

On isolating each problem and its foundation thereof, a teleology is thus identified, in which proximal causes lead themselves to whether cooperation may take place or not, as opposed to an isolation of its being true, which thus conforms with the axiom that *absolute certainty cannot therefore exist*. It's yet again evocative of Nick

The Epistemological Quest

Effingham's *epistemic contextualism* in which foreign knowledge is innocuously condoned until the severity of such contexts increase. A collage of authority, experience, emotions, knowledge, and most of all, trust, manifest themselves, which isn't so far away from positions authorities have taken advantage of over their subjects for hundreds to thousands of years.[66] Many today endure an amalgam of such problems that can't be reduced to anything binary, simple, or non-oscillating.

A Post-Cambrian Diffusion

While reason alleviates itself of such authority, thus allowing scepticism to reconcile such teleologies, what remains is an *epistemological purchase* one ought to cede distinguishing scientific communities that successfully evolved.[67] Be that as it may, for understanding knowledge on a microscopic level that hark back millions of years in time, distinct methods are yet advanced within Bayesian molecular clock dating processes, through the analyses of fossils endogenous to the fieldwork of genomics alike. Thus a practicable methodology analyses what genomics refer to as 'divergence time estimation', allowing for the compartmentalisation of species within a phylogenetic tree of life in an attempt to calibrate fossils and molecules via the accumulation of genetic sequence data, in order to understand a 'macroevolutionary process of speciation and extinction to estimating a timescale for life on Earth' (Mario dos Reis et al., abstract, 2016).

As a result of such advancing methodologies, successful mapping of complex functions within organelles in prokaryotes (the first forms of life), and eukaryotes (evolved across the two Archean and Proterozoic aeons), have ensued. Both cellular forms of life that entail distinguishing differences such as a membrane-bound nucleus in eukaryotes, as opposed to such prokaryotic progenitors, are respectively invoked, contributing to a better understanding of how multicellular forms had precisely evolved throughout the three timely eras of the Proterozoic aeon; the Paleoproterozoic, the Mesoproterozoic, and the Neoproterozoic eras (A.H. Knoll et al., 2006). With a phylogenetic map of Proterozoic cells successfully deliberated, vast periods of geological *deep time* that once bioengineered our pre- and post-Cambrian phyla have been epistemically diffused. Such diffusion that consists of that which we call the cnidarian, the annelid, the chordate,[68] and the arthropod are then closely studied.[69]

Be that as it may, conspiratorial narratives akin to the Cambrian explosion have certainly ensued, expounding an undercurrent of biblical *belief bias* ensconced within an offsetting theory of evolution, such that the notion of intelligent design has been used by Christian academics such as Jonathan Wells who inverts the Cambrian tree of life in order to confirm a creationist bias (Gregg, 2007). Whilst claiming the Cambrian explosion to be a *cause*[70] of a hierarchical classification of species into the manifold taxonomical groups understood as genera, families, orders, classes, phyla, kingdoms, and domains, Wells purports such groups to be manifest acts of an *intelligent design* without further need of a burden of proof. So, as I end this introduction with a verse from the book of Ecclesiastes, and on ending a brief quest of knowledge that may sit on life, time, and the universe, hopefully convincing the reader of the ways in which knowledge may present itself to the recipient, we shall start with what I believe is of a significant part of the life sciences, that of a theory one may say has been far too often neglected, the theory of sexual selection.

'He has made everything beautiful in its time. He has also set eternity in the hearts of men; yet they cannot fathom what God has done from beginning to end' (New International Version, 2000, Ecclesiastes. 3:11).

The Epistemological Quest

**Part I
Of Sexual Selection**

'Peafowl' (Mansur, ca. 1610)

Chapter 3

What If I Wanted to Get a Yoghurt on the Way Home, Would You Pay for That?

'The two fundamental conditions which must be fulfilled if an evolutionary change is to be ascribed to sexual selection are (i) the existence of sexual preference at least in one sex, and (ii) bionomic conditions in which such preference shall confer a reproductive advantage' (Fisher, 2018, p. 134).

I was recently audience to a discussion on a rather popular application called Clubhouse, an exclusive application for people to discuss various topics and issues with others around the world in a pass-the-mic style of conversation. This particular conversation was a rather heated discussion about whether 25 dollars was enough for a man to spend on his first date, predicated on what was heterosexual courtship for those involved in the discussion. Interestingly enough, a vociferous proportion of the men in the discussion claimed to be opposed to spending too much money on a woman, founded on how they believed chivalry had 'died out', and attributable to how faith had somewhat subsided over time. I recall an instance of a man who claimed that, today, there are many dates he may go on, and with a pervasive feminist movement compelling women to capitalise on their own options, it may turn out more *costly* for him in the long-run, and so cutting his budget is best for him. Another man said he was already responsible for children to other women, and so spending too much money on a woman on a first date would cut back on the welfare of his children. There were women in the conversation unsurprisingly disconcerted with the parsimonious demeanour of some of the men involved in the discussion, and even presented the question, 'What if I wanted to get a yoghurt on the way home, would you pay for that?' So, is 25 dollars too much for a man to spend on his first date today? Should he buy her a yoghurt on gentlemanly taking her back to her home, or will such a delicacy abet an already sunk cost? Part I endeavours to shed light on some of these compelling questions...[71]

On the Working Power of Sexual Selection

Spermatogenetic systems are of a *telos* that roots itself in the very first RNA cycles of life. It thus compels one to consider the colossal differences in the size the two male and female gonads produce in meiosis, notwithstanding such gametes in sperm competition, limitation, and its distinguishing phases (Parker and Lehtonen, 2014).[72] Such economics are germane to the Quaternary two and a half million years that superseded a *great leap forward* all but ten thousand years ago, and inasmuch as the nomenclature may vary on what concerns the evolutionary highway between our sub-tribe and genus, namely *australopithecine* and *homo*, such economics in behaviour tend to follow, giving ethologists something rather fascinating to unpack. Meiotic haploid cells, gametes, stringently abide by the laws of natural selection in phases that come to form the eukaryotic and diploid cell, the zygote. But part I does not aim to focus on the ways of meiotic genetic variation, crossing over, centromeres, recombinant chromosomes, or the ways in which spindle fibres distribute variegating chromosomes, as not only is this best done by the work of said biologists, but would not answer the question of what, or why such differences are best expressed throughout human behaviour, be it ethological or an eventual sociology. We shall ponder on how the produce of other cells, like nerve cells, are governed by the laws of meiotic cell division originated from a process that carries its anisogamous data through to those last X and Y chromosomes dividing the sexes, in so far as sexual dimorphism willingly takes the baton.

In the introduction to knowledge we looked at how the origin of life may have necessitated a congenial environment for the purpose of carbon-based chemicals to self-replicate, and so it may therefore compel us to ponder on how deoxynucleic information in the eukaryotic nucleus evolved cellular *functions* that likewise found its way to self-replicate under the slow timeframe of evolution. And so as it would appear, appertains to the final round of abounding millivolts action potential bestows upon the neuronal headquarters, a cerebrum, governed by ever more workable organelles that perform for the cell, which by the molecular biologist may be filtered down to the molecule, and even by the quantum mechanic filtered down to the atom.

As far back as Mesozoic life, around 200 million years ago, sexual selection conferred upon the organism sharing genotypes within a single monophyletic taxon (group of organisms).[73] Paraphyletic taxa in order to sexually select will share cistrons common to its ancestors, save the one or more monophyletic groups carrying the gene to be likewise phenotypically expressed. Sexually selective features find themselves duly ensconced within *analogous structures*, which in *convergent evolution* express characteristics of polyphyletic variation that stretch far beyond the kingdom's phyla.[74] When turned to strategies against the oviparous egg-laying species, the viviparous, economics of egg-laying and concomitant strategies thus unfurl such essential *spatiotemporal* essence. Albeit endogenous to mammalian species, other vertebrates, such as sharks, batoids, teleosts, caecilians, anurans, salamanders, snakes, and lizards may both endure and revel in such intersubphyletic reproductive strategies. So while Blackburn suggests that the 'viviparous female can protect her developing eggs from many sources of mortality, both biotic and abiotic, while buffering her eggs from environmental fluctuations and maintaining them physiologically' (Blackburn, 1999, page 6/998); in order to plausibly contend with the questions posed (of that which may sate a lady's appetite for a yoghurt at a reasonable cost to the male, or if 25 dollars truly outweighs its benefits), it may yet suffice to dig a little deeper into the theory of sexual selection.

Hatching biota and fauna of enduring parturition overlap in animals embryologically reproducing, as lecithotrophic stages provide the embryo with the salutary yolk of the ovum, while matrotrophy provides the cornucopia of nutrients both placental tissues and oviductal secretions most value, nonetheless not mutually exclusive, as between squamates, monotremes, and therians, oviparous lecithotrophy, viviparous lecithotrophy, oviparous matrotrophy, and viviparous matrotrophy occur (Blackburn, 1999). An instance would be of the elevated egg-yolk protein found in the redtail splitfin, vitellogenin, that occurs throughout developmental stages in female oogenesis (Iida et al., 2019).

The Epistemological Quest

Motile flagella of genetically variated spermatozoa toil against a teleogenetic tide, as parsimonious reserves snug tautly within the ovum. Gametes thus fuse, forming a zygote for what would be propitious embryological growth. As is notable of our sexually reproducing plants of the East Asian monoecious cherry blossom genus, *prunus*, male gametes; pollen grains, on approaching the season, are seasonally pollinated as hymenopteran bees in such teleogenesis mount the stigma, releasing to the ovule gametes ready for embryological development of the later fruit bearing seed. But moving back to our kingdom, it is viviparity that adheres to a productive *evolutionarily stable strategy*, for the *domestic bliss strategy* that Dawkins noted so ingeniously subsequently ensues for viviparous species. Such a strategy sees a female *inducing* male behaviours that will thus safeguard life's onerous embryo. A *coy* female forestalls copulation in what primatologists call an enduring *prospective phase* for the male. She is thus preoccupied with the *fast* female, who allows for an *acceptive phase* with little circumspection, destabilising a market for a want of assuaging of the deleterious *philanderer*:

'For simplicity I have talked as though a male were either purely honest or thoroughly deceitful. In reality it is more probable that all males, indeed all individuals, are a little bit deceitful, in that they are programmed to take advantage of opportunities to exploit their mates. Natural selection, by sharpening up the ability of each partner to detect dishonesty in the other, has kept largescale deceit down to a fairly low level. Males have more to gain from dishonesty than females, and we must expect that, even in those species where males show considerable parental altruism, they will usually tend to do a bit less work than the females, and be a bit more ready to abscond.[75] In birds and mammals this is certainly normally the case' (Dawkins, 1976, p. 202).[76]

Computational algorithms that play within evolutionary dichotomies of both passion and disgust are oscillated, as 'Natural Selection evolved passion and disgust as quick algorithms for evaluating reproduction odds' (Harari, 2017, p. 100).[77] However, memory in such complex species will hardly dispel of sexually neuroanatomical dimorphisms, just as an inquisitor will successfully question the endeavour of a poor young bushbuck in her unfortunate predicament. As she browses the woodlands of the savannah, she is all but one of thirty-five, when the group are suddenly approached by a lion. The lion leaps in pursuit of a bushbuck, in which the inquisitor is then effaced with two questions of significant outcomes:

1. Does the calf inhere the danger of the lion, and thus flees?
2. The calf had learned the dangers of the lion, and thus flees?

A nature of heuristics and biological predisposition for the calf most certainly ensues. An endocrinology of glucocorticoid secretion sees thoracic and lumbar regions of a sympathetic nervous system steadily release a flow of both epinephrine and norepinephrine to the fleeing vertebrate, one that sees a classic fight or flight response.[78] Over the vast animal kingdom, natural selection favours the eye for all its environmental utility. A cup-structured eye carrying light-sensitive cells evolves in gradients that slowly curve as a retina assimilates light from its multiple directions. The light-gathering lens thus focuses. Its data now focused with a cup-structured form is grounded in interminable cycles. Such RNA cycles prove profitable to the allele in both the maritime and terrestrial animal as they both hunts their food and seek a more than suitable mate. Across a vertebral subphylum, and even of the extraneous arthropod, notwithstanding its dearth of locomotion and cellular motility, evolves photoreceptive rod cells and cone cells that transfer light to the brain. Such rod cells and cone cells perform against the force of the electromagnetic field in which the organism is propelled into its intraspecific plight. The calf calibrates its *time* to move from A to B against the ever more dominant calibration of a lion's maturity, thus calibrating *space* within a timeframe to successfully abscond as sensorimotor reflexes viciously project signals to the soma. A neurological catalyst; that of long-term potentiation (LTP) in our species, excites neurotransmitters bound to postsynaptic receptors of the ever expectant dendritic spines of neuron B, or its myriad neuron Bs. Neural synaptic circuits thus strengthen as non NDMA inboxes flood with chemical correspondence, and 'smidgens' of calcium agglomerate in channels germane to a further explosion of excitation, activating a then worthy NDMA neurotransmitter. A deluge of calcium impels a glutamate receptor to multiply, and thus engenders a sensitivity representative of a pre-and-post-synaptic fortifying event, that which the neurologist will call the function of the hippocampus, and what the laity may call the art of *remembering* (Nicoll and Roche, 2013; Sapolsky, 2017).[79]

~

For the South African paleoanthropologist, Phillip Tobias, the once outspoken activist of the apartheid, saw four departures of character complexes of the early Pleistocene: the brain, the tooth, the hand, and the foot. Between 643 and 723 cc, a cranial volume far below any other species of *homo* but larger than the *australopithecine*, allowed for ventured savannahs and subtropical dry forests of a Tanzanian terrestrial biome, otherwise induced from Olduvai depositories in which tools controlled for such an ever inimitable fate, *Homo Faber*. Oldowan lithics encompass widespread stones that flake and chip sediments, lithic reduction that cracked open nuts and bones in which woodwork would hollow and slice hides allowing for camouflaged shelters. An Acheulian industry, principally of the *Homo erectus*, saw stone technology dexterously crack, slice, chip, flake, and pound while ameliorated digging, catching, and butchering of the unlucky mammal or fish soon quickly developed (Corbey, 2012). Cycles akin to such timely ice ages, such as that of the late Pleistocene 115 to 10 thousand years ago, are of Milankovitch cycles,[80] which, in a trajectory of an orbit, see eccentricity governed by gravitational forces in the highest of our system's masses, Jupiter and Saturn, revamped in cycles of every 100,000 years. Apsides of perihelion and aphelion in windows of 72 to 120 hours move in both prompt and tardy permutations per annum. The earth's tilt, obliquity, tacitly 23.5°, in which a greater tilt calls for extreme conditions in seasonal fluctuations,[81] render the higher intake of radiation in summers, and conversely, a lower that renders airs cooler in winters. Ice sheets reflect heat otherwise more steadily absorbed by waters keeping sea levels hundreds of feet high, in which obliquity occurs in cycles of every forty-one thousand years. Distal and proximal stars control for relative astronomical forces of our orbiting satellite and ever dependent star, thus contributing to the ever increasing seasonal vagaries in an axial precession lapsing every twenty-six thousand years (Buis, 2020).

Organs of Milankovitch produce, like that of the insular cortex in the mammal, is activated by a mistakenly rancid food source. As well as for a sexually selective telos, in so far as a selected assortment of the environment ensues, an insular cortex activates on those of subjective distaste (in as much as subjectivity quickly dissipates). Decisions found on emotions the VMPFC (ventral medial prefrontal cortex) harnesses[82] while a sceptically logic DLPFC (dorsal lateral prefrontal cortex) yields to pleasantries thus evocative of his most ardent sister (Sapolsky, 2017).[83] While isogamy slowly evolves into orders, anisogamous dimorphism sees pair-bonded species unwittingly predispose its behaviours:

Pair-bonded	Tournament
Extensive male parental behaviour	Minimal male parental behaviour
Male mating pickiness	Lower male mating pickiness
Lower variability in male reproductive success	Higher variability in male reproductive success
Smaller testes size	Larger size in testes
Lower sperm count	Higher sperm count
Lower male intrasexual aggression	Higher levels of male intrasexual aggression
Lower sexual dimorphism in weight	Higher sexual dimorphism in body weight
Lower sexual dimorphism in physiology	Higher sexual dimorphism in physiology
Lower sexual dimorphism in coloration	Higher sexual dimorphism in coloration
Lower sexual dimorphism in lifespan	Higher sexual dimorphism in life span

Figure 1.1

Parental responsibilities and increasing rates of cuckoldry thus levy on both the male and female, while such behaviours in tournament species are noted.

A telos for profitable genes are sought while cuckoldry precipitately plummets. The marmoset, tamarin, and owl monkey comprise those of the pair-bonding ethos, while the baboon, mandrill, rhesus, vervet, and chimpanzee comprise a battling tournament species. The swan, jackal, beaver, and prairie vole bond in ardent affinities, while the gazelle, lion, sheep, peacock, and elephant seal are reduced to a lifetime of skirmishes. (Sapolsky, 2017).

Of a tournament kind, the male vervet, *Chlorocebus pygerythrus*, in cost-efficient 'Red-White and Blue displays' (RWB),[84] delegates for an oncoming *coup d'état*. Levels of blue scrota signal stress in the belligerent as delegates level through loci in spectra. Swift calibrations of the propitious of moments sees meretriciously devised strategies played (Young et al., 2020). Will he come, go, stay, attack, retreat, or flex? Apropos of a risk of lost genes, an ESS is now present in the vervet compelled to take part in significant games.

From von Neumann economics to a Crafoord prize in game theory, the British biologist, John Maynard Smith, saw a practical matrix of such Hawk and Dove strategies played (Michod, 2005).[85] Animals bite, scratch, kick, retreat, and flex on what will be successful escalations of games. Hawks of a pertinacious nature genetically intersperse while the Dove may endure a gratuitous escalation of sorts:

The Epistemological Quest

The Payoff Matrix

	Hawk	Dove		Hawk	Dove
Hawk	V/2 − C/2	V		−1	2
Dove	0	V/2		0	1
			V=2 C=4		

From left to right: Figure 1.2 (algebra) and 1.2 (arithmetic); Population Model = ESS; Units:
* V = The value of what two animals are fighting over; the ***resource***.
* -C = The cost of injury for the animal playing the game.

Game 1
I am Hawk, and you are Dove

I am the Hawk and thus escalate. As you are the Dove, you flee, evading a cost of injury (-C). I consequently take the whole resource (V); a territory, an item of food, or a female.

Game 2
I am Dove, and you are Hawk

I flee to avoid injury (-C) and concede the resource (V), thus leave with nothing.

Game 3
Both are Hawk

One gets injured (-C), one gets the resource (V). We face the risk of injury half the time or successfully acquire the resource half the time, averaging itself out for both playing the game.

Game 4
Both are Dove

Neither bears the cost of injury (-C) and we both reach concessions in sharing the resource (V). (Maynard Smith, 1982)[86]

For a battling vervet, action is founded on a completion of such information:

"Thus suppose two people, *A* and *B*, are debating how a sum of £20 should be divided between them. Each might start by proposing that he receive £19 and his opponent £1, but the other would not agree. If there is no time limit, and no cost to continual negotiation, there is no reason why either should alter his proposal, and hence no way in which the argument can be settled. Breakdown occurs if neither player is prepared to alter his bid. If the result of a breakdown is that neither gets a share of the £20, or that both are involved in an expensive escalated contest, there is a reason for not being too intransigent." (Maynard Smith, 1982, p. 152).

And founded on incomplete information:

"If *A* and *B* are arguing about £20, and both know the sum at stake, this is a game of complete information. Compare this with the following imaginary example of wage bargaining. The management would prefer to give no rise at all, but would pay 10% rather than face a strike. The union would like as big a rise as possible, but would be willing to settle for 5% rather than strike. Clearly, a settlement would be welcomed by both sides at some point between 5 and 10%. The union, however, does not know that the management will go to 10%, and the management does not know that the union would settle for 5%. Further, it would not pay the union to announce right away that it would settle for 5%, because if it did, that is all it would get, and it is hoping for more. This, then, is a game of incomplete information; each side knows something that the other does not" (Maynard Smith, 1982, p. 152).

As Dominey (1984) demonstrates, of alternative mating tactics and evolutionarily stable strategies, one finds nuanced definitions of strategic and tactical significance...

Strategy
'A set of rules stipulating which alternative behavioral pattern, of several stated options, will be adopted (or with what probability) in any situation throughout life. With respect to the stated options, each individual must have one and only one strategy, and different strategies must represent differences in genotype'.

Tactic
'One of several stated behavioral options (phenotypes)'.

ESS (Evolutionarily Stable Strategy)
'A strategy such that if a critical proportion of the population adopt the ESS then no different strategy can produce, on average, a higher fitness'.

Mixed (Stochastic) ESS
'A strategy in which the tactics are stochastically assigned. Either individuals adopt several tactics probabilistically, e.g., '"guard" with probability p, 'sneak' with probability q," or individuals are randomly assigned permanently adopted tactics ("guard" or "sneak") with probabilities p and q. A strategy specifying a mixture of tactics is not automatically a "mixed strategy"'.

Pure Strategy
'A strategy which contains no probabilistic statement. A pure strategy is not necessarily a strategy in which only one tactic is expressed. For example, "in situation A 'guard'; in situation B 'sneak'" is a pure strategy, not a stochastic "mixed" strategy'.

Conditional Strategy
'A pure strategy specifying two or more condition-dependent tactics' (pp. 385-386).

Ethology observes strategic intermittence as variables expound in *asymmetries*.[87] The ornamented male, in such adorned iridescence, is only produce of female evolutionary scepticism, in which nature delegates through Fisherian selection the true secondary characteristics we refer to as sexual dimorphisms.[88] Of the two occurring outcomes for a runaway selection, stable equilibria see its long spans of evolutionary *tactics* level itself out from the once precariousness of its strategy. She is thus compelled to cut her expenses sure of endured predation as the male serves for haphazard arrays of a costly invasion. Manifest in sexual selection, a second outcome destabilises, or sees a definitively 'semi-stable exaggeration of preference for a male ornament, followed by a slow decline in preference due to the cost of choice' (Pomiankowski and Iwasa, 1998).[89]

Then of the great instances of exaggerated dimorphisms we look no further than to our birds-of-paradise family of the Southeast Asian tropics. The twelve-wired bird-of-paradise, *Seleucidis melanoleucus*, with his twelve wire-like filaments, strokes a face and hindquarter of his female conspecific. In the courting displays of a Fisherian selection motif she thus calibrates her chances of long-filament acquisition against an ignominious cost of failing to suitably reproduce.[90] Be that as it may, in so far as he flaunts in such ostentatious displays,[91] she thus assimilates such 'signals' as his handicaps are affluently *borne* in a suitably adapted environment (Zahavi, 1999; Zahavi, 2007).[92]

A heron incubates necessitous nestlings as he displays such costs as *affordable*, while the babbler in his custodian servitude will likewise display what is abounded in fitness. In such expensive displays, the gull, the heron, or babbler must portray such superfluity in fitness, as selection for survival finds itself somewhat ensconced between Darwinian and Wallaconian beauty and survival. In so far as fitness inclines itself to the dangers of life's most menacing fauna, a Hamilton-Zuk hypothesis concerns itself with selection and fitness against the smallest instances of life, that which may also be referred to as life's coevolving parasites (Hamliton and Zuk, 1982; Smyth, 1995).[93]

A Commentary

As we continue to deliberate whether twenty-five dollars is a worthy investment for the male (with a possible supplementary yoghurt on the way home if lucky), it may be useful to taxonomically retrograde from the vertebrate classes and briefly touch on both the phenotypes and large sex cell of an animal class of distinct phyla. Natural selection sees sexes as profitable for multicellular organisms that may isolate its data through the germline of a single sex (for instance levels of mitochondria in females) (Dawkins, 2016).[94] But what more should be said of a sexual selection that may guide us through the currents of an epistemological quest?

Of arthropods we find the insect class, *insecta*, evolving approximately three hundred and ninety-six million years ago germane to the respectively oldest Devonian fossil, *Rhyniognatha hirsti* (Haug and Haug, 2017). A class of organisms Charles Darwin found of particular interest ultimately compelled him to further suspicions of sexes that selected for fitness. For sociobiologists, insects may be dubbed significant for the following reasons: They...

- predate mammals by one hundred megannums or more.
- are products of the late Palaeozoic-Cambrian explosion.
- had survived both the Permian-Triassic and Cretaceous-Paleogene extinction events.[95]
- are eusocial instances of supporting kin selection and supergene theories.

Of the earliest disputed Devonian insect fossils, we find:
1. *Rhyniella praecursor* – A 400-million-year-old hexapod fossil of the entognathan springtails (albeit no longer considered insects).
2. *Rhyniognatha hirsti* – Both a possible myriapod and flying insect with fragmentary parts of its mandibles and head preserved. The species is now used to calibrate molecular clocks.
3. *Eopteran devonicum/Eopteridium striatum* – Originally thought to be insect wings, but later research suggests possible crustaceans that may be closely related to the mantis shrimp.
4. *Archaeognathans* – Known as the jumping bristletails, fragments were originally presumed to be comparable to the *Rhyniella praecursor* as the earliest evidence of insect fossils.
5. Arthropod cuticles of the 380-million-year-old Gilboa Fossil Forest (New York) – such remains are relatives of Chilopoda, Acari, Ricinulei, and Trigonotarbida, while of presumed archaeognathan and zygentoman affinities.
6. *Devonhexapodus bocksbergensis* – a possible early representative of the lineage towards insects.
7. *Leverhulmia mariae* – Originally thought to be a myriapod, but later speculatively interpreted as a possible archaeognathan or zygentoman insect.
8. *Strudiella devonica* – a latest addition to the presumed possible insect or non-insect Devonian arthropod (Haug and Haug, 2017).

Like many of our Holocene insects, lepidopteran species of butterflies and moths undergo the evolutionary spectacle of metamorphic change. Its ovular stages move from egg, larva, pupa, then matured and often winged imago.[96] Such species display proclivities in behavioural courtship as they battle, tick, vary in colour, and are observed to engage mimicry. Orthopteran males, which include species of crickets and grasshoppers, are unparalleled for their worthy saltatorial jump, stridulation, and musical abilities. Dipteran males include species of flies, mosquitoes, and gnats. Dimorphism is such that peculiar hairs on antennae is observed, while branched and palmated horns thus allow for humming abilities for a lengthy courtship with females. Of remarkable hymenopterans we see bees, ants, and wasps. Like that of their orthopteran cousins, they also display a flamboyant repertoire for stridulation, thus making vociferously distinguishing calls.[97] Coleopteran species include the varying beetles with the fine-polished appearance. They are thus displayed with varying colours and sizes of both a chest-like thorax and horns, while display such varying phenotypes. Hemipterans, including aphid, bed bug, and planthopper species, show the also remarkable stridulating ability to female conspecifics. Hemipterans may even take turns in musical contests and vary in songs and attraction-levels of voice. Neuropterans, which include both the mantis-fly and antlion species, are observed to display abdominal spots and varied colours, while have distinctly netted wings in sexual dimorphic morphologies. They hold distinguishing phenotypes from as early as its pupal state (Darwin, 2017).

But it seems, in as far as cladistics may show for insects that evolve in its distinct trajectory of a vertebrate phylum, natural selection favoured the sessile ovum for profitably successive replication:

'A somatic cell with a diameter of 10–20 μm typically takes about 24 hours to double its mass in preparation for cell division. At this rate of biosynthesis, such a cell would take a very long time to reach the thousand-fold greater mass of a mammalian egg with a diameter of 100 μm. It would take even longer to reach the million-fold greater mass of an insect egg with a diameter of 1000 μm. Yet some insects live only a few days and manage to produce eggs with diameters even greater than 1000 μm. It is clear that eggs must have special mechanisms for achieving their large size' (Alberts et al., 2002).

Of distinguishing phenotypes, which Darwin observably collected in writings, gear towards a compelling theory of sexual selection (for both the Wallace and Darwin dichotomy). So of the first Darwinian phenotypes in 1871 to a *Fisherian runaway*, *Hamilton-Zuk hypothesis*, *evolutionarily stable* and *domestic bliss strategy*, and then both the *extended phenotype* and a *handicap principle* in subsequent lustrums of the 80s, we may be momentarily consoled with a brief timeline of a sexual selection theory.[98]

~

Observed through secondary sexual characteristics of the lower classes of our kingdom, is what was discerned, sifted, and classified in subsequent generations of a more detailed entomology. As his early work made note of in the *Origin of Species*, in a carefully crafted theory of natural selection, was of the complex systems of hymenopteran eusociality. And so it serves us to deliberate on how eusocial sterility in working castes of insects evolve.

For Darwin, without the tools of what would have been Mendelian inheritance, the nature of alleles in sterile individuals was certainly difficult to ascertain. Both the performance of dominant and recessive alleles, and in how an individual organism could be so different from progenitors was amiss. Neuter insects, or sterile insects (such as in the working ant), greatly differ from parental progenitors while also differ from their progeny, thus becoming of distinct castes. Differences are such that it could have been easily elided by both a formatively linear theory of natural selection or Lamarckian heritability, as characteristics are as distinct as two species of the same genera, and even to the lengths of two genera of the same family. Distinct morphologies evolved under natural selection as morphologies had proved profitable to the parent, as castes engendered range in workers phenotypically advantageous for the colony, and hence had the impetus to evolve under the respective natural laws.

In the distinguishing roles of a hymenopteran colony, workers may guard and cut fruit, others tend fungus gardens, while others transport cut leaf fragments from one area to another. Over the successive generations of a supergene theory, distinct characteristics for size, body structure, jaw structure, and phenotypic behaviours are duly selected for as it proves profitable for the community at large, while compelling adaption to its respective environments.

Interestingly enough, the yet curiously crafted wax cells in the honeybee's honeycomb had troubled entomologists until the very final year of the 20th century. So much so that geometers were summoned to shed light on a honeybee's proclivity to form them, as we learn hexagons are the best way to divide a surface into equal areas under the 'honeycomb conjecture', calling for what are still hitherto remarkable phenotypic displays (Morgan, 1999).[99]

Of Waters

Of phenotypic displays we attribute to natural selection an adaptational *telos*, as *a priori* and *a posteriori* judgement structure all phenomena. Before pondering on what would be left of libertarian freewill, we must anchor at the significance of planetary waters. As briefly perused earlier, we know adaptation for living populations experience distal cycles of eccentricity, obliquity, and axial precession. We see how subtle changes in stellar distances affect lapses of *deep time* for the perceiving species on earth, while all types of evolutionary performance must find practicable ways of subsisting. A period of Devonian extinction events saw a struggle for existence in genera and families, which, subsequently led to vertebrate classes that would go on to survive upcoming extinction events.

Namibian fossils of *Otavia antiqua*, in a rifting supercontinent as long as 760 million years ago, found sponge-like fossils that first emerged and acquire its energy independent of the sun. As the lithosphere would rift across the Tonian period and stretch across a range of over 3 billion years, planetary history saw life's energic forms first hitherto flee from a first-hand source of the system's solar nest (Brain et al., 2012). Bilaterians would evolve 130 million years later that would see to subtle changes in the Australian Ediacaran, while a Cambrian explosion gifts the animal kingdom the finest of armaments as sea levels rose 89 million years later. Such weaponry proving essential for the next half-a-billion years, calcium, phosphate, and other significant chemicals thus dissolved into alkaline seas, and animals hunted and defended with teeth, claws, armour, spicules, spines, shells, and skeletons.[100]

Compelled to capitalise on terrestrial earth and its atmospheres as they would spend more and more time away from the seas, our amphibious ancestors were also coerced to ensure the germinal subsistence of the zygote. As a zygote proceeded to its worthwhile embryo, a blueprint for cellular germ line division developed. Those ancestral tetrapods would then give birth to the significant clade of our first sexually reproducing terrestrial amniotes, which, would then branch off into distinguishing clades of our genealogically sauropsid-synapsid pastime.[101]

Slowly achieving independence from the oceans and seas, and watertight skin cells thus gradually produced, cyanobacteria became a proximal cause in oxygenous fuel for both further and farther independence. The amniotic sac thus enriched itself with nutrients as contents of yolk became of vast significance.[102] As David Attenborough wrote best:

'Later, microscopic organisms called cyanobacteria began to photosynthesise, using energy from the Sun's rays to build their tissues. The exhaust gas of the process – oxygen – caused a revolution. It became the standard fuel for a much more efficient way of extracting energy from food, and so paved the way for the establishment of all complex life. Cyanobacteria still constitute a significant part of the phytoplankton that floats today in the upper levels of the ocean, You and I, and all the animals with

A Chip Away at Freewill; a Continued Critique on the *Social Constructionism* Fallacy (Formally Known as the *Divine Fallacy*)

While Milankovitch cycles may be compelling evidence of how earlier species withstood geophysical aberrations in *deep time*, an amniotic sac and its fluids made the subtlest of changes for the parity of the ever-expensive zygote. Questions must be fairly invoked to deliberate spaces between both genotype and phenotype (epigenetics), embryological development (ontogeny), and the environment. Irrespective of the thinker's philosophical locus between determinism and environmental factors thereof, it is still incumbent upon the thinker to truly fathom how important the phenotypic behaviour of our ancestors may have precluded our species from extinction in such a large and comprehensively intricate gene pool. Could behaviour be solely attributable to the by-product of consciousness? Or should it be ascribed to free will? Which thus leads to the next question. Can billions of neural cells that work in concert, making rational decisions in a possible DLPFC (dorsal lateral prefrontal cortex) against the billions of neurons that work in concert to make more emotionally centred decisions in a possible VMPFC (ventral medial prefrontal cortex), be imputed to natural selection in game theories? Or is it easily imputed to the by-product of consciousness that makes us anomalous to all the millions of other species that share the same planet? Altruism tends to be the recursive objection to zoological 'reductionism' and is usually paired with humanitarianism to portray how we differ as a species/genus to the rest of the animal kingdom. Meanwhile, it may be slightly more impartial to express that this objection lies within the scarcity of *knowing* exactly how other species within the same genus, *homo*, had behaved, as all other human species are, quite evidently, extinct.

However, when we fairly deliberate on the phenotypic behaviour of honeybees in producing the wonderfully hexagonal wax cells of the well-crafted honeycomb, do we truly believe honeybees logically calibrated areas and surfaces, or would we express that it is the instinct of the bee? In the same way, and also as an evolved social species that worked in concert over millions of years to espouse the benefits of sociality, in order to compensate for its relatively slow locomotion, bland teeth, and weaker prehensility (i.e. compared to even other mammals within the same class), did cellular structures compel themselves to opt for intraspecific emotional response? Is our sociality hidden behind traits within phenotypes, or is it simply a product of consciousness? Or is it a concoction of both? Is it plausible to face our bestial proclivities or model an intermediary species between our kind and the distant *pan* genus. But let's suppose for a moment that the common chimpanzee didn't exist, or that other humans were extant, would we still confer upon us that 'all human beings are born free and equal in dignity and rights that are endowed with reason and conscience and should act towards one another in a spirit of brotherhood' (Roosevelt, 2001)? Did not Jane Goodall unveil what archetypically stood *In the Shadow of Man* when pushed to think of what may lurk in the shadows of our ancestors (Goodall, 2010)? While we come back to the swinging pendulum between Darwin and Wallace and the theory of sexual selection, let's look at how the evolution of language may meddle between natural selection, sexual selection, and the theory of games.

The Epistemological Quest

A Working Telos of the Vocal Tract

Essential to *agree upon* in conferment of the vocal tract is that it thus coevolved with the brain. This should seem rather obvious. But, if it can at first be admitted via clear concessions, the vocal tract must also be adduced to the very same *telos* of those preliminary carbon-based chemicals evolving all but three billion years ago. Simply, if the gene or trait is profitable enough, it thus works its way into the gene pool. That's to say constituent organs won't be mutually exclusive in its rudimentary form. The concession would mean that a teleology of the gene *finding ways to replicate itself* via progressive mutations over long periods of *deep time* cannot be entirely elided, most especially when we think about the anatomy of the vocal tract. Given the colossal profit language has had over all other species, we know the working of language divides itself into grammar, phonology, semantics, and pragmatics.

In order to think about how the vocal tract may be a product of natural selection (again akin to the dictum in which profitable characteristics increase an organism's chances of survival and possible reproduction in a population) it's useful to first look at how a tract physiologically comes to produce something as complex as speech. The necessary tools the numerous amniotes came to both utilise and weaponize effectually, was a *Goldilocks* balance of oxygen, O_2, and carbon dioxide, CO_2, both essential atoms and molecules for life on earth. Against a congenial number of nitrogen and other such chemicals in an atmosphere, we find those precise molecules to be useful. Martin A. Nowak and David C. Krakauer suggest the evolution of language more than likely evolved under the ESS blueprint of game theory:

'Hence, we assume that both speaker and listener receive a reward for mutual understanding. If for example only the listener receives a benefit, then the evolution of language requires cooperation. In each round of the game, every individual communicates with every other individual, and the accumulated payoffs are summed up. The total payoff for each player represents the ability of this player to communicate information with other individuals of the community. Following the central assumption of evolutionary game theory (38), the payoff from the game is interpreted as fitness: individuals with a higher payoff have a higher survival chance and leave more offspring who learn the language of their parents by sampling their responses to individual objects' (Nowak and Krakauer, 1999, p. 8028).

As the fittest were naturally selected, rudimentary organs would soon evolve into the functions that the epiglottis and both ventricular and vocal folds would play out. The palate, cavities, tongue, windpipe (trachea), and oesophagus are thus adapted to chemicals emanated in a terrestrial troposphere. Before an utterance is made, both the lungs and the vocal tract precisely calibrate needed levels of breath. And so in order to increase the volume of lungs, a diaphragm would first need to find ways to contract.[103] Airflow rushes into the lobes (via bronchi and bronchiolar tubes) and into an alveolar sac (thus exchange the gasses that equalise the pressure offsetting for the lower numbers in the other now necessitous lobe) (Chaudhry and Bordoni, 2021). As a breath is singly released, a diaphragm relaxes, reducing the volume but yet increasing pressure in both of the lobes. As air pushes its way to the tract, air then needs to pass an array of what we refer to as speech-producing organs.

Now through and across a glottis and vocal folds, air passes both the epiglottis and pharynx. The uvula flaps as air passes through to the hollow oral and nasal cavities. Meanwhile, a soft palate (the velum) allows for a steady flow of the liquids and gasses. If the velum is down, opening a steady flow of air into a nasal cavity, air then travels through a cavity and exits the nose. That notwithstanding, whether the velum is up or down, air will then flow through the oral cavity and out of the mouth. Akin to the anatomy of the vocal tract, the hard palate (connected to the velum), likewise opens and closes for air to travel through a nasal cavity. In front of the hard palate and behind the teeth (again separating both cavities) we find the alveolar ridge. So the tongue (the essential specimen of linguistic evolution), can, for this instance, be divided into four sections. Moving away from the farthest section of the pharynx and into the cavity towards the mouth, of the sections of the tongue we find:

- The root, (the farthest away from the mouth)
- The body
- The blade
- The tip of the tongue

Thus we find the oral tract to encompass both upper and lower teeth and the upper and lower lip.

So it may not be so costly to dwell on the question: What actually happens when one chooses to speak? And thus what would be the foundational *telos* behind a simple act of an utterance? In terms of speaking, our organs evolved for a purpose: of manipulating sounds to be intraspecifically profitable for our species to communicate both effectively and quickly. So can speech be ascribed to the by-product of consciousness or is it simply a corollary of the evolutionary features for humanitarian altruism? We can say that as air passes back up out of the lungs and into the trachea (windpipe) and larynx (voice box),[104] we *de facto* necessitate the question on what our essential organs may be for producing speech from the neuron-bound organ, the brain, as we pave our way into the junction between anatomy and cognitive science. As Leiberman precisely explains:

'The key biological mechanisms that are necessary for human speech are (1) the supralaryngeal vocal tract and 'matching' neural mechanisms that (2) govern the complex articulatory maneuvers that underly speech, and (3) decode the acoustic cues for linguistic information in the speech signal' (Lieberman, 1986, p. 702).

Acoustic cues in archaic structures of communication manifest themselves through both manifold and complicated means. They tie with a physiological language that at previous times could have reasonably been a matter of a survivorship machine's demise (while hold the compelling argument to be profitable in the environment). As Steven Pinker's *The Language Instinct* detailed, the structures of grammar are latent before a language is learned, thus stressing that the earliest forms of the organs of the vocal tract would pass on to the next generation given its use of helping the organism to survive (Pinker, 1994).[105]

Picture a hunter gather of a band somewhat compelled to communicate an animal that weighs over sixty kilograms was approaching, or a rival band surreptitiously undergoes a local usurpation in what could appear as a quasi-devised plan. Thus we see language in its rudimentary form.

So back to sexual selection. The question on how much of neural linguistic areas mediate between reproduction and survival still circles the debacle between a Darwin-Wallace dichotomy, as Lieberman states, 'I propose that these neural mechanisms evolved by means of the Darwinian process of preadaptation from the part of the human brain that first evolved to facilitate the production of speech' (Lieberman, 1986, p. 702).

The follow-up on Darwin's claim that language came in instinctual states was fruitfully elucidated in Steven Pinker's *The language Instinct*, 1994 (after Noam Chomsky). As Pinker explained, the Broca's area of our brain sits on the left side of the frontal lobe and is responsible for the intricate craft of speech production. Conversely, the Wernicke's area of the brain, sitting on the left side of the brain in the temporal lobe, is responsible for the intricate craft of speech comprehension. Surprisingly, studies in which damage to either parts of the brain, namely, Broca's or Wernicke's aphasia, produce remarkable effects on speech, conveying difficulties in the coherence of speech when the Broca's area is damaged, while conversely showing no difficulty in comprehension. Inability is displayed in speech-comprehension when Wernicke's areas are damaged, meanwhile show no real signs of difficulty in speech production (with often vast circumlocution amidst subjects utterly irrelevant with the subject in conversation) (Pinker, 1994).

Interestingly enough, and honing language down to an epistemological perspective, what rises to the surface is why and how both the production and comprehension of speech could be profitable for the species in Broca's and Wernicke's areas. To the extent of espousing something as wondersome as an instinct for language, in which organisms wilfully avail themselves of the ability to speak to an extreme of garrulously expressing vast arrays of ideas exclusive to perimeters of the same species. As we may observe with our hymenopteran species (specifically bees and wasps), it begs the question: why would humming and buzzing be insufficient for the perimeters of our intraspecific communication? Or as we observed with our orthopterans (such as crickets and locusts): why wouldn't stridulation be enough for communication? Or why wouldn't the acoustic cues of quacking suffice? I'm sure a waterfowl would get the general gist of what needs to be communicated throughout intrafamilial signalling.[106] Instead, we are a species found to compartmentalise language into nouns, verbs, adjectives, adverbs, prepositions, consonants, vowels, diphthongs, glottal stops, and a sea of more curious features of language, while the young yet share an intraspecific proclivity to speak more than one rather effortlessly. The chasm in the genealogical tree between *Homo sapiens* and the not-so-distant genus *Pan* (which scientists and historians still ponder on in terms of speciation undergone over the last thirty-three thousand years since the extinction of the Neanderthals), endure the fragmentary elision of intermediary species that would have allowed for more sophisticated means to understand Broca's and Wernicke's areas.[107] Intermediary data would allow for scientists to observe how those areas of the brain precisely evolve, which is what I thus believe does lead to significant questions of the human psyche, and thus what I call the subtleties of intersexual influence. As spiders make webs, beavers build dams, monkeys swing, birds fly, and fish swim, humans display an unparalleled instinct to produce language across what could be observed as a product of the possible *Great Leap Forward*.

As long ago as the Mesozoic era (between a temporal range of two hundred and fifty-two to sixty-six million years ago) that ends at a K-T boundary, research in paleoecology and paleobiology endeavoured to precisely map out morphological differences in our distant dinosaurs that may show for the pressure of sexual selection:

'As a working hypothesis, we assume that those individuals of a particular dinosaur species showing the most conspicuously developed potential display structures are males, based on what is commonly seen in living animals. Possible dimorphism of this kind has been inferred for the horns and frills of ceratopsians, the thickened and ornamented skull domes of pachycephalosaurs, the cranial crests of lambeosaurine hadrosaurs, and more subtly in the cranial rugosities of therapods and the body armor of ankylosaurs (Farlow et al., 1995, p. 453).

The Epistemological Quest

Since the coinage of the term 'dinosaur' by the Lancastrian, Sir Richard Owen (Wessels and Taylor, 2017), this all instantiates upon the notion that vertebral reptiles, (i.e., tyrannosaurs, triceratops, velociraptors, stegosaurs, spinosauruses, archeopteryxes, brachiosauruses, allosauruses, apatosauruses, dilophosauruses, cryolophosauruses), just to name a few, are animals that ventured the same planet and did not run afoul of the very same laws of natural selection endured today, not being extraneous to the domain of consensual belief. The importance of understanding those sexually dimorphic morphologies, even of life a deep time ago, lies in allowing oneself to deliberate on how much asymmetries compensate for organisms (*the survivorship machine*), to subsist in a given population against adaptational laws. For instance, as we looked at earlier, *Homo sapiens* as Synapsid mammals endure the *telos* as viviparous species. Known well to academics found to dabble in paleobiology, is that almost every dinosaur is oviparous as birth transpires extraneous to the female herself. For ethologists that study extant oviparous species, it subsequently allows for researchers to reach a more founded conclusion on such game theories. Between a possible evolutionary game that's played by a mammal for over three hundred million years, such a game would be played across dimorphism, viviparity, oviparity, gametes, and polygamy, as even odour may play in sexual selection within mammalian behaviour (Blaustein, 1981).[108]

A Freudian, Jungian or Piagetian Perspective?

In 1899, Sigmund Freud and his newfound psychoanalysis interspersed within the scientific community in what became published as *The Interpretation of Dreams*. Termed 'founder' of psychoanalysis, Freud intrepidly suggests ideas within the dream endure *wish-fulfilment*, and thus a wilful substrate in what Freud defines '*cathexis*', battles against three nemeses of the psyche the organism finds itself few and far between. The **id**, defined as the instinctual embodiment of the mind, encompasses both aggressive and sexually, or an impetuously instinctual perception of the world. A **super**-ego (and yet an *id's* worst nemesis) is an embodiment of a moral conscience of the vitriol yet instinctual impetuosity of the id. The **ego** plays intermediary between both the vitriol of the id and stringency of the super-ego, which constantly mediates between both powers and allows for reconciliatory pathways between stark differences between each and the other, espousing the everlasting war in the psyche.
The clinical psychologist and public intellectual, Professor Jordan B Peterson, compares the id, ego, and superego as a system akin to 'executive, legislative, and judicial branches of a modern government' (Peterson, 2021, p. 4). Freud thus conveys a deeper transpiring process within the psyche demonstrably, which both *compress* and *condense* conceptual units within the psyche when experiencing dream-thought. As a feature of the process he believes a colossal significance dovetails, rendering '*cathecting energy*' both capable of condensation and discharge within vast dream-thought processes. Freud's 'salacious' *Libido*, placed between latent content of dream-thought processes, is demonstrative of how such libidinous is both latent and manifest through somnolent states, as he spoke profusely of his patients in clinical practice. There were systems of distinguishing conscious states which modify themselves through dream content and gave the famous instance of the legend of King Oedipus and the Oedipus Rex of Sophocles:

'Oedipus, the son of Laius, king of Thebes, and Jocasta, is exposed as a suckling, because an oracle had informed the father that his son, who was still unborn, would be his murderer. He is rescued, and grows up as a king's son at a foreign court, until,

being uncertain of his origin, he, too, consults the oracle, and is warned to avoid his native place, for he is destined to become the murderer of his father and the husband of his mother. On the road leading away from his supposed home he meets King Laius, and in a sudden quarrel strikes him dead. He comes to Thebes, where he solves the riddle of the Sphinx, who is barring the way to the city, whereupon he is elected king by the grateful Thebans, and is rewarded with the hand of Jocasta. He reigns for many years in peace and honour, and begets two sons and two daughters upon his unknown mother, until at last a plague breaks out – which causes the Thebans to consult the oracle anew. Here Sophocles' tragedy begins. The messengers bring the reply that the plague will stop as soon as the Murderer of Laius is driven from the country. But where is he?
Where shall be found,
Faint, and hard to be known, the trace of the ancient guilt?
The action of the play consists simply in the disclosure, approached step by step and artistically delayed (and comparable to the work of a psychoanalysis) that Oedipus himself is the murderer of Laius, and that he is the son of the murdered man and Jocasta. Shocked by the abominable crime which he has unwittingly committed, Oedipus blinds himself, and departs from his native city. The prophecy of the oracle has been fulfilled' (Freud, 1997, pp. 155-156).

As perverse as the Oedipus exemplar may be, and as licentiously branded by scholars alike, his work does not run so injuriously afoul on the onus of the ruminating psychoanalyst today.[109] Freud imputes symbols of somnolent states to a sex motif in clinical practice. It may yet be necessary to meditate on the essence, purpose, or teleological means for the human psyche to incorporate, or at least assimilate anything related to properties of sexual reproduction. In so far as sex dissimulates in academia, germane to its evidently practicable and sensitive nature, it should not fall too short of the *weltanschauung* intrepidly dismantled by psychoanalysis today.
Through analytical psychology Carl Jung necessitated the features of archetypes (universal motifs of the symbols and concepts of human experience) (Jung et al., 1968; Snowden, 2019). Thus, the question follows, 'what are the ontological properties of an archetype?' We look at principally tens, hundreds, thousands, millions, then billions of years of archetypal bugs and features of the human psyche that must be fairly deliberated across the vast span of 'deep time'. Thus it's only successfully achieved by making the concession that as sexually reproducing organisms, the psyche interprets *synthetically a posteriori* matter. The invocation of sexual objects in the human psyche cannot be refuted hastily. A further subsequent and yet necessary question for the ruminator follows: 'what is the feasibility of impartially ruminating on archetypal features across the long span of *deep time?*' As we looked at earlier, anisogamy and morphologies of isogametes across evolutionary history relate to the brain, all while natural selection works on populations. The evolution of the brain is somewhat nebulous when contrasted with the unknown of sexual selection in palaeolithic progenitors. Behaviour of other primates has been observed, as we perused briefly with a look at Sapolsky's both pair-bonded and tournament species. It seems to be quite clear sex games come with a rubric of different rules for other primates when females sexually select.[110] Under a Freudian perspective, the gradual process of encephalisation and the proliferation of brain cells inundate the very latent content circulating the id which many other animals unwittingly live by as a manifest survival strategy. Then rising to the surface of what appears to be the alternative of the id, essentially the replacement of the id under order and immutable despotism, is the superego, merely predicated on rationale and so playing out as a subset of the logos that helps the organism survive

in an intelligently sacrificial and intraspecific environment. Within this process of encephalisation (and also akin to the contraction of reason) inasmuch as the psyche can be perceived, the ego would have undergone the laborious task of restructuring a sexual domain, mediating between the order of a stringent superego, whilst aiding and abetting the teleology of the id, harnessed by what could be defined as the vitriolic selfishness and teleology of the gene.

Jean Piaget, the once Swiss psychologist known for his craft of unveiling the earliest of our operations, believed *episteme* lied between the spaces of *Play, Dreams and Imitation in Childhood* (Piaget, 1962). As we think about processes of the human psyche in developmental psychology, Piaget calls to the structure and restructure of epistemic spaces within early sensorimotor structure and assimilation. Such spaces he defined as *schemas*, i.e., elementary structures in what a child comes to *know* before restructuring the thought. Throughout distinguished stages of childhood development, an equilibrium takes place between restructuring processes he terms *assimilation* and *accommodation*. Assimilating activity of a child tends to incorporate external objects and *assimilate* such objects into *schemas*, while balancing those newly found and assimilated objects between what comes to then comprehensively accommodate. Accommodation, which lies between the restructure of those external objects and then a reapplication, or ontological properties, thus becomes of an external world. The well-known stages of the development of a child comprise of:

- a sensorimotor stage (0-24 months)
- a preoperational stage (2-7 years)
- a concrete operational stage (7-12 years)
- a formal operational stage (12 years+)

Play and imitation Piaget observed as analogous to structuring processes of accommodation. Such that a child restructures and applies *schemas* to an external world. And so accommodation, which harnesses schemas, has the psyche assimilating such objects and their representations. The teleological means for accommodation is for *the child to survive in the environment* eventually endured (and does so independently) upon completion of all operational stages once leaving adolescence and thus 'fleeing his nest'. Piaget gives the instance of a two-year-old child in the inception of his preoperational stage encountering a man with frizzy hair on the side of his head but yet bald at the top. The two-year-old shouts, 'clown', as the concept of a clown was prior assimilated. After the father explains to the child that he isn't wearing a costume nor is making people laugh, the child further accommodates the idea of the clown to the standard *episteme* and concept of a clown (Piaget, 1962).

In Kantian terms, such assimilation of the chair (in preoperational stages) falls into one of two intuitions. As opposed to a *pure* intuition, which would be *a priori*, such as space and time and devoid of experience, assimilation would be an *empirical* intuition founded on sensibility or experience alone. The properties of a chair that the child assimilates is that the chair has four legs. But yet tables also have four legs, so what would be added? What's added is that the chair has four legs, and you can sit on it. Meanwhile, a horse also has four legs, and one sits on a horse. All that lies between the empirical intuitions of a horse and the chair is transition into *schemas* that manifest themselves under the restructuring necessary to apply them to the external world (all within the process of what's accommodated).

As the issue of sexual selection further develops, the idea may thus be applied experimentally. If we were to render an encapsulation of archetypes into a theory of games, what may be found? Ludic activities and imitation in childhood would be rehearsals for games succeeding operational stages. Such archetypal networks would then run adrift of amalgams of genes both for the profitable phenotypic displays and their response to act out in the environment (and does so intraspecifically). As stated by Darwin of the abrasiveness we find in both intraspecific and interspecific competition:

'And we have seen in the chapter of the Struggle for Existence that it is the most closely-allied forms – varieties of the same species, and species of the same genus or of related genera – which, from having nearly the same structure, constitution, and habits, generally come into the severest competition with each other.' (Darwin, 2017, 125).

When distilling the organism down to its genome and consider the plight of intraspecific competition, it may suffice to look from the point of view of the gene. Germane to the dictum that 'the probability of a gene reappearing for the organism under dominant and recessive alleles may pass on through generations', we thus use a male to rehearse the very idea. If a male is of a certain height, shorter than the tallest male in his environment yet slightly taller than the average male, when juxtaposed with the tallest male in a dominance hierarchy, for example (or in that case within sexual selection), something interesting does occur. We know that the male inherits fifty percent of his father's genes and fifty percent of his mother's (compared to that of what the rest of the population inherits), who both inherited fifty percent of *their* father's genes and fifty percent of their mother's; the boy's grandparents. Genes equally inherited within the generational fifty percent for profitable phenotypic traits, whether they be physiological or behavioural, may be inherited, standing in as profitable for the organism within his environment. As we know the male is shorter than the taller male in this competition, how does he intend to compensate for his dearth in height (given the assumption that an approximation of the range of taller males is a feature for sexual selection)? Is this offset by his beauty, strength, intelligence? Well, akin to the four Piagetian operational stages of childhood, there is a rooted archetypal network at play. This is extended away from a physiological phenotype and plays itself out behaviourally, offsetting those asymmetries via optimal behaviours for survival both assimilated and accommodated by the child, which then becomes subsequently adduced to his ontological security post-formal operational stages of childhood.[111] In such accommodated schemas of assimilated phenomena, i.e. in the restructuring of assimilated phenomena mapped out through ludic activity and imitation thereof, extant archetypal behaviours level themselves out which allow the organism to offset given asymmetries in the manifold games played throughout sexual selection, relative to his fluctuating environments. Through genetic inheritance within a probabilistic set of extant alleles, passing themselves on for millions of years, the most optimal genes for both the female and male play themselves out within ludic and imitative processes during operational stages, thus undergoing both necessary assimilation and accommodation proved profitable in his environment, leading to a net *evolving* of our species across the gene pool.

The archetypal network rooted in the organism for both the male and the female child must be able to detect, restructure, or accommodate the invariable *schemas* which render partial subtleties towards powers of female selection within the species. Therefore, the male will assimilate and accommodate what conveys itself as the most optimal or *powerful* characteristics for the selectable qualities of a male in his environment, via an archetypal embodiment of the best, or most selectable male body throughout his operational stages (germane to the tools he has). Likewise, females archetypally embody characteristics assimilated, and subsequently accommodate *schemas* into optimal characteristics for her as the selecting body which wields the power to select, akin to a compensation of asymmetries experienced within intrasexual competition. Generally speaking, mothers and fathers are most *powerful* and present figures across operational stages, and so in order for a gene to be successfully expressed, neuronal cells must be able to restructure its phenomena in order to fruitfully encounter its means to replicate within a population. Given a congenial setting for ludic, oneiric, and imitative freedom during operational stages, those with the highest of intellectual faculties generally accommodate much better, accounting for asymmetries that can readily be applied to an ontological position in which the individual capitalises on either their power to sexually select or be rendered selectable following the formal operational stages of childhood.

So, to what extent should phenomena be sexual in the human psyche? It might be profitable for *synthetic a posteriori* schemas to discern the sexes and successfully accommodate those properties hitherto to her reaching puberty (given that our sexes are similar in so many ways). If the psyche both assimilates and accommodates the earliest of sex differences of mother, father, brother, sister, grandfather, grandmother, male, and female in both physiology and behaviour, it might be profitable for her accommodating power to then apply the newly stratified *schemas* to the external world, thus profiting her. This leads to what I truly profess to be *intraspecific* and *intersexual influence* at play. The concept of *intraspecific influence* should hopefully appear quite evident (being the free will lost conducive to such intraspecific presence). For instance, if you were to accidentally tumble and fall over when surrounded by one hundred ants, I'm sure you wouldn't feel as embarrassed as falling over in a city centre surrounded by one hundred of your own species! The ludic, oneiric and imitative schemas of childhood can then only be imputed to accommodation throughout childhood, selecting for the most optimal behaviours for the organism once superseding the formal operational stages intraspecifically, consequentially reducing free will. The same idea can be applied to the psyche for *intersexual influence*, (being predicated on free will lost by the presence of the opposite sex) evidently under heterosexually centred individuals. Intersexual activity, exaggerating for the individual around puberty, may manifest itself in ludic activities, imitation, or mild to complete withdrawal which as a corollary reduces free will for the individual (in so far as the individual may be aware of this). However, it is but impossible to be aware of the comprehensive degree of assimilated and accommodated schema through operational stages.

The Oedipus story is only indicative of early representations of how both assimilation and accommodation play out through could-be incipient stages of sexual selection, rooted in evolutionary changes over periods of time. With that being said, it is more than necessary to adumbrate the powers of sexual selection. To see exactly how they have come to blossom in manifold and relative forms in contemporary societies, thus rendering themselves observable to the astute anthropologist, is why it deserves its undivided attention.

When playing with the idea of such powers of sexual selection (sociologically speaking), while also considering the myriad endeavours of misarticulating such theories as socially constructed, misguided approaches have only shrouded what those relative powers manifest themselves to be. With the means to protect and safeguard the sociological imperative that the primary issue grounded in contemporary western democracy today is that we contain a dominance patriarchy today, it only proves prejudicial to the idea of the philosophical logos we share in reaching (or for the means of weltanschauung). It is also notwithstanding the evidence for these types of spurious premises. If you are a social constructionist who does believe that there is only ever an underlying patriarchal dominance hierarchy at play, transpiring for thousands of years, I would ask, 'to what degree have women been elided in this dominance hierarchy in our history? Do we honestly believe that women have played no part in our societies' incredulous progress? Or is it that you believe for the most part women have been oppressed?' What does the data that represents a patriarchal dominance hierarchy suggest about such a complicated empirical claim through history?[112] And so in terms of dominance, what does it mean to *have power* and how exactly does an individual or a group come to *dominate*?[113] Before I elucidate what the ontic subsets of the power of sexual selection may be, let's look at one of the first significant yet intrepid attempts to apply sexual selection to the human species.

Geoffrey Miller

In 1998, as evolutionary psychology was slowly arising, Geoffrey Miller, the evolutionary psychologist, published both an insightful and important review on how the theory of sexual selection can be applied, quite directly, on to human nature. The caveat with the time of what seemed to be an inopportune review, was that, as displayed earlier with Fisher, Hamilton, Maynard-Smith, and Zahavi, many biologists still unfortunately objected to the idea of sexual selection since Darwin's claim in 1871. As opposed to the theories of cultural relativism, evading the realm of sexual selection at play within evolutionary games and predicating itself on a pretext of our species being conscious organisms, Geoffrey miller approached sexual selection evolutionarily. Miller professes sexual selection to be grounded under working fragments of evolutionary psychology, being much more implicated within the biological faculties of human behaviour, which he explained as a manifestation of our physiology, primarily grounded in endocrine networks that effectively shape such human behaviour:

'Sexual selection does not stop when copulation begins. Indeed, gonads and genitals are the clearest expressions of sexual selection, because they are most directly responsible for fertilization, and they typically serve no survival functions. The traditional view that 'primary sexual characters' such as penises are "necessary for breeding hence are favored by natural selection" (Anderson, 1994, p.14) is misleading. If sexual competition and mate choice can affect genitals, then genitals can be shaped by sexual selection' (Miller, 1998, p. 10).

It only invokes the teleological substrate for the everlasting gene to survive in its means for immortality, and not that of the misconstrued idea that sexual selection can only be ascribed and reduced to the contours of copulation, especially for something as significant as for creating life. Miller explains how culture should not be solely imputed to individual behaviour, but, should be perceived as emerging from sexual competition within the myriad numbers of individuals pursuing the myriad numbers of mating strategies, all performed within myriad but yet distinct arenas. This means that cultural dimorphism should be ascribed to reflecting a difference in teleological motivation and sexual strategy, rather than a difference in basic mental capacity, which goes back to the more Darwinian sense of features being profitable for the species, rather than rudimentary, vestigial, or a bug, which would then run afoul of the laws of natural selection.

The 5 Ontic Subsets of the Power of Sexual Selection

Firstly, and most importantly, in separating the wheat from the chaff, I vehemently believe that Darwin's anthropological accounts of sexual selection in *The Descent of Man and Selection in Relation to Sex*, are, for the most part, tenuous. The first reason

The Epistemological Quest

I impute tenability to such poorly written accounts on the anthropological domain of sexual selection, while also exculpating the endeavour, is because of how unfathomably complicated our species is compared to all other species hitherto of him then writing in his lifetime, which would have been something scientists would have invariably come to endure. The second reason, which is unspeakably incumbent upon the theory, is that the armamentarium academics have today in such manifold populations, hitherto unprecedented due to the dearth of migration in our species and conducive to a scarce feasibility to migrate through the dearth of technological advancements, allow schools of sociology, anthropology, and psychology to give much more of a reliable account, without completely dispelling of the necessary wheat from Darwin's chaff.[114] Before pondering on the human power of sexual selection, and discerning how these ontic subsets may work as interwoven faculties within what we should call contemporary western societies (or their environments), let's firstly dissect the power of sexual selection into the five ontic subsets that come to weave themselves into one relatively working power, submerging as both latent and manifest biological advantages to reproduce for the male, compensating for asymmetries in what is possibly played out in even deeper instantiated evolutionary games.

1. Beauty

Starting with the first subset, as traditionally ontic, is the powerful subset of beauty, which we can say has been a power extant for time immemorially unknown, spanning millions to billions of years, predicated on whatever the sexually reproducing female considered and considers beauty to be, today appearing in its phenomenologically psychological and contiguous form. For the female, and, in so far as females disagree on their depiction of beautiful males, it proves to be a fervent and relatively contiguous power for the male. Beauty for the female ocular experience founds itself on areas such as height, head size, mandibular structures, brow ridge, eyebrow shape, pigmentation, ratios of size, length and protrusion in physiognomies. For eyes; eye shape, protrusion, and fold. For lips; lip size, protrusion, and pigmentation. For noses; size, protrusion, and pigmentation, and so on. The beautiful male will not have to be physically superior to his male counterparts, only that he may only comprise features that may be founded on ratios of symmetries and asymmetries in a particularly fusiform physiognomy.

2. Physicality

The second subset I call physicality. Empirically axiomatic in most academic communities that may at least contend with anthropological history know reasonably well that our ancestors were violent (at least in our genus spanning approximately 2.5 million years ago). The evolution and physiology of the skull, chest, arms, legs, feet, and essential bone structures were features for violence and played out as features for enabling males to fight in and out of hunter gathering bands (let alone for hunting prowess or defence). In turn, a proclivity to pick out and select for characteristics which are indicative of a male being stronger than others are certainly dominant within a process of selection for many females, as it plays out as conclusively profitable for her survival.

3. Resource

The third subset, and rather important, is resource. The significant and sensitive question of resource has been recursively portrayed in our *a priori* structures of stories that hark back thousands of years for our species. How important are resources for a member of the human species today? Without the ponderance on economics, we can stick to a simple benefit-cost model on resource that renders females necessitating *qualities* for a male to be resourcefully stable rather than simply how much capital a male may have in contemporary western societies. Within the Hobbesian 'laws of nature', or 'articles of peace' we cede to capitalism, everything that is profitable or costly is solely predicated on time. As a female who selects, notwithstanding her own capital, the temporal advantage a male can offer as a payoff for the female to allow her labour to subside and endure the temporally taxing occurrence of pregnancy and raising her altricial offspring, evidently may lead to how resource displays itself as a subset within contemporary sexual selection.

4. Status

The fourth subset I call status. Status, as opposed to fame, accounts for all males in its multiplicities of environments. Well understood by sociologists alike, becoming the most social species to have ever existed, sociability accounts for so much. Females may algorithmically assimilate and calibrate a male's value in a given environment, and in turn necessitate the assurance of his value via the response of his ontological position by such intraspecific organisms in his myriad vicinities. Again, and well understood by psychologists alike, women showing more traits inclined to sociability than men is often the case, and so indicators of status may emerge as profitable for a female in her selecting power in speculatively maintaining her position in a dominance hierarchy, ensuring her own survivorship and the stable custodianship of her offspring.

5. Intelligence

The fifth subset, and what many will term the most ontic subset of the power of sexual selection, is intelligence. In as far as all subsets run contiguously, intelligence may be by far the most profitable and essential subset to the power of sexual selection (so much so that it is difficult to recognise). It can be ascribed to the temporal success of our species, conducive to the likes of allopatric speciation, sympatric speciation, and the extinction of intermediary species. Meanwhile, and in so far as intelligence is disputed in academic circles, an understanding of how intelligence can play out as a subset of sexual selection, playing out profitably for the male in becoming as selectable as he possibly can, most certainly has to be ventured.[115]

~

How the subsets of an effectual power manifest themselves into something that can be profitably ontic for sexually selective advantage is unspeakably paramount when we endeavour to reify the metonyms of such biological faculties organising themselves into all complex sociological phenomena. As Miller stated quite implicitly, cultural dimorphism should be perceived as effects led by causes of biological faculties in sexual selection rather than be merely confounded as their causes. Nevertheless, the key *a priori* biophysics of space and time that intersperse themselves in a recipe for sexual selection, approximating those necessary ontic subsets, accounting for a single working power of sexual selection will be germane to senescence and age. We can quite easily ponder on this momentarily when we hark back to the physical problem of space and time. Succeeding the age of consent,

The Epistemological Quest

how significant is the spatial-temporal problem of age within any lasting courtship? Age plays itself out as quintessentially instantiated within sexual selection, instantiating itself throughout the spatial-temporal attributes germane to a cornucopia of evolutionary games being played. In turn, it may cause for both symmetries and asymmetries played out in an effect of cultural dimorphism led by either advantages or gratuitous biological burdens both revelled and endured by the sexes (being viviparity, puberty, intellectual prowess, and menopause), rather than being egregiously confounded as such superficially professed causes.

Our biological repertoire for 'deferred gratification' (or sacrifice) could also be conducing cultural dimorphism in the sexes. The teleological means for the male organism to become as selectable as he possibly can for the gene may be rooted in an incognizant and long-term evolutionary strategy to acquire the female, rendered through an approximation of archetypically accommodated ontic subsets of a single working power. The reproductive awards for all forms of creativity tantalise the male as the sacrifice for increasing reproductive success (or fitness) is only achieved by inducing him, and by no means would induce the female (as her gene bears no profit).) Time sacrificed for the ontic subsets to empower the male reproductively may begin to amass for the male, germane to his intellectual faculties in his ability to calibrate his asymmetries and advantages and synchronise the levels of sacrifice to achieve the levels of creativity he is willing to acquire. This is all akin to the male competition he adapts to in his environment as he aims to acquire the female, and so may evolve akin to male competition over centuries, millennia, or megannums.

As we impartially meditate on subjective qualia, dissecting and discerning each and every instance within a classical phenomenology, what may be paramount in how ontic subsets of a single power of sexual selection can be played, essential to the epistemological contours of this theory, and as noted earlier, is that we hold an unfathomably complex instinct for language (as noted, in the Wechsler IQ test, a whole fifty percent of the test tests for verbal communicative ability). Ruminating further, when interposing the compatibilist philosophy of free will, and all the circumstances in which the free will of an organism is reduced, the proclivities to utter must be somewhat inferential when we summon the teleological means to speak, or utter. One most certainly errs when arguing that a will to speak, or utter, could be characterised by such a libertarian dictum that one truly holds the inalienable freedom of his or her own sovereignty to act. Nonetheless, an arithmetician or a skilled bookkeeper may reasonably object and contend that one could infer such finite thresholds of what an organism most certainly doesn't speak, or utter, in so far as we ever understand the infinite possibilities of what could be conferred! If at any instance, speech could be ascribed to sexual selection in such finite utterances, or a teleological ground for purposeful action could be ascribed to sexual selection, this would only diminish the limits on free will respectively. The oneiric, ludic, imitative, and sexual grounds both assimilated and accommodated mediate between the evolutionary games that structure the faculties of asymmetries in mechanical levels of biological restructuring, allowing the organism to compensate throughout such ontic subsets of sexual selection. The question of dominance and power should then be revisited when we deliberate sexual selection. For the female who selects (relative to parallel ontologies), she may meander the psychological trajectories in coming to fathom her value in such power of selecting. As we endure the force of spacetime and continue to be fettered by a biophysically temporal senescence, the female can only make a reckoning that may evolve fruitfully throughout the first twenty-five to thirty-five years of her life, calibrating the most selectable male possible within properties that induce her to select, throughout such complicated algorithms all against the odds of losing her acquisition. Intellectual faculties calibrate those proclivities that endeavour to deceive in the male, relative to his own ontological ability to endure such struggle for existence, testing for the worthwhile domains in both sacrificing and investing for the smartest of females, indicative of her proclivities in evolutionary games. Many may object to while ponder on the subsets that cause the inclination of the male, which stand in as two ontic subsets of the selecting power of the female:

1. Beauty

'All of these reflections prompted a profound realization, albeit one which she was not consciously aware of, that her heart's desire was to keep for herself, yet at the same time she reminded herself that she could not and might not keep him; her pure and beautiful nature, which at other times was so lighthearted and readily found a way out of predicaments, sensed the oppressive power of melancholy. Banishing the prospect of happiness. Her heart was heavy, and her vision was clouded by sadness'
(von Goeth, 1989, p. 119).

The first ontic subset for the female is beauty. Beauty has and will always manifest itself as a power for the female to select. In as much as beauty variegates for the male throughout the variegating environments, beauty yet again proves as a manifestation of contiguous forces for a quasi-axiomatic definition of how beauty appears to the heterosexual qualia of a male. For the female, again, this could be predicated on areas such as height, head size, brow ridge, eyebrow shape and pigmentation; eye shape, protrusion, and fold, mandibular structures, lip size, protrusion, and pigmentation; nose size, protrusion and pigmentation; ratios of size, length and protrusion in physiognomies, and so on. As opposed to male beauty in the species, neoteny proves highly advantageous for the female ontologically, thus increases power to select, augmented by the advantage of a gracile *anthropic* form. Extending beauty to corporeal areas then predicates beauty on ratios of breast size, protrusion, and shape; waist size, protrusion and shape, legs, bottom, steatopygous qualities, and so on. Significantly involved is the vocal tract, dictating the sound of the female's voice, all germane to the shape and sizes of both pharyngeal and laryngeal organs that render her sounding as she will, for instance.

2. Intelligence

*"'no one can be really esteemed accomplished, who does not greatly surpass what is usually met with. A woman must have a thorough knowledge of music, singing, drawing, dancing, and the modern languages, to deserve the world; and besides all this, she must possess a certain something in her air and manner of walking, the tone of her voice, her address and expressions, or the word will be but half deserved."
"All this she must possess," added Darcy, "and to all this she must yet add something more substantial, in the improvement of her mind by extensive reading"'* (Austen, 2014, p. 39).

The second ontic subset for the female is intelligence. A better lens to look through concerning intelligence (notwithstanding the degree to which IQ tests tell for intelligence) in so far as it may dissipate in our understanding for intelligence, is the lens of biology, namely, of our own species. For the female, sapience, or cognizance, may be necessary. For both sexes, it may be necessary to be relatively aware of one's faculties that stand in for asymmetries, or an ontic armamentarium of the powers and degree to which one's subsets play throughout intrasexual competition. This is typically rehearsed through oneiric experience, principally throughout her formative stages of childhood that stand in for the length of how long she chooses to allow courtship to be displayed, in as much as courtship does not become courtship, which will be a part of her selecting advantage.

Such sociological phenomena weave both ontic subsets, and such powers for sexual selection may be firmly implicated in all axioms that come to be engendered as a precise definition of what can be accurately defined as courtship. This issue incessantly obtrudes in the labour-force for the sociological domain approximating itself on a teleology of sexual selection, instantiated in physiological and endocrine processes that amount to unpremeditated action. Amidst primatological circles, the malleable polarisation of courtship oscillates between two significant phases: the proceptive phase, and the acceptive phase. As briefly mentioned, a proceptive phase merely founds itself on receptivity, on how receptive the female *chooses* to be and what she is willing to *receive* from her courter, which, fundamentally, will be predicated on his ontically selectable status. Nevertheless (and quite importantly), an acceptive phase is founded on her position to either veto or accept, namely, copulation, which is entirely predicated on how long and what she condoned in previous proceptive phases. Germane to an epistemic biology (and notwithstanding incest), I will be audacious in contending that all heterosexual males are *always* in prospective phases, and that the female selects, and so the female may maintain the tacit psychological advantage to stipulate exactly to the degree of what is or isn't courtship in the species![116]

Space and Time

As puberty is endured and the female succeeds her formative stages, progressing her latter adolescence and continuing to vicenarian years, intellectual faculties and a capacity to experience and apprehend *spacetime* most certainly matures in adapting to her environment. Ontic subsets the male wields will most certainly be detected by a newfound inclination to select for what will be most profitable through her endeavour in both space and time. Closest to the lower forms of our taxonomical order, or class, will be the ontic subset beauty. For the intellectually premature faculties in the female, her propensities will be most impetuously inclined to necessitating beauty,[117] conducive to the dearth of apprehending its ephemerally profitable nature, and in how little useful it will be in profiting her in her taxingly viviparous state before parturition, as she is subsequently amenable to the burden of nurturing her offspring. Moving away from the lower forms of our order, or class, and approximating the most supreme faculties of intellect for the female, we are met with the ontic subset of physicality, in as far as physicality does not become of utility for the premature female in her capacity to select for what is ephemeral in its nature, and hence least profitable. As a species rejected of the physical prowess of many distinct mammals, and distinct primates, physicality as a predicate for increased survivorship and reproduction, or survival of an immortal gene, is of little utility as a selective advantage for the matured intellectual reckoning within space and time for the female. Once reaching the latter stages of her vicenarian years, or for late bloomers in the former stages of her tricenarian years, algorithms detecting for likeability, manifesting itself in social status, irrespective of the size of the social group, proves ever more profitable in its sense of being indicative of both investment and monogamous longevity in her teleological calibration in the gene, attributable to her increased social status. Resource is usually misconstrued within its local utility within sexual selection. It could be better understood to polemically proposition, for the purpose of being understood, that shortcut selection for the richest of males is akin to both intellectually premature faculties of the female, or simply lower intellectual prowess, against the odds of him being selected by another female that he most certainly may avail himself of the opportunity to capitalise on. For the female with either matured or higher faculties of intellect (and germane to other ontic subsets),[118] anything behaviourally indicative of financial stability proves profitable for the female, in respect of what should be understood as economic competition within the species harbouring sufficient resources for her survival and eventually her offspring. This is successfully achieved by the male if he can wield the ontic power of wealth against the primitive proclivity in parsimoniously harbouring his resources, which enters the sociological domain for those atavistic predispositions instantiated in dominance hierarchies. The ontic subset of intelligence can be played out as anything and everything that assimilates and works sagaciously with the data counting for all other ontic subsets of the power of sexual selection. Intelligence will be an adaptational feature that runs contiguously, extending and becoming larger in size in its spherical sense, being profitable for all social areas that prove the male to perform illustriously in space and time. Once addressing the anatomy of the brain and all functions heretofore never dismantled, before one hundred and fifty years or so, intelligence has manifested itself in such expansively comprehensive modes that IQ testing can only detect for intelligence by its mantra, that 'if one performs well on one domain, then one should invariably perform well on others', therefore detecting intelligence to its feasible degree. One could therefore induce that sexual selection within evolutionary games detects for as much time she acquires the best possible mate, within the smallest space expended in her quest, germane to the value of what she will come to select.

Taking matters further, and elucidating on a biophysical approach of sexual selection, is a fruitful perspective on the essence of *spacetime* in human sexual selection. David A Puts, an anthropologist of the Pennsylvania State University, believes quite rightly (unlike many who meddle and distort the true nature of the sexual selection theory) that scientific literature unfurling sexual selection approximates male contests, as opposed to females solely selecting for male traits. Puts elaborates by pointing to thresholds within biophysical dimensions accounting for male contests, as opposed to Hobbesian 'articles of peace' that would be absent in early hominids tens of thousands of years ago. As opposed to two-dimensional spaces for male contest to possess the female, three-dimensional spaces (such as air, water, and trees), make the possession and surveillance of a female much more of an arduous task, as opposed to two-dimensional spaces (such as land), or one-dimensional (such as tunnels). In a three-dimensional space, an organism may move *forwards*, *backwards*, *left*, *right*, *up*, and *down*, as opposed to two-dimensional, moving either *backwards*, *forwards*, *left*, and *right*, or one-dimensional, moving *backwards* or *forwards*. Puts divided three-dimensional spaces into hemispheric or spherical regions as most airborne species do not often travel down into the ground or spend too much time in water. Dung beetles competing for mates cajole competitors, as surveillance is easier, and contests are more likely to transpire in one-dimensional tunnels. For male dung beetles to survive and continue to be naturally selected, dung beetles must go *through* male competitors (as opposed to *around* them) to access the female. For the male dung beetles to find a mate and reproduce such incessant contests must reoccur in one-dimensional spaces. Likewise, fur seals competing for mates on land are also subject to the thresholds that limit the feasibility to survey, harness female movements, or allow competitors to abscond, as terrestrial dimensions compel contest. For male fur seals to have any reproductive success is only through male contest, as the feasibility for the dominant male to guard his harem thus doesn't prove so arduous. Blue headed wrasses that guard coral reefs are both compelled to survey and contest, as three-dimensional spaces bounded by a territory limits activity, as one cannot easily travel *through* the reef! Blue headed wrasses travel in *all* directions, save the reef, so for males, dimensions make it reasonably taxing to guard and preclude females from leaving, or from a male absconding from a contest and then surreptitiously returning from distinct directions. Male bottlenose dolphins competing for mates are borne with the hindrance of guarding females from all open dimensions, allowing evolutionary games to play themselves out through variables ulterior to cajolery male contests (Puts, 2010).

Apropos of necessary ontic subsets, a reason the very subsets may be sutured in the form of subsets, and not powers, is that, as cultural dimorphism is presented as a manifestation of sexual dimorphism, appearing on the surface as sociologically motivated, its utility is only found in the form of a sexually selectable and ontic organism, consequently allowing for cultural bifurcation to transpire. Power or dominance for the female differ in her teleological mission for the gene (as opposed to the onus of the male). Her ontic value paralleled with such teleological will as the female, only thus diminishes the libertarian dictum that any free will may exist for her, and that's notwithstanding sexual predilections. Likewise, for the male subjected to his own teleological missions and trajectories, may only empower himself where natural selection sees fit, in as much as the ontic subsets that necessitate the survival of the species may fluctuate few and far between sexual selection, and all facets concerning natural selection.[119] In making a carefully drafted concession, and likewise drafting a reasonably evolutionary proposition, it may be suffice to say power may be anything that may equip an organism of any species across all domains with the necessary armamentarium to either **increase**, or at least **harness** one's own survivorship and reproduction, in as far as power may be vicariously inhered by other organisms of the same species. For the working power of sexual selection to successfully dovetail variegated subsets of sexual selection, or ontic subsets to do so empowering the female, power then only adduces to this very evolutionary proposition, which in sociological spaces should not be taken lightly by any stretch, form, or sense of the critical imagination.

Concerning space, time, benefit, and cost, while returning to a *domestic-bliss strategy* precisely promulgated by Dawkins, females may only increase such intellectual capacities through taxing processes of adaptation in engendering a worthwhile ESS in any given environment. Adaptational capacities may mediate between philanderer and faithful males, accounting for anything that can be assimilated as data pertaining to both time and space. Her highest faculties will apprehend the variegating strategies devised for her ability to test the faithful, achieved by extensive probationary periods that will thus compute all relevant data into algorithms, telling for faithful countenance and behaviour through the calculating probabilities. As briefly mentioned pertaining to the archetypal realm with all that may be inculcated in such complex psychoanalysis, data concomitant to his familial affairs will be beneficial for the coy strategist to assimilate any profitable behaviour indicative of his own familial stability and what she may expect. Data will then compute itself into revelatory algorithms aiding the female to make judicious and prudent decisions, which will account for availability of the best she can possibly acquire in limited space and time. The necessary data will continue to compute itself calibrating the ontic benefits and costs of the suturing subsets of a working power, which is also conducive to her own ontic status of what she will have inhered as the *power* to select for in her environment. These very ontic subsets germane to their given environments have become so much more intricate in twenty-first century phenomena, as technology has only multiplied and amalgamated evolutionary games played in enclosed two-dimensional environments, as it has most certainly added its own dimension of what could be referred to as *technological spacetime*. For the feasibility of the male to avail himself of the opportunity to court in this newfound dimension of *technological spacetime* has most certainly hitherto been played without precedent, and the female's expanded capacities to select is most latent in her environment also, and this is notwithstanding an individual's disaffected decision to reject *technological spacetime* as 'a', or 'the' new phenomenon! For the *fast* female, her feasibility to expedite any courtship with the philandering male is certainly profitable in her strategy, which may adversely stress *coy* females in a given non-technological spacetime (as their environment continues to be somewhat adulterated). Likewise, the *philanderer* may only capitalise on such a titillating dimension, which may prove to become either profitable or injurious for him in his quest to acquire the best of the *coy* females in space and time (which will not be facile whatsoever). In so far as the *philandering* male may deceive the *coy* female of his 'faithfulness', will be in so far as the *fast* female may deceive the *philandering* male of her 'coyness'. Likewise, in so far as the *faithful* male may deceive the *coy* female of his **ability** to be philanderer (or the worthwhile lothario in an environment which he has allegedly 'opted out of)',[120] will be in so far as the *coy* female may deceive the *faithful* male that she has no interest in the philanderer whatsoever!

A Brief Conclusion on the Theory of Sexual Selection

On closing part I of the book, and on closing a brief trajectory of sexual selection, it may be necessary for us to conclusively venture yet again the epistemological representations that aid us in understanding the very nature of complicated qualia, and all imbrications laid out across all culturally identified residue. Evolution is by no means a simple matter, nor is anything that displays itself as pertinent to the domains within life and all Kantian *analytical a priori* judgements of space and time endured as organisms. As natural selection symmetrically populates sexes in its environment (Fisher, 2018), and as idealistic philosophies desire *infinite progress* of our species (as opposed to an *infinite regress*), I believe that a true biological comprehension of the sexes allows one to evade the trap of committing an *appeal to social constructionism (divine)* fallacy, as Albert Borgmann stated quite usefully:

'Training these questions on gender as a construction, one wants to know what the construction is imposed upon. On genderless human beings? If so, is not the genderless human being a construction as well? What is that design imposed upon? Primates? Animals? Or featureless stuff?' (Borgmann, 1999, p. 132).

Intermediary forms of our sexes and understanding variations, has been, and interestingly enough, best represented in endocrine circles. The debate on what makes a female a female, and conversely, a male a male, still permeates yet again our sociological institutions in an endeavour to fathom the sexes. Meanwhile, like with the progressive evolution of neuroscience, the brain is best understood when things go wrong!

It's quite often misconstrued in many institutions of education approximating the sciences, even of well-educated individuals, that the division of the sexes can be binarily divided into the XX and XY chromosomes for what we fundamentally define as a male, or as a female. Others also egregiously argue the XX or XY chromosomes could also fundamentally dictate how sexes can be divided phenotypically. Nonetheless, summoning the fields of genomics, ontogeny, and epigenetics, specialists show the data presented is so much more complicated in understanding how genes interact prenatally, and how labile trajectories of molecules may be in travelling where they are 'supposed' to arrive. According to two cases that demonstrate complex systems of sex differences, the simplification of binary XX vs XY chromosomes may not be so useful in telling the entire story of the sexes (albeit *extremely* exceptional).

For instance, congenital adrenal hyperplasia (CAH), a syndrome of androgen activity, affect both males and females. Typically studied are females, attributable to the subversive nature in phenotypic behaviour. CAH, quite frankly, is an issue akin to cortisol production. A mutation in an enzyme within adrenal glands causes CAH to occur, and so instead of the adrenal glands creating necessary glucocorticoids, they manufacture other androgens and testosterone prenatally, consisting of an XX chromosome. The brain detects low levels of glucocorticoids in the blood, and so compels the adrenal glands to work harder in engendering the necessary glucocorticoids, thus creating overproduction. The result, i.e. such overproduction, causes 'hyperplasia', meaning that the adrenal glands that sit on top of both kidneys grow larger in size. Such overflow of testosterone, which are known as the sex hormones and so should be noted in its purpose for sexual selection, paralleled with the scarcity of corticosteroids produced by the adrenal glands, led many to estimate hormones such as testosterone to be a primary factor in dictating sex differences. The corollaries of these prenatal occurrences result in both interesting traits and behaviours. Females exposed to higher-than-average levels of androgens were less tender-minded, less interested in infants, and were more physically aggressive in response. Conversely, and galvanising the work of bioethicists were the resulting effects of CAH treatment. Today, prenatal screening for CAH can be undergone, and the prenatal virilisation of the foetus can be somewhat precluded by compensating corticosteroids, which in turn increases the chances of the child being heterosexual, again, raising flags and questions for its locality in both sexual selection and bioethics (Mathews et al., 2009; Sapolsky, 2017).

Added to such a conundrum of the female and male chromosomes is a most fascinating syndrome, androgen insensitivity syndrome (AIS), which likewise plays out in an endocrine system, particular to individuals who entail of the XY chromosome. Unlike CAH girls enduring overproduction of androgens instead of the necessary corticosteroids, AIS is conducive to a defective androgen receptor (AR), insensitive to the androgens that the AR is fundamental for receiving when processed in the testes. In turn, testosterone is thus not assimilated for the virilisation of the foetus, resulting in a female phenotype for the foetus. It's only when the child experiences puberty when specialists discover that the child, or the girl, actually holds an XY chromosome, as the child endures puberty save menstruation, attributable to the far misplacement of her testes. AIS differs in individuals who experience the syndrome in that distinct levels have been purposefully categorised. The syndrome of androgen insensitivity may lead to ambiguity in genitalia, ranging from 1, being a normal masculinisation of the utero, to a 7, which would be phenotypically expressed female genitalia. Grade 3 or 4 would be the severe ambiguity the syndrome typically causes as growth in specific areas combining both sexual phenotypes are expressed. In unfurling the necessary genetics, while akin to the uncanny behaviour of genes such as the AR, endocrinologists find that genes, such as that of the AR gene determining sexual and reproductive fitness, occupy a vital role in the maintenance of our species as they find testosterone to be a major hormone for the sexual dimorphism of nonreproductive tissues (as opposed to the reproductive). The AR gene may not be significant in testicular production, meanwhile, does remain essential for spermatogenesis during stages of mitosis and meiosis. Productive forms are essential for the meditator to reconcile vacillating vagaries that play out between sexual dimorphism and sexual selection, while discerning the ramifications for genes in failing to undergo the necessary androgenisation throughout evolution.

Likewise, the necessary detection and DNA binding on androgen receptors are mutations that may be essential for sexually dimorphic fitness, granting sexual selection as the biological mechanism that continues and maintains the species. It seems that RNA splicing (essentially the expression of exons and omitted introns), also play out fundamentally for genes to then express themselves on to foetal stages in transferring recessive and dominant alleles, which are then finally expressed in organisms which then speculatively flourish in such *working powers* (Quigley et al., 1995).

When taking syndromes of such into necessary consideration, scientific discoveries discerning epigenetics, ontogeny, and endocrine systems only invoke questions on both intrasexual and intraspecific influence (more than the question of what a male and a female are). As adumbrated in individuals comprising the CAH syndrome in particular, corticosteroids are signals received from the brain via transportation of sanguineous systems, and one can only ponder on the simple question in this case, 'on how much libertarian control can an individual hold on corticoid secretion concerning one's own blood', notwithstanding the lack of control one may have in harnessing all other scientific variables.

So it may be prudent to make concessions on scientifically epistemic grounds, and in doing so we may successfully necessitate the existence of our extant species leading to gradual educational reform, whether the species be as big of an organism as an elephant or giraffe, or microscopic in its biological nature.

When peering through the lens of a theory on sexual selection, and on how closely I may expound this theory on such ontic subsets germane to spatial and temporal entities, what should be implicated through a historical prism, best represented by historians such as Yuval Noah Harari, is a technological revolution we continue to unprecedentedly experience and endure (Harari, 2017; Harari, 2018). Thus far the spaces and times both sexes have had to venture to select for and be selected by have been limited by the dearth of technology continuously capitalised on today, which may not prove to be as profitable for our species, in so far as it procures the very fruitful experiences as never experienced before. In its sense, long-term monogamy, again a facet of space and time cohered in such a Maynard Smith perspective, may be a feature for both sexes evolved from algorithms that play out for the feasibility of selecting the fittest within enclosed societies. Meanwhile, the rise of dating applications, in which those very algorithms for females to select in hitherto enclosed environments are now being doctored by programmers, thus remap the user's capacity to select for (or be selected by) mates falling outside of those very perimeters, which will have psychological ramifications for the worse in the species. Even when turning to infidelity, restructured spaces and times technology offsets deleteriously endangers the canons of traditional monogamy, as the feasibility to deceive is multiplied by the viable networks of communication, compensating for space and time historically lost, attributable to the complete omission of the type of communication that has thus far not existed until today. As we continue our epistemological quest, and as for a brief and more than contemporary perspective, there it goes for the theory of sexual selection.

**Part II
Of Good and Evil**

'Heaven and Hell' (Tassaert, ca. 1850)

The Epistemological Quest

This part concerns knowledge and morality in its quest. It's a larger perusal of some of the compelling modules of knowledge, as opposed to any real claims. So, I will get my claim out of the way in its simplest sense:

All we are given is belief about the world. In its strictest sense, knowledge can be considered as a separate entity, and thus it must be reliable for sustenance in the world. Both the separated entity, knowledge, and your belief, have relationships with he who holds the exact same set. How all knowledge doesn't perish is only achieved through the tools of evidence, logic, scepticism, and probability; that's nothing else withstanding these tools to how we know about the world...

Chapter 4

Of Contractualism and Objectionable Conditions

As for the first part of this book, the adornment of the natural world led critics to object to scientific theorists and their theories. Such objections contend scientific epistemological propositions that concern the natural world to be merely conducive to contumelies, guilty of a naturalistic fallacy, in that just because something may *be*, does not mean that it *ought*. So it may be necessary for us to revisit our concessions by moving back to the drawing board, unfurl the epistemological contours and see how one comes to actually *know* a thing to be a thing and yet exist in the world. The solipsists holds that knowledge extraneous to the mind is uncertain, thus the external world and all other minds cannot be *known* subjectively. The philosophical drawing board used in epistemological discussions pertaining to supernatural entities may be forever incessant in those often acrimonious debates (in as far as the state of the world does not change by a commission of anything as philosophically significant as extra-terrestrial life).[121] But for where we arrive in such concessions, these very types of debates unwittingly emanate into a contractual philosophy concerning the infinite progress we logically agreed to take precedence.[122]
For instance, the solipsist's reasonable contention that one can never be indubitably certain of extraneous thoughts quickly inculcates the supernatural. This then ties itself to the doctrine of *methodological naturalism*, in that one can never indubitably be certain of supernatural causation, and we are thus limited to exploring all naturalistic explanations (i.e. the phenomenal world), which, again, is a reasonable proposition, in so far as two thinkers warrant cooperation. To offer scope to a methodologically naturalist claim, would mean, that one may invariably approximate the possible eschewal of an antithetical *philosophical naturalism*, holding the natural world to be all that therefore exists. In as far as it appears reasonable to placate the naturalist, heretofore we have only been free from apodictic dogma concerning the natural world, and hence the natural world being all that is consequentially **proven**. If we progress to a 'grand and unified *concessional* epistemology', must we decry a theory of philosophical naturalism given the are limits to our methods?
The capacity to deduce nests in epistemological act. The mechanism of interpreting knowledge thus ventures vast ontological spaces external to the subject, essentially deducing ontological faculties of all other minds to think as the subjective, and thus experiences the natural world as the subject may do so, notwithstanding certainty, essentially germane to the solipsist that the external world will most always continue unbeknownst to the subject. Founded on a serialisation of justified beliefs, which, when fundamentally speaking, negates that of a belief being entirely self-justified, is a foundationalist's claim. All that is known in the world; what logicians term 'alethic' (as opposed to what is conversely known by the subject; 'epistemic'), means one could never compartmentalise knowledge into compartments of such vacuous properties, thus making the concession that a case for ontology manifests those very invocations. For instance, in the case of classical Euclidean geometry, the sum of all angles of a triangle are 180°. Such justification rests on previous beliefs, which rests on other previous beliefs, then on other previous beliefs, *ad infinitum*. The foundationalist contends that knowledge must then be finite, as there can be no knowledge that is self-justified other than knowledge that derives extraneous to the domain of knowledge, which leads many to a doctrine of methodological naturalism in that supernatural causation will most always remain nebulous in its **methods** of justification. Coherentists thus challenge the domain of knowledge and challenge the

The Epistemological Quest

tenets of foundationalist epistemological belief. The Coherentist sees foundations of those beliefs will be either non-foundational, or merely 'unjustified' belief:

- 1. Certain Belief ~ Belief = Non-foundational
- 2. Certain Belief ~ Belief = Unjustified

If the foundationalist's belief is *justified* by another belief, it must be that certain belief isn't certain, in that it results to certain belief being a regular belief. In the second instance, if the foundationalist has a certain belief *unjustified*, then there should be no *reason* for the foundationalist to believe the certain belief to be true, it would simply mean the belief remains unjustified. For instance, if I am cold, it must rest on the belief that it is snowing, which then rests on another belief *justifying* itself (i.e. that the angle of the earth relative to where I am situated on the planet is at a particular locus). An instance of the second objection may hold that God does not exist. I am now borne with the burden of proof. I cannot justify that God does not exist, nor that he or she does exist, so the belief remains to be untrue as it does not yet rest on another belief, such as that the angles of a triangle add up to 180°. For coherentists, knowledge may rest on a coherence of tenets that nest themselves in logic, maths, and other forms of truth.

It seems, especially as noted in preliminary stages of this book, that a precursor for anything contentious is that knowledge should rest on the duality of scepticism, namely, on tenets that oscillate between philosophies of denialism and credulity. Irrespective of where an individual purports to sit on the scepticism position, it is both consensus and scientific experimentation that edified, is edifying, and will edify epistemological knowledge to any infinite progress. Methodological naturalism that has cherry-picked its methods on the pursuit of knowledge must give way to the naturalist world for all it has contributed against the *burden of proof*; or the burden of the naturalist's experimentation.

Logic and its Laws

'Continually and, if possible, on the occasion of every imagination, test it by natural science, by psychology, by logic' (Aurelius, 2020, p. 121).

An onus to prove, or in logical terms, 'the burden of proof', persuades the dogma of atheism (or moves those to agnosticism). Reason has been conducive to the empirical claim that agnosticism can be hardly deemed a relationship with the supernatural world, but a relationship with *knowledge*, meaning it morphs around contours of epistemology. A relationship with knowledge instead of the supernatural would yet approximate absolutism and certainty, in which the agnostic will claim that they are maximally certain that a supernatural entity does not exist (instead of claiming they could be absolutely certain), hence transposed as an epistemological claim. The epistemological claim is thus germane to 'degrees of which one thing becomes another', and 'another thus does not become the other', which aims to eschew philosophical naturalism and any religious presupposition that the supernatural should conclusively exist. Deconstructing those laws of logic, is a perspective necessitating the significance of continuity and the degree in which a thing becomes a thing, in so far as it fails to become another thing, as noted by the philosopher, Stephen Clark, at the University of Liverpool. The three laws of logic propose that:

- If p, then p [*The law of identity*]

The Epistemological Quest

o   Not both p and not -p [*The law of non-contradiction*]
o   Either p or not -p [*The law of the excluded middle*]

Clark, accounting for both space and time within logical absolutes, dismantles them by accentuating the caveat of such a dilemma: with 'I' being the law of identity, 'NC', the law of non-contradiction, and 'EM', the law of the excluded middle:

'Consider any such difference, whether in time or space: from L1 to L2, we'll say, an object x is A, and from L2 to L3, that object x is not-A (by not being A). Consider what is true at L2 itself. Is x at L2 A or not-A? If it is both then NC fails; if it is neither EM fails' (Clark, 2008, p. 4).

In the theory of evolution, in the popular dilemma of what came first, 'the chicken or the egg?', it seems that the method of language (profitable for the species used to expedite a process of intraspecific communication) obtrudes as an essential problem of the before, the continuous, and 'afters', rather than anything that may maximally present itself as certain of what lies beyond the logician. Clark's claim to continuity is that 'our reality is ineradicably *continuous*, and there are therefore no abrupt changes of the kind that language and logic might lead us to suppose' (Clark, 2008, p. 4). Conflicted with the laws of logic that not even the supernatural objects to, Clark invokes the more complicated question via a supplementary logical absolute, i,e. the law of double negation:

o   If p then not not -p [*The law of double negation*]

The inversion of anything one may perceive in a proposition of what may or may not be reality, whether supernatural, or anything that may exist or not exist simultaneously, is in its most logical sense what Clark accentuates as an ontic, yet epistemic, contradiction. Discourse germane to the logical absolutes most always engenders the question of, 'would two rocks be two rocks if all human minds ceased to exist?' It seems that the paradox in a perception of space and time amplifies throughout the method we use (language), being that *continuity* may challenge the very absoluteness of its logical sphere.

The issue of effectual scepticism for anything pertaining to reliable epistemology may be instantiated in what I previously mentioned as 'epistemological purchase', in its score for what may be earned as epistemological points, which is most fundamentally an ontologically subjective relationship to *knowledge*. When suturing the philosophies of the logical absolutes with solipsist assertions, it may be incumbent upon the reader, hopefully not revelatory, to share tacit understanding that one cannot bear the temerity to express any claims about the external world or anything concomitant. For anything to engender infinite progress in the epistemological playing field with knowledge, I believe to be solely achieved via consensus. Epistemological purchase is a method inhered unbeknownst to free will, being inhered in a sense of intersubjective reality, and most definitely in the Hararian sense of imagined orders (Harari, 2015).[123] In so far as it becomes ever more complicated when we approach the larger domains of thought, data suggest we are doing rather well as a species. It means that scepticism, for what can be argued contextually, also becomes useful in its propitious *times* and *spaces*.

The Epistemological Quest

A Sceptical Sceptic on Scepticism

Let's look at how the clinical pharmacologist at the University of Oxford, Jeffrey K Aronson, as his experience attests to such large-scale issues in medicine, uses five types of scepticism conveying its take on knowledge.

o **Philosophical scepticism** is predicated on its classical sense of doubting the very existence of knowledge, being a Pyrrhonian question concerning itself with the possibility of knowledge. Kant ventures his way to antinomies in its noumenal sense of the breakdown in reason, in which the impossibility of knowledge is a problem if two opposing logical arguments aren't necessarily *non sequiturs*.

o **Voltairian Scepticism** being scepticism that postulates doubt from knowledge, in which any proposition should be principally effaced with doubt, which most certainly variegates in degrees of sensitivity.

o **Scientific scepticism** necessitates reasonable doubt. The cultivation of scientific scepticism is essential for a form of scepticism to work effectually as experimentation, interpretation and conclusion should work cogently and rigorously before knowledge can be justly criticised, as new ideas can most often be unjustly cast away, as the English surgeon Wilfred Trotter's famous aphorism held, that 'The most powerful antigen known to man is a new idea' (Aronson, 2015, p. 1).

o **Dogmatic scepticism**, or **negative dogmatic scepticism**, postulates *a priori* knowledge in impossibility. This is an assertion that allows the worst possible situation to take precedence, hence leading to the negative sceptical doubts as a truth claim pertaining to impossibility in essence.

o **Nihilistic scepticism** can be best understood through instances of the new and old. It arrests credulity and approximates credulity to denialism through the discernment of incredulity, in which something can never be truly believed, or for what it's worth, trusted (Aronson, 2015).[124]

The weaponization of scepticism against the rise of both literacy and population over the last two centuries most certainly ensued, and thus a domain of epistemology invariably expanded, contracted, and distorted within spaces of propagated data. Approximating philosophies of contractualism, a delineation was necessary for both the devils of dogmatism and nihilism running asunder, as an adulteration of its natural form could have certainly ensued. As the Harvard philosopher Gisela Striker had shown, the first definitions were of 'philosophers who suspend judgement, refrain from making any assertions, either about philosophical problems or about anything whatsoever, including everyday statements or facts' (Striker, 2001, p. 113). Sextus Empiricus, the Phyrronian philosopher (second-to-third century), portrayed scepticism as both a form of investigation and search, instead of mere philosophy of doubt (which is scepticism in its often misconstrued form today). It seems that one cannot continue to be inimically opposed to the true art and mastery in scepticism today. In as much as inexperience of information society endures, rendering epistemic knowledge thus difficult to discern, it seems that a turn to contractualism is what may profit the trajectory of our epistemological quest.

The Epistemological Quest

Cessions, Concessions, and *Blind* Privileges of Society

'The passions that incline men to peace, are fear of death; desire of such things as are necessary to commodious living; and a hope by their industry to obtain them. And reason suggesteth convenient articles of peace, upon which men maybe drawn to agreement' (Hobbes, 1998, p. 86).

Civilisation as a term, in its most common sense, is a philosophy of tract between citizen, state, government, or superior body. In which the citizen is burdened with his own proclivity to violence (most especially of the most trivial of reasons), he thus frees himself of his will. Superior bodies may not oblige the citizen to fraternise with others (in as far as the state would not for 'respect' of others), and as a result of the tract on ceded proclivities and an act of disquietude, acquire anything necessary from such superior bodies (all of which is for the sole purpose of subsisting).[125] One may cede more of what she agrees or disagrees upon to the state (time), in order to flourish in *civilisation*. As much as is ceded to the state, the more disconcerted one feels about what's yielded in belief, accounts for all benefits that compel the agreement in the cessation. Albeit you may secede from the state, abscond and live in other confined communities of the world (as many have done so for millennia). Parallel to the benefit of what's acquired from the state is the costly cession of what's agreed and disagreed upon, which passes on to the realm of what we then call *politics*.[126] Before we touch on foundations of political philosophy, we must locate such origins of what's studied, *ethics*. Before the emergence of contemporary *realpolitik*, it's ethics in classical antiquity are rooted as a colossal proximal cause, which harks back far beyond the proximal cause of the agricultural revolution. But to start with the inimical point of contention, we will start from the other side of the coin first and make our way backwards to religiosity.

A Moral Landscape and its Epistemological Purchase

Published by the moral philosopher and public intellectual, Sam Harris, was his polemical work, *The Moral Landscape*, gaining traction in the philosophical world and beyond (Harris, 2012). Antithetical to religion, and believing to be deleterious to progress, he recursively unfurls his philosophical prism. Harris grounds his claims under the canopy of science which would lead to invariable human progress. He believes one to primarily ground facts and values from science without the yield from religion. Harris outlines the very nature of those caveats which substantiate the notion of what I call *epistemological purchase*. In what the world languidly reveals itself as true, or epistemic in knowledge, proliferates what seems to have evolved in the problems of a twenty-first century:

'Conversely, those who are more knowledgeable about a subject tend to be acutely aware of the great expertise of others. This creates a rather unlovely asymmetry in public discourse – one that is generally on display whenever scientists debate religious apologists. For instance, when a scientist speaks with appropriate circumspection about controversies in his field, or about the limits of his own understanding, his opponent will often make wildly unjustified assertions about just which religious doctrines can be inserted into the space provided. Thus, one often finds people with no scientific training speaking with apparent certainty about the theological implications of quantum mechanics, cosmology, or molecular biology' (Harris, 2012, p. 161).

The Epistemological Quest

The ramifications of arrogated authority by intellectuals alike, who elucidate on such phenomena that naturalistic explanations for the external world support in naturalistic philosophies, only fortify the bulwarks of irreconcilability between religion and science. If the most preeminent of intellectuals obstinately disagree on fundamental dicta, then, according to Harris, it both incessantly thwarts and undermines those that consecrate lives to specialise (in *scientific* experimentation) in the natural world. That being said, and what must be reverberated (in as much as Harris pithily commits to *scientific* experimentation in the phenomenal world), is that philosophical naturalism is all that must be necessitated and methodological naturalism **can only be** a private affair. Here I can only put forward those instances of methodological naturalism to support Harris' claim. Even as Steven Pinker substantiated in his book on a history of violence, *The Better Angels of Our Nature*, if we count the innumerable times of human sacrifice, witch hunting, and deaths without *natural* and *scientific* trial, *methodological* naturalism would be responsible for deaths in hundreds of millions, if not billions (Pinker, 2021).

Public vs Private

The case of public and private proves necessary for the question in our epistemological quest, notwithstanding purchase, nature, or a supernatural world. Incumbent upon the meditator, is to impartially agree, or at least merely understand what methodological naturalism *is*, rather than incessantly misconstrue the very essence of its doctrine. Tiddy Smith deconstructs the misconception of the method of the form, as he states, 'methodological naturalism does not obligate science to reject supernatural entities, but to reject supernatural methods of acquiring evidence' (Smith, 2017, p. 3). Smith breaks methodological naturalism into both 'intrinsic methodological naturalism', and 'pragmatic methodological naturalism'. He does so for the purpose of discerning the caveats of vitriol in a supernatural world (as opposed to advocates of the natural). *Intrinsic methodological naturalism* holds that science is conclusively irrelevant to all explanations of the supernatural. *Pragmatic methodological naturalism* holds that science accounts for supernatural entities and dispelled of them as the mere non-existence is justified, or false. Smith believes the problem situates itself within an epistemological approach, in which the problem becomes instantiated in *justifications* of knowledge, as opposed to *explanations* thereof. Essentially, it's a matter of 'method', not 'conclusion'. In a natural world, played out through the essence of philosophical naturalism, the method of justification is wholly predicated on evidence, proof, or experimentation, pointed out as evolving along the emergence of medieval philosophy, crafted in the thinking of those such as Duns Scotus, Adelard of Bath, William of Ockham, Thomas Aquinas, Siger of Brabant, Nicole Oresme, Boethius of Dacia and John Buridan. Repudiation of insufficient supernatural explanations had never been ascribed to explanations as such, but the manner in which they are *justified*, which medieval philosophers believed naturalistic explanations couldn't account for. The method in which 'pragmatic methodological naturalism' is situated, is held in the classical belief that phenomena can only be justified by *faith*, which may be the very method that frustrates natural philosophers alike (Smith, 2017). Notwithstanding the natural philosopher's disquietude to this choice of method, and irrespective of natural philosophy's accomplishments for its own method of justification, natural philosophy continues to be unable to justify all phenomena (nor does it claim to justify all phenomena), in so far as *noumena* dissimulates in the natural world. And so, I believe there should be a case between **public** and **private**.

A public domain would mean necessitating natural philosophy for all it's worth, increasing what Harris defines as human **wellbeing**. This can be justly achieved irrespective of one's own religious affiliation or belief, allowing epistemological purchase in science to lay precedence to phenomena, which would continue untainted, even coinciding with the most religious achieved in societies today. By no means is this a simplification of the political philosophy that continues as complex. The private domain would mean what one believes privately will continue to be *inalienable* in one's right as a citizen, by contractual concession. The private sphere would continue to be areas where citizens can practise privately, the religious freedom to practise, act, and be, with unabated impunity, a cause for liberal religious ontology and freedom. In which an essence of any such phenomena would approximate a public (such as institutions concerning citizens working in **altruistic** spheres most especially), natural philosophy continues to take precedence in a rule of thumb. This would a) simplify matters of human wellbeing, and b) take into account all-natural contingencies in its endeavour of remaining altruistic, which would merely entail of cooperation with those necessary specialists in their variegating fields of study. The public sphere would continue to be untainted by phenomena (ulterior to its form of justification), which will clearly continue to be natural, as opposed to a possible amalgamation of vindicating methods, which Smith implicitly highlighted.

Red Houses

As philosophical naturalism yields to the *private* and natural philosophy is thus invoked *publicly*, it proves itself sociologically practicable for the masses in terms of science, reasoning, and even humanism. Instantiated in the postulates of inductive, deductive, and abductive reasoning is what concerns a natural or supernatural world. If one drives past nine hundred and ninety-nine houses out of one thousand, the driver induces the next, or last house as red. I would add, in so far as the next house may be induced as red, the possibility of the last house changing colour may also be inferred. Essentially, it's just a little paint. The question could be asked, 'how do you *know* there are only one thousand houses and what is the method in which you came to justify this belief, without being sceptical of there being a thousand more?' Now the driver's induction is still somewhat predicated on the belief that a thousand more houses are merely possible, and that distinguished colours of paint can be found in myriad stores. This is a cause of the natural world in which induction plays out, in that the benefit-cost relation empowers the subject where one sees fit, which is often the substrate for all sports played in the species. The issue emerges when deductive reasoning becomes *conferred*. As the benefit-cost relation of induction does not prove injurious to our daily activities, we understand induction continues innocuously. Nevertheless, we deduce that every house falls under the laws of nature, so, in as far as the next house isn't red, or in as far as there is an additional one thousand houses when passing the thousandth house, we may deduce natural laws as the same prior to the driver's present (for example, the road being of matter and the laws of gravity continuing to be held), otherwise the driver would have never got in the car in the first place, against the odds of falling off the edge of the earth!

Such a form of deduction is quintessential in fields of science to be free in their methods and modes of justification. As Harris points out, matters such as stem cell research should be a nominally public issue. Germane to his philosophy of wellbeing, and irrespective of the varying anomalous cases, living and human flourishing should take precedence in contractual philosophies of civilisation.[127] In public spaces and institutions such as scholar education and universities, the freedom to propagate private phenomena of that in which is justified by the *method* of **faith** should be continued to be freely professed as *opinion/belief* (doxa), but by no means *knowledge* (episteme)!

For instance, naturalism (or science) in times of stress obtrude in the issue of epistemological purchase when effaced with life-or-death conundrums, which may be a novel approach the twenty first century is unwittingly reconciling with in the emergence of science, as this very instance conveys in the public success of the covid-19 pandemic:

'The catholic church instructs the faithful to stay away from the churches. Israel has closed down its synagogues. The Islamic Republic of Iran is discouraging people from visiting mosques. Temples and sects of all kinds have suspended public ceremonies. And all because scientists have made calculations, and recommended closing down these holy places' (Harari, 2020).

Notwithstanding the ramifications of radicalism on both sides of the atheist religious-coin, science is playing a more central figure in the humanist spectrum (as Harris rightly points out). It thus expands a natural world which religious fundamentalism is more and more anathema to, in which epistemological purchase may brandish discoveries in the hope of pushing religious dogmatism out of the epistemological spectrum. Nevertheless, and as many believe, if naturalism succeeds in that goal, by no means whatsoever can disdain suffice in inadvertently dispelling the baby from its bathwater, which is why I aim to focus in part on the moral philosophy of a Judeo-Christian world and its spectrum, and in order to achieve this, one must hark back to a time I believe is significant to ponder on, what I lay claim to be in the underpinning of the west as we perceive it today. So, to finish demarcating the epistemological contours, vetted out in the circles concerning knowledge and a natural world, and before venturing the contours of what may morally distinguish as good and evil in the most fragile of the human histories, we must turn to the essence of modality.

Modality

Pertaining to certainty and *what is in the world* is what philosophers esoterically dispute as possibility, thus coming in variegating forms as **modalities**. Modality alleviates the antipathy one may hold in a religious debate, thus embedding itself in the progressive political state in which religion may play its role throughout ongoing civilisations, whether dilapidating attributable to the vigour of philosophical naturalism, or conversely doing so attributable to the methods of methodological naturalism. The four essences that unfurl possibility of any such thing lies in a *proposition*, or *entity/object* being either Possible, Impossible, Necessary, or Contingent; alethic modalities.

Alethic modal **necessity** is said to be predicated on knowing such things that must be possible in *all* possible worlds, respective of its form of modality. Whether truncating a piece of this world, or invoking other worlds, necessity thus occurs through all possible worlds. Conversely, alethic modal **impossibility** does not occur in *any* world, whether truncating a fraction of the possible world we experience or invoking other such worlds. Alethic modal **possibility** would be merely congruous with a possible world in which if the proposition transpires in at least one possible world, it is thus proven to be alethically possible. An alethically *contingent* world would be converse if the proposition were to transpire throughout some of the worlds instead of all or none of them.

Alethic, what's true in the world, and epistemic, what's true in one's mind, is what philosophers have often pondered on in the quest for ultimate truth, and hence alethic modalities must thus be unfurled to challenge naturalism whether it appears to be naturally methodological or philosophical in essence. The essentialist theory thus draws the distinction between alethic and epistemic modalities through philological foundations of all possibility from the world and of the individual mind, as the British mathematician and philosopher, Bob Hale wrote:

'By the *nature*, or *essence*, of a thing, I mean *what it is to be that thing*. This is what is given by a definition of the thing. For example, the definition of *circle* is: set of points in a plane equidistant from some given point. The definition of *mammal* is: air-breathing animal with a backbone and, if female, mammary glands. This is definition in an Aristotelian sense – what is defined is *not*, or not primarily, a *word* for the thing, but *the thing itself*. A correct definition may serve to state what the word means – as with the previous examples – but it need not: for example, gold is the element with 79 protons per atom, but this is not what the world 'gold' means' (Hale, 2012, p. 129).

Axiomatic is the course of modality, that one thing may be possible given one type of modality, but yet impossible given another. Logical possibility, or narrow logical possibility, as mentioned previously, predicates itself on the logic of non-contradiction, in that nothing can possibly be itself and another at the same time, hence is illogical, or logically impossible. So, as believed through alethic possibility and logical modalities, Barack Obama could not both be Barack Obama and not Barack Obama at the same time; the law of non-contradiction.

The second from the three most prominent modalities would be nomological possibility, or physical and natural possibility. The laws of nature must coincide with the laws of philosophical naturalism and laws of nature. For instance, if someone is to proposition that the possibility for Barack Obama to be Barack Obama and not be Barack Obama at the same time, stemming from the logically absolute and modal claim, would mean for the nomological possibility that Barack Obama must thus be a monozygotic twin, hold the same passport, along with a 100% heritability score (although one could still name them different things)! Now what could be the possibility of such an event transpiring given his inimitably illustrious career and that no two presidents can take office at the same time?[128] Meanwhile the more contentious idea of nomological possibility is that the laws of nature are fixed, in terms of the relationship of masses, the speed of light, and others of the extreme instances of our laws of nature.

Thus we turn to metaphysical possibility, or broadly logical possibility. Notwithstanding the severity of logical and nomological modality, metaphysical possibility enforces more stringent laws than other modalities. The possibility of

The Epistemological Quest

anything being merely possible may be in how God intended for things to be, germane to ideas of intelligent design and contingent religious philosophies of how the laws of nature could have been (or be) different from now.

This book addresses the most significant modality adducing to what should be the isolated philosophy for human progress and flourishing; 'epistemic possibility'. An evidence-based 'epistemic possibility' dissects modalities for all we *could* know, and thus *know*! On progressively necessitating *what we know*, thus leads us to contractual philosophy evading Hobbesian 'states of war', and thus approximating more conciliatory 'articles of peace'. On the invocation of biological possibility as a significant modality, what if we cannot make the necessary concession on the apodictically alethic claim that we are all animals, and thus an individual claims that they must not be an animal due to what he epistemically knows? Well, as uncertain, and irrespectively uncertain as a solipsist stance may be on all epistemic claims, one cannot make plausible nor viable progressive concessions. On reaching *epistemic possibility*, let's now turn to the prism on knowledge and what is known to be good or evil in the world, given one of the biggest events that would shape the Western world in unfathomable degrees. To do this we will hark back to a time of classical antiquity. After years of Christian persecution and the awkward change from Republic to Empire, under the first Augustus (Octavian), we precisely found Constantinian, Byzantine, and Western Rome (particularly the years preceding 192 CE) on good and evil. But on the more than possible epistemic concessional cause for Christianity, we will start with Constantine the Great.[129]

'It would probably have surprised both Pliny and Trajan to discover that 2,000 years later the most famous of their exchanges is to do with an apparently insignificant, but awkward and time-consuming, new religious group: the Christians' (Beard, 2016, p. 476).

Constantine the Great

In the year 312 CE, the Roman emperor Constantine, born in Niš, today the third largest city in Serbia, experienced what some argue to be the most significant vision of possibly the last two thousand years. In his preparation before the 'Battle of the Milvian bridge', a gruesome battle which was to take place against the inexorability of the Roman emperor Maxentius, a 'remarkable sign appeared in the heavens above the Sun' (Nicholson, 2000, p. 310). As the historian Eusebius of Caesarea accounts, on gathering his troops in what was also witnessed in such mystified amazement, Constantine's dream in which Christ had appeared to him with the same sign (which many record as the Chi Rho and became a part of imperial insignia), led to Constantine's sympathy for Christianity; then eventual conversion. Incumbent upon the story of Constantine's vision, in as much as historians and theologians have assiduously collated archives pertaining to the events that occurred amid the Battle of the Milvian Bridge, was the demeanour, countenance and character of such an inimitable character, given the idiosyncrasies and essence of such a figure. And to account for the endeavour of merely touching upon the enigma of a figure as Constantine, compared to other contemporaries subsequent to 192 CE (the end of Commodus' tenure), and considering he initiated what came to be a long and robust Eastern empire, one has to begin with a brief account of what played out as the preliminary politics of Constantine and the Byzantine Empire.

In a rather chronological approach, the British military historian, Charles Oman, laid out a sequence of the Byzantine events that led to the fall of the empire in 1453, at the death of Constantine XI and Mehmed II walks into the famous Hagia Sophia to

The Epistemological Quest

eventually put an end to the empire, on its second siege and its turn to the mosque. According to Oman, Constantine I, on unwittingly founding a successful empire, achieved the unthinkably ingenious task of sailing the shores of the Bosphorus straight and estuary of the golden horn, concluding, what would become the city of Constantinople would be a most congenial location to build and govern its citizens (Oman, 2008). Constantine, akin to the geography of the shores, land, and estuary, saw it impenetrable, and that all entrances into the empire would almost coruscate visibility, thus rendering it difficult for oncoming sieges and snap attacks attributable to the dimensions engendered by altitude, precipitation, and geometrical spaces. Fundamentally, the city would be self-fortified.
His ingenuity could be ascribed to a conflation and susceptibility of rare talent, but most importantly what had been vicariously assimilated from his father, Constantius Chlorus, in what would be the armamentarium to rule an empire, hence leading to what had become the battle to become the single ruling emperor of both the east and west in an endeavour to expand and become arguably the most canonised emperor of Rome.

The Tetrarchy

Notwithstanding the myriad accounts given on Constantine I by authors and historians (such as Eusebius and Lactantius), the vacillation between the claim that Constantine had usurped his power or took what was rightfully his has been justly accentuated in academic circles. In order to understand this, one has to unfurl what was meted out by Diocletian. Diocletian, notoriously infamous to the Christian world for the *Great Persecution* of Christians, had formed the imperious system of the tetrarchy, which was to preclude inherited rights to the throne and instantiate meritocracy as the prelude to imperial power across the Roman empire. And so, as it stood, the tetrarchy would entail of two senior emperors who would encompass that nominal title *Augustus*, while two prospective emperors would encompass the nominal title *Caesar*, in which the demise of the Augustus would invariably lead to the succession of Caesar to Augustus, and what Diocletian would learn, even before his death, is that the tetrarchy would be indubitably ephemeral. Given the colossal geography of the Roman empire, the east was ruled by Diocletian (as Augustus), and the west by Maximian. Constantine's father, Constantius Chlorus, had been somewhat anointed and promoted by Maximian and chosen as the Caesar of the west, while Galerius had chosen for Caesar to Diocletian on the East. As nepotism, partisanship, politics and power would inevitably play a part in such a debile practice of an imperial judiciary, this was about as far as the tetrarchy would appear successful, which barely succeeded a generation. Prior to around 303 CE, a tacit agreement was made that Maximian's son, Maxentius, would become the new Caesar, and that Constantius' son, Constantine, would also prospectively take subordinate position. The Caesar of the east, Galerius, however had ulterior motives with Diocletian in the east. This was that Maximinus Daia and Severus would become the new Caesars to rule the empire.
On the 1st May 305 CE, on Diocletian and Maximian abdicating their position as co-emperors, Maximian handed his purple cloak to Severus, proclaiming Severus as new Caesar, while proclaiming Constantine's father, Constantius, new Augustus. in which Constantine had stormed out of the court through mere indignation. in the eastern realm of the empire, precisely in Nicomedia, while Constantine was soothing his own disconcertion, Galerius was also proclaimed Augustus by Diocletian, rendering the two new Augusti of the Roman empire powerful *imperators*, essentially holding the wanted keys of the kingdom. In so far as Constantius and

Galerius nominally held the same position, it must be tacitly agreed Constantius was seen as more of a senior authority, attributable to his partisanship with Maximian and his cordial senior relations with the courts, in which Constantine would eventually come to surreptitiously capitalise on.

As Constantius crossed over to British soil in a military expedition against the Picts, which would eventually lead to his death in 306, Constantine remained apprehensive about the precarious future he would obtain throughout the empire. All this was until he received news of his father's state and thus travelled across to Britain before his father were to die. Nonetheless, it was a Germanic ex-tribal king of the Alamanni, Crocus, who would be a driving force for Constantine's testimony as Constantius had proclaimed Constantine *Augustus*, in so far as it would prove inimical to the tetrarchy, which was merely nested on meritocracy, instead of a system in which imperial power would be duly inhered. As Crocus accompanied Constantius to York during the time before his death, while possibly commanding a robust auxiliary unit of between 3,000 to 6,000 men, Constantine obtained the necessary hubris to see out his plans coalescing and embarking on his claim as the new Augustus in the West. Albeit Constantine knew Severus would be 'the'/'a' rightful successor, this only presented Galerius with three options once receiving such information while fulfilling duties on the east: either 'to try to displace Constantine by force, to ignore him, or to recognize him as the new *caesar* in the West (Doležal, 2019, p. 28)'. As Maximinus Daia had been proclaimed *caesar* in the west, and on Galerius calibrating the best stance to evade civil war or gratuitous skirmishes with Constantine's claim, Galerius took the judicious approach and proclaimed Constantine *caesar*; and so he sent him his purple robe. Consequently, on Constantine having his position to the imperial inculcation legitimised in the 'third Tetrarchy', Maxentius, son of Maximian, only arrogated his authority across the land of Italy and reprised Constantine's act of endeavouring to inhere his rights to the third **Tetrarchy (Doležal, 2019)**. This is what would consequently lead to the 'Battle of the Milvian Bridge'; and subsequently, Constantine's vision…

Constantine, being son of senior *Augustus*, trained in administration, finance, and experienced in military expedition, rendered Constantine an inimitably preeminent force, holding a demeanour and sense of courage unparalleled with those of his fellow emperors. Notwithstanding Constantine's cross with Christianity, and how perspective was so distinguished from the former Diocletian, he was the force behind his own teleological motive to become as powerful as he did, change the fate for the cult, and eventually form one of the longest and most vigorous empires witnessed in human history. For Christianity, and for many historians, it was all instantiated in the *edict of Milan*. An edict which retrieved the possessions to the Christians, banned further pernicious persecutions, emancipated them from prisons, and would form the foundation for the religion to expand congruously with the Roman empire.

Bresheit

On assiduously auditing the very first verses, chapters, and pages of *the beginning* with a keen and sceptical eye, it may be rather onerous in avoiding the temptation to parse and discern such evolutionary flaws in what God may have duly created in that first curious week:

'Then God said, "Let the land produce vegetation: seed-bearing plants and trees on the land that bear fruit with seed in it, according to their various kinds." And it was so. The

land produced vegetation: plants bearing seed according to their kinds and trees bearing fruit with seed in it according to their kinds. And God saw that it was good. And there was evening, and there was morning – the third day' (New International Version, 2000, Genesis. 1:11-13).

...

'And there was evening, and there was morning- the fifth day. And God said, "Let the land produce living creatures according to their kinds: livestock, creatures that move along the ground, and wild animals, each according to its kind." And it was so. God made the wild animals according to their kinds, the livestock according to their kinds, and all the creatures that move along the ground according to their kinds. And God saw that it was good' (New International Version, 2000, Genesis. 1:23-25).

In God's creation of the Garden of Eden, in its literal sense, recursively spurning objections point out creationist verses such as, in that, if evolution didn't exist, carnivorous plants were essentially starved (or fasted) for the first few days of creation until they could eventually consume some type of energy. Objections that follow could be, how would an explanation of complex animals such as Cnidarians suffice without evolution considering such peculiar intermediary complex forms? and so on. As of the third decade of the twenty first century, it may be politically safe to allow these epistemological frustrations to subside, and endeavour to pull apart the *articles* that, despite one's antipathy with religious documents as such, generate *peace* as per the theory of methodological naturalism alone. And so, being necessary to fathom the context of such scripture in its mere psychoanalytical sense and on peering through the historical lens, what's often elided in the critical approach of such time is just how existentially strenuous life had been, in a society amiss of the unspeakable privileges we do not vacillate in capitalising on almost every day. Essentially, it helps in meditating on humanist phenomena parsing the vagaries within ontological existence over two hundred years ago, firstly in perceiving those humans as philosophically distinct animals faced with an unfathomable amount of death, toil, and importantly, sacrifice. To achieve this without appropriating the responsibilities of a historian, we should attempt to undergo an important exercise which will be revisited throughout the very juxtaposition of *good* and *evil* in this part of the book. This exercise should allow us to be carefully critical of every possibility that will be mapped out with clear rules as a philosophical subject, which fundamentally, should mediate between what's *manifest* and *latent* as a deeper form of meditation. Contrary to the laws of space and time briefly perused throughout the essence of part I, we must break those laws and enter, for the sake of simplicity, a meditative time chamber. Played out in many scholarly activities, it proves effectual in so far as the student, many times of adolescents and early vicenarians, is stopped at the vulgarity of the extreme measures concomitant with the foundations of such meditative exercise. For the purpose of epistemology, we will transpose the hinges and traverse the spectrum of comprehended space and time germane to what will appear merely foreign to the common ontologies of today, which render for those approximating atheism an inverse **intraspecific epistemology**. Bear with me on the story, as there is a significant point I aim to make from this.

The Epistemological Cul-de-sac

Robert

As Robert glances through the window and scarcely assimilates the iridescent contours of the meadow, he speculates on his evening activities and rejoices in another day thus passed. He finds that tomorrow is yet Sunday. Robert, in such somnolent rumination, ponders on as Alfred continues to count what was given in fee as he gruntles loquaciously on account of his wife's dyspeptic caprices amid his social relations in the manorial court. Robert contemplates the awaited rest he will gladly relish tomorrow by waking up an hour later, feeling the recursive exhaustion he so often endures as early as Tuesday evening, but takes its toll so onerously by Saturday afternoon, which he avails himself of the propitious time to initiate his excitement for his Sunday repose before service. Both Lucy and Sarah know not to bother, and that, in as much as they sporadically and peevishly bicker, and are found to laugh so raucously in their ludic activities, tacitly understand Robert must not be awoken and such reconciliation should transpire before breakfast, before Mother Margery's not so innocuous castigations result in such painful tolls. The porridge will be hot and served with tea, which will be ready approximating 08:15 am. Meanwhile, Robert has become so inured with the taxing impatience of service, so Robert casually finishes relieving himself and one hears those convalescent stretches and yawns, which sits exactly between 08:15 and 08:20 am without failure, compensating for those such strenuous minutes of tardiness in breakfast being served as Margery removes her apron and takes her seat, as the table decrepitly rocks and tumbles in its overused murmurs.

Alfred's wife's dyspepsia is somewhat akin to the discomfort Robert feels towards Alfred in having to pay his monthly feud. In fact, Robert doesn't mind watching the men in their robes as they enter the manorial court, the only thing irritating is the coercion in being held up by Alfred's garrulous talk of his spouse and the affairs of the manorial court, which, to be rather honest, Robert feels he already came to fathom autodidactically throughout his late vicenarian years. Not that Robert is so disaffected, Robert is good with numbers and is probably the best reader of his four brothers. But it's just that, to put it frankly, Alfred would be the runt of the manorial litter that Robert could possibly have to confer monthly to pay his fief, but maybe one would all be the same once known in such cordial affairs.

As it stands, in such a small district of Northumbria, Robert feels as blessed as he possibly could be by the hand of the Lord. His two daughters, Lucy, now eight years old, and Sarah, celebrating her seventh birthday almost three weeks ago, now, are nothing shy of the blessing the lord has graced Robert and his family with. Margery had lost two children to Robert coming on almost five to seven years ago now, and the next child, growing graciously in the womb that will bear the dearness of the Lord's will, will be baptised with the Levitical honour as 'Eleazar'. Robert's inclination to the stories of the Tanakh are the edifying properties in which Robert encounters his ontological status and security, which Margery also found so endearing at her betrothal all of them years ago. In the book of Leviticus, like Aaron, Robert's two children were placed into the hands of the lord for reasons that belie the justifications germane to reason, and so lie on the extraneous perimeters of such supernature, which Robert will once achieve closure when weeping in the arms of the Cherubim while repenting for such sins and imploring absolution from our Lord and Saviour Jesus Christ. As per Exodus:

'These are the names of the sons of Iseael who went to Egypt with Jacob, each with his family: Reuben, Simeon, Levi and Judah; Issachar, Zebulun and Benjamin; Dan and Naphtali; Gad and Asher. The descendents of Jacob numbered seventy in all; Joseph was

The Epistemological Quest

already in Egypt. Now Joseph and all his brothers and all that generation died, but the Israelites were fruitful and multiplied greatly and became exceedingly numerous, so that the land was filled with them' (New International Version, 2000, Exodus. 1:1-7). Aaron, the brother of Moses and spokesman of the Lord, bore four sons, the first born Nadab, Abihu, Eleazar, and Ithamar. The fate of Nadab and Abihu, descendants of the twelves tribes of Israel, Jacob, had unfurled their misfortune as they were effaced by the wrath of the Lord in the tabernacle, in which their unauthorised access in such sacred occurrence in the Lords perfection cursed the lives of the first two sons of Eleazar, putting a blemish on the Levitical priesthood of Aaron's very first two sons.

'Aaron's sons Nadab and Abihu took their censers, put fire in them and added incense; and they offered unauthorised fire before the Lord, contrary to his command. So fire came out of from the presence of the Lord and consumed them, and they died before the Lord. Moses then said to Aaron, "This is what the Lord spoke of when he said: 'Among those who approach me I will show myself holy; in the sight of all the people I will be honoured.' (New International Version, 2000, Leviticus. 10:1-2).

Eleazar was the anointed answer to the morose within the hearts of dear Margery and Robert. And the hearts of those were merely compensated by the blessing of their two daughters as they both survived their fifth birthday. The blessings extend to the poison inflicted upon infidelity and heresy that lurks eminently and beyond the demarcations of Northumbria. Robert duly pays his feud and insouciantly casts away anything indicative of the ingratitude he dare not bear on his Christian heart. His father, of the unfree serfs, demonstrably raised a family and did so well. Robert's success could be ascribed to his intellectual prowess, freeing his fate from the peasantry his father had endured and enabled him to trade the Gestum with the feud he pays to his Lord, giving his family the protection of the Lord's hand in all affairs and inciting the quest for righteousness to inculcate him with the affairs he will allow his son Eleazar to eventually bear for all of his worth. The sutures that rest on the hearts of men alike will be borne by the parturition of the very nature of what he has been instructed to fulfil in his dreams of what Eleazar will become and will come to rest on the calamities endured by his two loving daughters.

The Lord, in his miraculous countenance mere flesh endeavours to understand, becomes painfully misconstrued in its imperfection. It is evident that the pain in Robert's heart for the loss of his sons be rendered irreparable. His consecration to the Lord is the only manner in which Robert is absolved from his fallibility. 'Blessed be the soul of Robert', says Robert's local priest.

Alfred

Now, Robert pays his Feud timely like clockwork to Alfred. And as Alfred continues to speak, which Robert deems imperceptibly garrulous and loquacious (in that Robert avails himself of the opportunity to ponder on such loquacity), Alfred thus alleviates himself of manorial duties by externalising such affairs on about the only person phenomenologically extraneous to the political disquietude of the court. Alfred was part of what the feudal state calls *Homage*, a body of subjects whose allegiance was allied to the manorial court, and what Alfred had so fortuitously evaded akin to the menial nature of most of his scope, was that of the *Host*, a feudal military service in the Lord's army. As Alfred's endeavour to commiserate on the vagaries of the female run acrimoniously amiss, Alfred's heart remained interminably vacuous, justly ascribed to the unfortunate passing of two of his closest siblings. He often ruminates on the fate of his long-lost brother's widow, and the nieces and nephews thereof, losing a husband and father in another war with the French. King Phillip II of France, in his reprisal, had promised to regain the Angevin from King John of England, resulting in the collapse of the ephemeral Angevin empire. Alfred's brothers, baptised as Mark, Albert, and Luke, were issued from their governance as knights to fight in the King's battle but with the protection of the lord at their hand. Alfred's warmth towards his brother's widows is empathic, albeit his heart was rendered vacuous attributable to the death of the closest of his brothers, Luke, who's spouse bore three children all dying at parturition, and successively it seems their hearts died alongside them. Alfred's estate still seeks the solace in such manorial distractions as he, who unlike Robert and the majority of the populous Britons, enjoys his time parsing the Gospel independently, along with the epistles of Saint Paul as Alfred's intentions lie with the nebulousness of other tomorrows.

'Now listen, you who say, "Today or tomorrow we will go to this or that city, spend a year there, carry on business and make money." Why, you do not even know what will happen tomorrow. What is your life? You are a mist that appears for a little while and then vanishes. Instead, you ought to say, "If it is the Lord's will, we will live and do this or that." As it is, you boast and brag. All such boasting is evil. Anyone, then, who knows the good he ought to do and doesn't do it, sins' (New International Version, 2000, James. 4:13-17).

As per the book of James, the *patience in suffering* sees how the farmer waits for the land to yield its valuable crops and is patient for the autumn and spring rains as the people await patiently for Christ's coming (New International Version, 2000, James. 5:7-8). Alfred's imperceptibility of the *telos* of life dawns on such moribund misfortunes he acquired of late, and so the epistles lay somewhat antidotal in intending to fathom the evanescence of his bereft heart, losing his siblings to the allegiance of His Majesty.

Margery

Margery awaits the man she most loves patiently, the father of her children, to join her as she lays the table while simultaneously chastising her children, Lucy and Sarah, as such hyperactivity yields to the echo of their father's footsteps entering the room to breakfast. Margery's righteousness and general predisposition, to care for her neighbour, yielded to the often villainous ramifications of her privileged estate, endured in the trauma when losing such children those very nebulous five years ago. Margery's alacritous affairs, in gradually consecrating her precepts and permuting those into a philosophy that became duly manifest, is what laudably delighted her community and those close by her in the vicinity of her such humble estate. Her privilege extended beyond the sociology of her estate as she was imminently honoured the responsibility of *Launder* in the manorial court, consolidating such a family position in the district and bringing the extra shilling to Robert's home, notwithstanding the paucity some would say is augmented to Robert's affair.

Be that as it may, on the previous day of Robert and his family beginning grace and poising themselves to breakfast, while Margery was engaged with daily gossip often revelled in, she would audit the Christianness of her fellow subjects (also revelled with fellow launders as they would finish laboriously folding linen). Margery's moral compass seemed to be somewhat unsynchronised due to the vitriol of the story she just so apparently heard within the course of that particular conversation.

As Margery would recount, it was a bondwoman in such an affair she heretofore wished to remain unmediated by the manorial court. Alfred, who we know receives the feud from Margery's husband, Robert, serving in the manorial court, mourns the loss of his younger brother, Luke, who died in the service of His Majesty against the victorious belligerents of King Phillip II. Ironically, Luke, in his fortunate career in his knighthood in a prominent district, often sought political affairs in the manorial

court within what has been merely suspected to be calumnious political support of a fellow vassal, who goes by the name of George. On such sordid and surreptitious machinations, in an avaricious attempt to abound to his estate, Luke thus availed himself of the prestigious position he accrued with a non-consensual courting of the said bondwoman, leading to her gratuitous conception. Margery's disquietude lies in hearing the bondwoman was consequently levied with a fine that not only she is unable to pay, but also that her masters are unwilling to pay, attributable to unchasteness on her account. While the intervention of which disseminates amidst the district as the latest cause for discussion, Luke's widow, who we know unfortunately had lost her three children at parturition, only continues to drown herself in the very melancholy that succeeded the loss of her dear children, all the while the bondwoman is charged with the crime of *Leywrite*.
The priest recites the words of Psalm:

'The Lord watches over the alien and sustains the fatherless and the widow, but he frustrates the ways of the wicked' (New International Version, 2000, Psalm. 146:9).

As Margery continues her rumination that besets a moral compass, Margery's estate is only shrouded by the gratitude often run afoul of the proclivity to incessant returns of ingratitude, which subject themselves to the human impurity of Eve enraging the Lord. As Margery seeks solace from her priest to cure her of recessive adulterations, he tells her the story of Jesus:

'Jesus sat down opposite the place where the offerings were put and watched the crowd putting their money into the temple treasury. Many rich people threw in large amounts. But a poor widow came and put in two very small copper coins, worth only a fraction of a penny. Calling his disciples to him, Jesus said, "I tell you the truth, this poor widow has put more into the treasury than all the others. They all gave out of their wealth; but she, out of her poverty, put in everything – all she has to live on"' (New International Version, 2000, Mark. 12:41-44).

On meditating the corollaries of such realities transpiring after Constantine's revelatory vision, in a devised system of feudalism as exemplarily successive to Constantine's vision and consequent power, it must be understood how Christianity had become so well instantiated in the very hearts and philosophies of ordinary society, interwoven with *death*, *injustice*, and *sacrifice*. The religion of Christianity some will argue could have been any other expansive sect. As it can be fairly understood, there were other sects that were a corollary of apparitions and those very religions still fruitfully and universally intersperse today, populating the planet in millions. This matter of fact I continue to ascribe to a *burden of proof* as mentioned earlier, interwoven in *methodological naturalism* as the failure to procure philosophical naturalism as the alpha philosophy. The rhetoric of the story of Robert, Alfred, and Margery is... to lay the following challenge.

~

While meditating on such myriad calamities endured in their lives, how would you, in travelling back eight hundred years to a time of an early thirteenth century, successfully brandish what you *know* about the world to convince the members of this feudal state otherwise, in that it suffices to allow their religion to wilfully subside. I'm sure it's reasonable to assume it wouldn't be so difficult a task given the eight hundred years of difference in time. Be that as it may, I would argue that probability runs against you. For instance (as you read these lines), I'd say the probability of you being a medical doctor is low, and even if you are a medical doctor, there is a low probably that what you *know* suffices in allowing such religion to subside, in that your specialised field is most probably insufficient in compensating for the sacrifices the characters would most continue to have to endure, or the death they would continue to attribute to such supernatural will. And this is all merely irrespective or what you assert or *know*. Then we should invoke the probability of you being one of the most honourable scientists, a theoretical physicist or any of the most illustrious cosmologists, biologists, ethologists, chemists, quantum mechanics, and so on. The probability of you being able to explicate what you know, evading the esoteric nature of such fields, to such lay estates of Robert, Alfred, and Margery approximates unlikeliness, principally, imputing the matter to Robert's dearth in literacy, and Margery's complete inability to both read and write, so anything written would appear to be somewhat futile.[130] The safest bet would be to speak to others associated with the sciences of the time, but one must run into the same cul-de-sac in the onus of the scientist having to explicate what would be speculatively understood back to the three members of this rural and feudal state.

It seems the highest probability in endeavouring to challenge such religious instantiations may then lie in philosophy (namely eschatology). As Trisel laid out the alternative perspective of death through the interrogative of, 'Does Death Give meaning to Life (Trisel, 2015)' it seems that a well ensconced, and unassailable philosophy would be needed to come to terms with life, germane to such senescence and vicissitudes of their time. In as much as one may disagree, it appears as though writings of the Old and New Testament were leading philosophies for the layperson, as it is embedded with what became a social philosophy of alleviation of the severity of death, a mechanism concomitant with space and time.

The Epistemological Cake

It may help to acquire a good gist on how I compartmentalise, or lay out, all human *knowledge* throughout the world. Importantly, I set out knowledge as the layers of a cake, the foundation of all knowledge being science (*Scientia*). So, I *know*, rather than need to *justify*, all knowledge to be scientific and part of the natural world as understood by the philosophical naturalist. The next layer embedded in a world of knowledge is philosophy (*Philosophia*), which mediates between what would be the top layer, the superficial layer as the surface; all social science (*Societas*). Along with the middle layer of the cake I would also inculcate psychology, psychoanalysis, and anything related to the psyche, in that what is merely propositioned must not have to be scientifically *tested*, or *justified*, but propositioned for the foundational sciences to thus investigate. The social sciences will ethereally proposition free of the 'burden of proof'; hence we find democracy on the surface of all knowledge. For the social sciences to reach the foundational sciences one must utilise philosophy, and for the foundational sciences to reach the social sciences one must also embark on the task of using philosophy. The layperson, as per democracy, stands at the surface of all knowledge and thus avails herself of her franchise in engaging politically (through the means of democracy). Albeit her democratic position holds no epistemological status, or purchase, in any of the realms of foundational science which I hope appears epistemically self-evident for the reader. Religion plays its role throughout all of the layers of the cake and claims that knowledge can be *justified* through other means, instead of such claims from the foundational scientists who will continue to be incessantly frustrated as *justification* should only be rendered through scientific means. The field of theology throughout institutional universities also capitalise on their inalienable franchise to invoke faith as a form of justification throughout this middle layer of the cake, which again should be kept clear from the form of *justification* used in foundational science. Hence, for knowledge, it seems *society*

regards those with a proclivity to the foundational sciences to be often highly regarded and even much more remunerated, compared to those subscribed to knowledge within the contours of the top layer, attributable to its method of justification which has evidently aided humanity in various and still yet unspeakable ways.
The conundrum for the epistemological cake lies in its foundation. A most remarkable fact is that the foundation of the cake is the *youngest* part of the cake, in so far as science played its utilitarian role throughout history. The oldest is the middle, in that the middle, engendering its archaic philosophy, has played out as the foundation for epistemological knowledge for time immemorial. This is merely conducive to the prehensible grip many still have on its utilitarian role, entailing a stronghold on such *narrative* philosophy as an antidote to human suffering. Ensconced is the predisposition to garner its invocations of meaning from the middle, which has been historically adulterated in a sense of spiritualism that was not only deleterious to progress, but sheerly barbaric! The birth of foundational science became conducive to even more homely cruelty within pseudoscientific theories such as phrenology, used across abolitionist politics in late eighteenth and early nineteenth century, and thus head on into Nazism, that would adduce any leading Darwinian theories to justify some of the most depraved behaviour in human history.
If one would object to *epistemological purchase* and its use in foundational science, let's peruse a familiar instance of how a political event has this structure shaped out in the epistemological cake. In 2002, when the 9/11 attacks on the World Trade Center by radical Wahabbi Islamist terrorists occurred, how would you know it had, or was, happening? The first source of knowledge you may have assimilated would have been via journalism, a TV station, or prolific gossip. On endeavouring to discover more about the real causes of the event after the initial shock and while completely transfixed by the nefarious nature of the attack, one was inclined to investigate more. The epistemological layer such *knowledge* is situated is the top, what journalism plays out as the geopolitical filter for many people throughout the world. TV stations compete with the best of specialists to procure information on the attack. The first with the necessary *epistemic purchase* for television were geopoliticians. Geopoliticians who successfully informed would be said to 'dabble' in the middle, whether it be the politics of the inculpated, or of subsequent actions of the United States. The most epistemological purchase regarding the attack was given to pilots, architects, bomb disposal officers, fire protection specialists, air force specialists, coroners, physicists, and the rest of the most preeminent scientists who fastidiously endeavoured to garner as much traceable evidence that could be adduced to the jigsaw of such a depraved attack.
As I will reverberate, as I continue in such an analogy, one cannot dispel of philosophy. Known tacitly well in the academic community, is that, with all the time subscribed by the scientist to compete effectually, and approximate anything indicative of a Nobel Laureate, the scientist thus yields to the art of articulating her prowess to those within the spaces of the top. The duty is thus transferred to philosophy to endeavour to meditate on such discoveries and possibly politicise those very discoveries. The humanities, those residing firmly in this top layer, will hold an inimitable and often flawless repertoire in appealing to the masses, those that principally reside close to the surface of this cake, using the consolatory gift of charisma, while amalgamated with its cadence and emotional displays.

The Epistemological Quest

A Philosophy of Sacrifice

Even today, from a merely mathematical perspective, one must be expansive of 13,800,000,000 years in meditation. Following the expansion would be a subsequent 9,200,000,000 years until the formation of the solar system until life most possibly came along (between such a time and 2,500,000,000 years ago). As I avoid continuing in such fashion, the point is that this is an unfathomable amount of time to logically endeavour to fathom. Less than 0.0001% of the 2,500,000,000 years ago, the start of the Proterozoic eon, would be enough time to extirpate all meaning in an individual's life today, which is as long as two and a half thousand years! That would be an extirpation of family, friends, music, education, literature, technology, sport, work, politics, science, religion, and all of the things that embed the meaning of daily life to give one the ontic security to sleep until tomorrow. If we look at life expectancy of the thirteenth century, which had been as comparatively short as sixty-five, how else could such a cognizant animal as *Homo sapiens* endeavour to cope with this Darwinian *struggle for existence*.

My suspicion is that it would have to predicate itself on a powerful and yet accessible philosophy of *death*, and this I believe has been inimitably achieved, all through the powers of Christianity. Espoused not through the entirety of the symbol of Jesus Christ, but his *death*. Concomitant with such powers are the powers of fable, lore, or narration. Narration, as it evolves with space and time, in that one could not find another patient enough to endure the impatience of such temporal structures on succeeding adolescence, is a feature of humanity that no other species has the intellectual faculties to rejoice in. This extends from lovers of TV series to the lovers of narrative philosophy. As I lay out some of the most powerful narratives dovetailed in the moral philosophy of such good and evil, a powerful and yet implicatory religion also perched upon the west; Islam.

As Charles Oman (2008) laid out variegated corollaries conducive to events throughout the vigour of the byzantine empire, the crusades of the Popes, skirmishes of the Seljuks, and the interminable fight for control of Jerusalem, such events espoused both sociological and geographical imputations for a multiculturalism that would see itself out. In which a slice of human will is taken up to five times a day, as acolytes face both al-Masjid al-Haram and the Kaaba (the qiblah), they direct their prayers. As one of the famous translations of the Quran is versed:

'And from whatsoever place thou comest forth turn thy face towards the Sacred Mosque. And wherever you are turn your faces towards it, so that people may have no plea against you except such of them as are unjust – so fear them not and fear Me – and that I may complete My favour to you and that you may go aright' (Holy Quran, 2002, 2:150).

As the death of the prophet in 632 C.E. became so poignantly indelible for much of a theological world, it was only the inception of a Rashidun Caliphate and its military expedition until 661 that led to the eventual fall of the Sasanian empire in 651 C.E. But such military success had come from prior military expansion of the Maghreb between 647 and 709 C.E. (across populated areas of the Sahel). It happened so that by 642 C.E. Mesopotamia, Syria, Egypt and Armenia were invaded, concluding a belligerent conquest of the Persian empire with the conclusive defeat of the Persian army at the battle of Nahavand (Blake, 1950). Ascribed to such commandeering of the great Caliph Umar, and the subsequent three Caliphs ossifying Islamic philosophy throughout the west, such success led to the bifurcation of the Sunni/Shia divide we observe throughout Islamic and religious credence today. And as we find of the vast majority of the Shia population, as of 2014, most reside in the historic and controversial Persian state we know as Iran.

Nation	Shia	Total Population of Muslims
Iran	90 – 95%	99%
Azerbaijan	65 – 75%	97%
Bahrain	65 – 75%	70%
Iraq	45 – 55%	99%
Lebanon	45 – 55%	61%
Syria	15 – 20%	93%
Afghanistan	10 – 15%	99%
Pakistan	10 – 15%	96%
Saudi Arabia	10 – 15%	93%
Turkey	10 – 15%	98%
Eqypt	< 1%	95%

Figure 1.3: Shia population (Lipka, 2014)

The Sunni/Shia religious credence somewhat bifurcated attributable to the belief in the Prophet's closest companion Abu Bakr, grounding such beliefs of the Sunni, in that one must endeavour to embody the prophet in all of his ways, hence consecrating oneself to the manners, demeanour and countenance of the Prophet (that of the Sunnah). Conversely, credence of the Shia becomes instantiated in the philosophy of the Iman, through such derivations leading back genealogically to the familial relations of prophet Muhammad instead of from extraneous tribes that followed his prophecy.

An Antithetical Islam and its Progressive Concessions in the West

Bill Warner, in so far as he's branded radical (and interestingly enough a former physics professor), staunchly contends a scientific approach onto the doctrine of Islam (and in exercising the freedom to render such a doctrine reprehensible). Warner conveys Islam to be founded under a trilogy played out as an apparatus of the doctrine; that of the Hadith, Sira and the Quran. The Hadith would be as much as 60% of the doctrine, the Sira 26%, and the Quran, 14%, anathema to the misconception that the Quran would be the principal religious text. As Warner points out in his analogy, 'is that the Sunna is water and the Sira and the Hadith are the glasses that hold the water' (Warner, 2010, p.10). Warner's turns to the *Dhimmitude* which the doctrine professes, which, fundamentally, would be non-Muslims under Muslim rule. Warner turns to more of a belligerent stance against the *Kafir*, which he believes *the apologist cannot build a bridge of compromise*, as he deems such endeavour logically impossible. Warner ascribes the very caveats to the issues that the doctrine of Islam professes, for instance, how a Meccan Koran

professes how one should treat the Kafir as equal, while the Medinan to profess treating the Kafir as enemy, a mere contradiction according to Warner.[131]
'Those who disbelieve and hinder (men) from Allah's way, We will add chastisement to their chastisement because they made mischief' (Holy Quran, 2002, 16: 88).
Maajid Nawaz, the prominent British activist and radio presenter, enduring much in a journey to politics leading to his counter-extremism stance, is a prominent figure for what I would call Islam's case of identity politics in the United Kingdom. From the plight of vicious racism throughout denarian years to advising the 43rd president of the United States, George W. Bush, in a Texan garden on the 'War on Terror', Maajid Nawaz tells his story of being radicalised throughout his subscription to the Hizb al-Tahrir, leading to his incarceration in Egypt for as long as five years. In such five taxing and stultifying years, Nawaz appeased the irreconcilability of his complicated British identity with the carefully assiduous study of the Quran, the Arabic language, and quite gracefully, classical English literature. The liberalisation of his dogma, initially incited by his own mother in a chapter he honourably names 'The womb that bore me', enlightened Nawaz in the then subsequent conquest of deradicalization, in what one could call a righteous effort to protect the amenable minds of such pernicious dogma as he ventured in a rather successful reign across Pakistan and the United Kingdom.

Chapter 5

A Balance of Good and Beyond...

It was just over four years ago, on the morning of 22nd of May 2017, on getting ready for work for what would be another regular working day, that I received messages from friends around the world asking if I was okay. I hastily switched on the news, and sadly heard of the twenty-two unfortunate victims of a heinous suicide bombing and hoped in these quasi-condolences I had not personally known any of the people who unfortunately lay victim to the attacker's insurgency. I could say, being relatively fortunate, that I didn't know any of the people who were murdered, however, and unfortunately, I more than ironically recognised the murderer. His name, Salman Ramadan Abedi, born of Libyan parents of the Salafi credence, had been a contemporary of mine in the very same boys-public school I attended in South Manchester. My high school, sporting a population of around 900 boys at the time, a relatively small school compared to others in the South Manchester region, was the school we had shared. The sudden and stricken disquietude of the murder evoked indelibly vivid experiences of my high school days. The high school happened to be situated in the unpropitiously rival area to where my brother and I were living, rendering us for the most part of the first two years of high school taciturn, merely attributable to gang affiliations and subsequent hostilities from 1990s South Manchester. Most importantly recalled of the experience, was the implicit fact that the school comprised of over 95% Muslims, with segregation dividing the school into what we knew as the 'Libyans', 'Somalians', 'Blacks', Indians (Sikhs), and then a vast majority of South Asian Pakistanis. Others would thus be interspersed Iranians, Turkish, Egyptians, Algerians, with less than 3% of the school white British. On enrolling on my first few days of high school, bearing in mind it was only two years after the 9/11 attacks (and my primary school being primarily white British with others forming minorities), most of the Muslim population in such novel experience were, quite frankly, indiscernible.

Like most things, becoming inured and acclimatising to the diversity of school life, I would remember the long conversations, disputes, and unsolicited cultural classes discerning such distinctions between religion, faith, belief, politics, geopolitics, and quite invariably, racism. Distinguished languages would inundate the school canteen, hallways, and social spaces while Friday prayer would be offered to the pious at Jummah. I felt that after leaving high school, I saw many Muslim ex-schoolmates somewhat stultified by the loss of hubris from such segregation throughout their denarian years, which was what I would come to experience two years later once leaving sixth form. Reality extraneous to the high-school experience did not entertain the Muslim by any stretch of the imagination, and as a result the inevitability of disaffection in the Muslim male was only heightened, thus expanding the susceptibility of radicalisation, which seemed to become the fate of Salman Ramadan Abedi, the boy who happened to share that same 22 bus ride home from school.

Salafis, Wahhabis, and Sacrifice

It was the year 1703, in Najd, Uyainah, present-day Saudi Arabia that Ibn Adb al-Wahhab was born. It seemed, just over a millennium since the prophet had died, that the ways of Islam were somewhat dilapidating and thus diverged from the original teachings of Islam. It became a case of definition, which somewhat inverted and amalgamated over time that distinguished the Salafist creed from the Whabbi. In reference to the most pious of predecessors, the Wahabbis preferred the term *Al-Muwahhidun* or *Salafiyuun* which referred to the Salafis who would consecrate their lives to the ways and teachings of Islam; praying, washing, and fasting properly; and most importantly, acting as the prophet would have desired (akin to the first three generations of Muslims). It was in the 1970s when the term 'Salafi' was associated with Wahabbism, and rightly so, as the Salafi creed is a subscription of honour to the Muslim, which one could argue ontologically would be the most pious of all Muslims. For the Salafi, the subscription to the *mazhab*, within Islamic jurisprudence would be instantiated in such creed, which would again be akin to the *Sunnah*, and the traditions of the prophet. The intercession between man and God, the *tawassul*, Salafis believe is a Muslim practice adulterating the puritanical sense of Islam, which they call the *Tawhid*. The common misconception is that all Salafi are Wahabbis due to the fervent ideological facets that combine them in such belief, but the matter is that Salafism is the much traditional sense of the doctrine of Islam, meaning not all Salafis would necessarily be Wahabbis (Bin Ali, Saful Alam Shah Bin Sudiman, 2016). When distilled, the misunderstanding of Salafism being a minatory doctrine should not be so, as the often long-bearded man would spend his days in long fasts and the edification of his being through parsing scripture and re-enacting the ways of the prophet, the *Sunnah*. The radicalised thus hijack the very alacritous and complicated religious philosophy to justify a vacuous sense of a *Shahadah*, martyrdom, and have thus stoked the sense of disaffection to be even equanimously professed in nations such as my own, the United Kingdom. Maajid Nawaz recalls his story of being surrounded by a group of racists who were poised to attack him and his friends being vastly outnumbered. His friend, on playing with such vitriolic ignorance of his belligerents, had warned the group that he was in the possession of a bomb he was willing to use in *the green rucksack* he was holding, thus willing to make such a sacrifice. On hearing the menacing threat, the asinine racists didn't vacillate and agreed to leave Nawaz and his friends in peace, in which Nawaz coined 'the green rucksack moment'. When effacing injustice and the reality of the evil that plagues the crevices in the world, the proclivity to supersede those very injustices seem to evolve in the deceitful face of sacrificial martyrdom, a *Shahadah*. The incompetent architect, once failing to construct the contraption that filters the evil they so confoundedly discover, is only left with the obsolescent issue that has plagued sociological reason for time immemorial, poor extrapolation. Innocent lives are lost, and the problem continues to perpetuate, or often exacerbate, without the martyrs sacrifice, and so those green rucksack must thus be spared in the inevitably meandering roads that lead to this ineffectual cul-de-sac of radicalism.

Harriet Beecher Stowe, the literary genius and political abolitionist on writing her novel almost a decade before the inception of the American Civil War, writes of Uncle Tom who toils the cotton fields of a cruel and villainous master, Simon Legree. Legree, foreboding administration of his plantation, bought his chattel in hope of promoting uncle Tom to the obsequious position of overseer, where he would capitalise on the arrogated perks procured by his master, at the cost of lashing other slaves to keep them in order. It was when he instructed Tom to lash two female slaves, that Tom vehemently refuses, leading to the vicious ramifications of Tom's recalcitrance. Stowe's invocation of John Newton's *amazing grace* is both the sacrificial and eschatological crux consoling him in his estate:

'How long Tom lay there, he knew not. When he came to himself, the fire was gone out, his clothes were wet with the chill and drenching dews, but the dread soul-crisis was past, and, in the joy that filled him, he no longer felt hunger, cold, degradation, disappointment, wretchedness. From his deepest soul, he that hour loosed and parted from every hope in the life that now is, and offered his own will an unquestioning sacrifice to the Infinite. Tom looked up to the silent, ever-living stars, types of the angelic hosts who ever look down on a man; and the solitude of the night rang with the triumphant words of a hymn, which he had sung often in happier days, but never with such feeling as now –

> *The earth shall be dissolved like snow,*
> *The sun shall cease to shine;*
> *But God, who called me here below,*
> *Shall be forever mine.*
>
> *And when this mortal life shall fail,*
> *And flesh and sense shall cease,*
> *I shall possess within the veil*
> *A life of joy and peace.*
>
> *When we've been there ten thousand years,*
> *Bright shining like the sun,*
> *We've no less days to sing God's praise*
> *Than when we first begun'* (Stowe, 1995, pp. 362-363).

Stowe embodies an evangelical narrative in the archetype of Christ in Uncle Tom, later killed by the pusillanimous nature of his master, Simon Legree, as Tom withholds information regarding the whereabouts of two escaped slaves, Emmeline and Cassy. The eschatological alternative to the darkest of human experience displays itself through Stowe's narrative, in how such doleful melancholy occurred throughout the horrors of an imbruted, and, according to Stowe, an unchristian political system of slavey.

Time and space was appropriated from the slaves on those plantations via villainously strategic use of terrestrial animals such as dogs and horses to ensure the evasion of escape, and the harsh deterrent of a dog's jaw, bullets, but most menacing, torture, to compensate for any scarcity of despondence on the plantation. Thus the only alternative was suicide, which would entail of the pusillanimous by-product of capitulation and mere acquiescence, so, 'Could there be another alternative?' For Stowe, yes there was. The alternative Stowe instantiates in Uncle Tom was the potence of grace, as Legree, in enforcing Tom to lash the two slaves reminds him that he bought him in *body* and *soul*. Tom recalcitrates his master's claim, making him aware that the only *One* who could own his *soul*, was, the Lord. The eschatological alternative of grace is practicably devoid of the pusillanimity of self-sacrifice, which would be the capitulating and acquiescing burden of suicide, as Tom 'shall possess within the veil a life of joy and peace'. The slave's heart is thus devoid of necessitating fear of such a nefarious master, the dog's jaw, being caught, tortured, and devoid of Legree's caprices, as for the psychopathological games the master would play was to gain arrogated and ultimate power of the subject. The slave could only win by invoking what was not earthly, as he ceded victory to his earthly fate, which would consequentially mean the slave slowly chipping away on the ignominy of his earthly defeat through the supernatural invocation of a stable philosophy, Christianity. As the verse in the hymn reminds the sufferer:

> *Twas grace that taught my heart to fear*
> *And grace my heart relieved*

The Epistemological Quest

*How precious did that grace appear
The hour I first believed.*

The eschatological alternate inscribed on the hearts of Stowe's readers allows the pain in her story to alleviate itself through a Christian and evangelical narrative in sacrifice. For a philosophy as ingenious as Stowe's, depicting the vitriol of slavery in the United States of America, it seems, that whether choosing to look at a philosophy of two thousand years ago, or of twenty, sacrifice will play out archetypically in empowering the senescent primate that evolved to the sagacious *Homo sapiens*, for it to be at least practicable or viable in any actionable sense.

Marriage, Responsibility, and an Illusion of Free Will: Before it Happens or After?

'You have heard that it was said, Do not commit adultery. 'But I tell you that anyone who looks at a woman lustfully has already committed adultery with her in his heart. If your right eye causes you to sin, gouge it out and throw it away. It is better for you to lose one part of your body than for your whole body to be thrown into hell. And if your right hand causes you to sin, cut it off and throw it away. It is better for you to lose one part of your body than for your whole body to go into hell' (New International Version, 2000, Matthews. 5:27-30).

As adumbrated in the first part of this book, in particular looking closely at sexual selection, and thus invoking the biological world and its alethic modal possibilities, the question thus obtrudes itself of the futility of marriage as we travel through the third decade of the twenty first century. Contraception has freed the viviparous female of the fetters of space and time and such suboptimal marriages; contraception has also freed the heterosexually viviparous female sexually. But for the injurious and unprecedented dangers of technological compensation for space and time, as mentioned in the first part of this book, what happens now or next?
Through the prism of biological modality, and keeping the argument purely ethological, I introduce my own, and understandably polemic concept of 'biological responsibility', to look at the situation in which two people become romantically involved. Before we venture biological responsibility, let's firstly give a true definition of what biological responsibility really could be. There is no better way to venture responsibility than to use the tool of the legal sciences; the Socratic method.[132] Responsibility according to the laws of space and time could look something like:

Since nobody can and has ever been able to travel back in time, we suffer the fate of being the mere fallible cousin of the chimpanzee, and so responsibility is, irrespective of the *depth* of the mistake, a mistake that will be avoided at all costs in and only in the direction of a *time arrow*. A mistake can be so grave the *soma* may espouse irrational behaviour; that which we call *trauma*.[133]

Unlike other modes of responsibility, the form of responsibility through the prism of biological modality necessitates two rules for the individual playing the game (the most special cases evidently notwithstanding):

1. I must keep myself alive; survive.
2. I must reproduce with the fittest I could possibly acquire as an organism; reproduce.

Without the confounded inclination of reductionism in such claims, leading to the *post hoc* fallacy that one should resort to living as those animals of the lower faculties of intellect, we must necessitate two rules within biological modality, given the context of such high faculties of human rationality we daily compete with. So, let's look at an example which would test the first rule.

~ I must keep myself alive: survive.

I am an inner city adolescent who will be turning eighteen in a couple of months in the year of 2009. My father, Nigerian, and my mother, both Jamaican and English, means that approximately 75% of my features are of sub-Saharan African descent while approximately 25% is northern European. One could argue that unfortunately, this may mean my ontological position in South Manchester could be counterproductive to an easy road in where I desire to go in my career. Gang violence through the 1990s in South Manchester was imbricated and conducive to the racial hostility that plagued the city since West Indians first disembarked the Windrush in the 1950s, and eventual sub-Saharan Africans and South Asians came to settle in the 1980s. To cut a complicated situation short, there are places for me that are suboptimal, what we know as 'rival territories'. Be that as it may, I should invoke the **first** article of Eleanor Roosevelt's draft on a Universal Declaration which holds my inalienable human right:

'All human beings are born free and equal in dignity and rights. They are endowed with reason and conscience and should act towards one another in a spirit of brotherhood' (Roosevelt, 2001).

In so far as I know there are areas that would *decrease my chances* of survival, must I mandate to the criminals I encounter that every human act towards one another in a spirit of brotherhood? Notwithstanding those idealistic philosophies, is it how human behaviour works? On the question of shirking responsibility on to those that must protect me is where we reach the fine line in political philosophy of how much authorities take responsibility in the event of a crime, both for the criminal and the victim, and both **before it happens and after.** As the British 19th century philosopher John Stuart Mill wrote succinctly:[134]

'It is one of the undisputed functions of government to take precautions against crime before it has been committed as well as to detect and punish it afterwards. The preventative function of government, however, is far more liable to be abused, to the prejudice of liberty, than the punitory function; for there is hardly any part of the legitimate freedom of action of a human being which would not admit of being represented, and fairly too, as increasing the facilities for some form or other of delinquency' (Mill, p. 81).

But when Mill says freedom, what could *freedom* possibly mean?

Hitler, Volition, the Pre-supplementary Motor Area, and Alleviated Responsibility

Between the months of November 20th, 1945, and October 1st, 1946, lawyers of the Nuremberg trials were effaced with an indiscernible problem. The war finally ended, Adolf Hitler had died, and lawyers were amassed with a group of the most villainous and high-ranking Nazi officials. Upon perusing the reasons of evil, and when asked why they did what they did, it was imperceptible whether responsibility was simply

assumed or shirked; if responsibility was plausibly or implausibly mitigated, which is thus a necessary pathway into the field of neuroscience.
Patrick Haggard, the neurologist of University College London, was one of the neuroscientists who requestioned volition in human will, and if, upon metrics of neuroscience, volition was measurable. Through the division of *intention* from *action* when making decisions, one avoids the problem of dualism in acting as if they were the same, and through neural circuits if one can find *who* it is who is actually *acting*. Let's say you contemplate on whether to raise your hand. What is it that allows *you* to raise it? Haggard displays four significant parts of the brain that allow for a decision to be eventually made. The frontopolar cortex (within the prefrontal cortex) would be responsible for the prior *intention* of deliberating whether or not your hand will be raised. The pre-supplementary motor area (PreSMA) would then be responsible for the *preparation* and 'readiness potential' of acting. The supplementary motor area (SMA and part of the brain shared with other primates), communicates with the primary motor cortex for body-specific preparation that moves to the spinal cord, on to the muscles, and then on to the movement. In patients with Parkinson's disease, reduced output to the preSMA from the basal ganglia is known to show (a) less frequent *actions* and (b) slower *actions* than healthy individuals. Thus Haggard attributes our 'urge to move' to direct simulation of the preSMA (Haggard, 2008).
Research from Chun Soon et al. (2008) looked at unconscious determinants of decisions in the human brain, 'where the earliest unconscious precursors of the motor decision originated in frontopolar cortex, from where they influenced the buildup of decision-related information in the precuneus and later in SMA, where it remained unconscious for up to a few seconds' (Soon et al., 2008, p. 545). Sitting between the temporal lobe and the parietal lobe in the cerebral cortex we find the temporoparietal junction. Two hypotheses are of a disruption to the right temporoparietal junction (RTPJ) with transcranial magnetic stimulation (TMS) in belief attribution, and that of the TMS to the RTPJ causing disruptions of other cognitive functions around those principal areas (Young et al., 2010). As both *intention* and *action* that urge us to act are merely products of physiology, the question is thus posed of those Nazis, who can truly take responsibility for all intention and action?

~

As I hope we have hitherto understood, all human behaviour is produced by physiology, and thus every production of physiology must have evolved through a means of natural selection. It means decisions that influence one's physiology is only the environment and its influences in it, which thus becomes a cause of natural selection. If the cortex is the principal part of the brain being responsible for decisions (of *intention*), then human volition through *space* and *time* is most especially flawed in the legal sciences. As Haggard explains:

'All known human cultures have the concept that an individual is responsible to society for their actions. This in turn rests on a concept of volition: individuals control their own actions, and their conscious knowledge of what they are doing should allow them to choose between right and wrong actions. Thus, in systems based on Roman Law, committing a crime generally requires both a physical action (*actus reus*) and the conscious experience of performing the action (*mens rea*). Views on the psychology of intention and action are thus engrained in the language that is used to discuss morality and responsibility. However, the links between conscious intention, physical action and responsibility are both problematic and highly important. Intention without action is sometimes sufficient for responsibility and sometimes not. For example, it is widely held that thought alone is not a crime. Someone who wants to hit another person but holds back at the last moment has not acted and therefore cannot be responsible. In other situations, preparation that is relevant to an action can be thought sufficient for responsibility, even if the planned action is prevented. Recent terrorism trials provide a topical example. Equally, action without intention is sometimes judged sufficient for responsibility and sometimes not. A person may be judged not guilty of an action that they clearly committed if they were not consciously aware of their actions, such as in sleepwalking assault. Although neuroscientific descriptions of the brain circuits that generate action and conscious awareness can contribute an evidence-based theory of responsibility, it is unclear whether they can capture all the nuances of our social and legal concepts of responsibility. This Review has resisted the traditional, philosophical idea that conscious thoughts cause voluntary actions, in favour of a neuroscientific model of decisions about action in relatively unconstrained situations. How might this model relate to responsibility? The initial 'whether decision', based on reasons and motivations for action, and the final check before action are both highly relevant to responsibility. By contrast, decisions regarding how and when an action is performed are less crucial. Responsibility might depend on the reason that triggered a neural process culminating in action, and on whether a final check should have stopped the action. Interestingly, both decisions have a strong normative element: although a person's brain decides the actions that they carry out, culture and education teach people what are acceptable reasons for action, what are not, and when a final predictive check should recommend withholding action. Culture and education therefore represent powerful learning signals for the brain's cognitive–motor circuits. A neuroscientific approach to responsibility may depend not only on the neural processes that underlie volition, but also on the brain systems that give an individual the general cognitive capacity to understand how society constrains volition, and how to adapt appropriately to those constraints. A basic level of functioning of the social brain, as well as the cognitive–motor brain, is essential for our conventional concept of responsibility for action' (Haggard, 2008, p. 944).

On shirking responsibility, are the Nazi officials truly *responsible* if they were simply taking orders? In so far as we are aware of the ramifications of both refusing orders, or of hiding a Jew, (which would lead to the accusation of defection of which one is shot in the head), there were Nazi's and Germans that found responsibility of humanity miraculously prioritised. But in this case the exception cannot make the rule. For the majority of those who were *responsible* for those heinous crimes, where would the locus of responsibility be when concerning *intention* or *action*? The action itself, muscles, the spinal cord, the primary motor area, the supplementary motor area, the pre-supplementary motor area, the prefrontal cortex, or its *influences* (the environment)? And if you accept that it's influence, how far back do we travel in a Nazi's germline that brought about their prefrontal cortex in the first place? As Stanley Milgram had shown in his famous experiment in the 60s, how far does *obedience* play in the make-up of the decisions on the PFC (Milgram, 1963).[135] As Zimbardo had shown in the 70s in terms of *influences* and those circumstantially wielding the powers as Nazi officials, how far does *power* play in the make-up of the decisions on the PFC (Haney et al., 1973)? Through the laws of causation and variability between the species, it's where responsibility seems to dissipate.

> ~ *I must reproduce with the fittest I could possibly acquire as an organism; reproduce.*

The Epistemological Quest

As we should understand responsibility cannot philosophically exist, I will now refer to responsibility as 'influence', or at least 'responsibility' as if it were an agreed influence. How much responsibility should the state have in who a citizen becomes romantically involved with?' As presciently articulated by John Stuart Mill, it seems the fine line between liberty/authority came to manifest itself best in his subsequent century. Before parsing those dangers and how they may be inhered in schools of political philosophy today, this second rule must first be addressed.

Without the necessity of the esoteric, we could start with a neurochemistry in the primate order. If we could make concessions that we have processes throughout our body that regulate behaviour, which we also agree is well documented in scientific literature, this would be a start. The next concessions are of classic instances that neurochemistry regulates; neuropeptides. The love hormones, such as levels of oxytocin, vasopressin, and serotonin are found to activate when some form of social bonding transpires. An example would be of the small tamarin monkeys purported to have high levels of oxytocin secretion in female members when grooming and in physical contact, while males secreted higher levels of oxytocin during sexual intercourse (Sapolsky, 2017). So, we could say that given the tacit understanding we have of a neurochemical, biochemical, and endocrine systems regulating for such behaviour in neural circuits, 'biological influence' comes to play in the arena.

Think of the neurochemical benefits you enjoy of keeping a person around who makes you laugh (which is always a wonderful feeling)... Now think of the neurochemical costs you may endure when keeping a person around who you fear? When in the context of romantic involvement, it appears to be rather simple? One would evade the latter and marry the former. Be that as it may, knowing the plight of domestic abuse suggests it is never so simple when we refer to responsibility. When we look closely at sexual dimorphism and pay attention to physical asymmetries, it's visibly obvious to find differences in physical prowess, and so it shouldn't be such a surprise for ethologists to observe the Dawkings' *coy* strategy rendered by the female, as there should be a timely process for the female to *observe* his behaviour and trace anything indicative of the male being less than congenial in an environment that would make her vulnerable against her physical asymmetries. Keeping the *coy* strategy purely mathematical, within the evolutionary game of benefit and cost, we can say that the more time the female *waits*, the higher the probability she has of assimilating a good *understanding* of his behaviour! Well, is that entirely true? Not quite. Let's say that the more time a female *spends* with a male, the higher the probability she has of assimilating a goof *understanding* of his behaviours as the process of assimilation organises itself into algorithms that suit the two rules. Through a subconscious teleological process that many women will be merely incognizant of, is what transpires in a pre-frontal cortex. This will transpire through long periods of observation. Quite simply, in adapting to her environment, and thus making the incisive decision to possibly reject the male after his long courting period, this is how biological influence may be *profitable*. But as one should note in this game, this patient process doesn't seem to be entirely fair for the male.

Respective of the *coy* strategy, we find that males have thus adapted to play against the *coy* strategy, or to put it frankly, learn how to deceive the female propitiously, which thus happens, likewise, teleologically. Incognizant of the *intersexual influence* being depleted from his free will, he will court the female commensurate to the extrapolations his sensical apparatus will make from the intercourse he has hitherto made with her, both verbally and non-verbally. He has, or is, thus adapting to the environment. As the survivorship machine for the gene, his role is to convince the female she has assimilated everything she needs to know, and that she can now thus proceed with an *acceptive* phase. John Maynard Smith's *evolutionarily stable strategy* yet occurs, which manifests itself into what would look like an average amount of time a female can make the male wait in an environment commensurate to the competition of the other females in how long they also make him wait. In so far as the *coy* strategy seems to benefit the female, it may be beneficial for the male to detect the female who plays a *coy* strategy, as it may be merely suboptimal for a male to accidentally raise a child that isn't his own, or at least be inclined to reassure that his child is actually his, which brings us back to parenting.

Rule 5: *'Do not let your kids do anything that makes you dislike them'* (Peterson, 2019)

As briefly mentioned in part I, perusing Piaget's accommodation and assimilation of the child in sensitive operational stages, the biological *influence* of the parent, male or female, must thus begin in courtship. Again, when we delve a little deeper into how epigenetics may play themselves out under such evolutionary processes that transpire over the millions of years of *deep time*, those very displays of fitness are duly ensconced within biological indicators and signals that may deem their partner fit during courtship. The sociological ramifications of biological influence thus mediate between where the individual may *assume responsibility*, as opposed to the **state**, which leads to the complicated phenomena of such political philosophy through the lens of both domestic violence and others of profusely sexual crimes.[136] As we focus through the lens of biological modality, that's the focus of all living organisms within the geological underpinnings of planet earth, one must see the miracle of attempting some type of marital agreement unscathed. Evolved to find that which titillates or gives pleasure in such romantic bonds, one must see the miracle endeavoured. Remembering that in so far as we share most of our DNA with the same species, we share, or optimally, are sharing less of our DNA with the strangers we end up romantically involved in. What then is *profitable* in such estranged affairs would be for the two strangers to be connected through 50% of their DNA that is shared with the child by both partners, and anything approximating this fact!

As I often retort in the discourse of biological influence, considering the myriad accessible ways of moving within the expanded networks of space and time, the question is thus invoked, 'who would be *politically* responsible for a male gratuitously impregnating a female on a consensually brief and casual encounter while he deliberately performs his act under an alias on vacation?' As unjust and insubordinate as his comport may be branded by many (or lauded), I can only see, in the biological sense, that we are thus left with two outstanding solutions. (1) Offer more power to the state to track down men, or (2), prevent such deleterious acts from occurring. As history and classical literature has well-informed us once parsing the events of Stalinist Russia, and disassembling the hypothetical of a highly surveyed state, Orwellian literature and the story of 1984 should hopefully be truly indicative of the solution we are left with; the latter.

As per the literature on child psychology, children are by no means organisms to be taken lightly in a civilised or a cooperating society, and so conducive to the existence of such children, the process of conception should not be taken so lightly either, that's both in the light of sexual crime and of the consenting process in itself. As the clinical psychologist Jordan Peterson tells us that it may be profitable to *not let our children do anything that makes us dislike them*, thus influence must thus start from somewhere. Jordan Peterson, along with another Canadian psychologist, Joseph L Flanders of McGill university, wrote an insightful article in 2005 unfurling the intricacies of play and the regulation of aggression in children. On pursuing how play and aggression are thus regulated in highlighting the activity of the brain (and also the possible desuetude that may occur), Peterson and Flanders oscillate a philosophical pendulum between the Hobbes and Rosseau debate, concluding that inhered motivation must thus be regulated, and brought under control via socialisation (Peterson and Flanders, 2005). As well argued by Sapolsky, when looking at the behaviour of a child, one must contextualise the child as a product of evolution, in that how the child responds to the world will be dictated by the relationship her *genes has with the environment* (which spans from seconds ago, minutes ago, hours ago, decades ago, thousands of years ago, and millions of years ago. It means, given the plight of survival in our species throughout history, that some children will be more susceptible to aggressive behaviour than others, and that where aggression has been useful in the trajectory of our species, some females will unwittingly find aggression profitable as instantiated in the subsets of sexual selection I mentioned in part I. This opens a whole argument on the subject of social constructionism I will not explore, which evidently leaves the scope of what I profess to be *biological influence*.

The conundrum of Dantes' affairs are constantly experienced by many in the western world today. The betrayal that Edmond Dantes became the resentful avatar of is experienced by so many, which I believe to biologically and thus mathematically makes sense through the lens of evolutionarily stable strategies and benefit/cost relations. What seems to adulterate such traditional systems I see is the sensitively antithetic relationship we seem to have with tradition and those who still expound the oldest and most obsolete fractions of traditional scripture, turning a reasonable debate into one politically conflicted. Thus the philosophy of traditional scripture is in danger of being completely dispelled of, meaning distinguished philosophies we find in scripture are being complicitly inculpated along with all other deleterious philosophies of contemporary society, which I find to be simply ludicrous. One of those distinguished philosophies that has been in unfortunate danger of being dispelled of, akin to the lens of biological influence when we think of the *essence* of rule II, is marriage.

Conversations emanating from the institution of marriage in a more contemporary lens have espoused new types of relationships such as polyamory, while conversations in feminist schools of thought discuss alternative systems that emancipate the female from childcare and other conjugal roles. Notwithstanding the vagaries some type of marriage has been found to exist in all cultures of the world. Even if sceptical about the claim, it could be sceptically reduced to most cultures in the world, which necessitates some type of tract in sexual relations across all of humanity. One must ask themselves the simple question, 'are ideas such as polyamory as unprecedented in thought hitherto to the 20th and 21st century? I think not.

Moving away from the stringent laws of marriage in protestant traditions of the Lutheran, Calvinist, and Anglican models of marriage did early enlightenment philosophy take, as the idea of marriage had been moderately liberated as philosophers recognised marriage was 'at once a natural, social, and contractual association, with a number of its basic terms pre-set by nature and society in order to protect the natural rights and duties of husbands and wives, parents and children' (Witte, 2015, p.16). What we had thus learned was a philosophy approximating the liberation of women from incommensurate and sanctimonious behaviour of a *minority* of men, working its way into the 20th century under the invariably perilous trajectory given 20th century politics and wars.[137]

> *"'Haven't you read," he replied, "that at the beginning the Creator 'made them male and female', and said, 'For this reason a man will leave his father and mother and be*

united to his wife, and the two will become one flesh'? So they are no longer two, but one. Therefore what God has joined together, let man not separate."
"Why then," they asked, "did Moses command that a man give his wife a certificate of divorce and send her away?"
Jesus replied, "Moses permitted you to divorce your wives because your hearts were hard. But it was not this way from the beginning. I tell you that anyone who divorces his wife, except for marital unfaithfulness, and marries another woman commits adultery."
The disciples said to him, "If this is the situation between a husband and wife, it is better not to marry."
Jesus replied, "not everyone can accept this word, but only those to whom it has been given. For some are eunuchs because they were born that way; others were made that way by men; and others have renounced marriage because of the kingdom of heaven. The one who can accept this should accept it' (New International Version, 2000, Matthews. 19:4-12).

Constraints of biological influence, save the minority of homosexual affairs and marriage, which thus lead to children, must instantiate itself in a cogent philosophy aiming to protect all parties involved. It seems *family law* may do well in protecting women, men do poorly in protecting themselves, and children go completely undemocratized in such affairs, leading to children dolefully free of fathers in the family, or a brigade of new stepfamily the child becomes compelled to acclimatise to. In so far as single-parent families and stepfamilies may come to function leaving little or no collateral damage, it still leaves the traces of immeasurability in the undemocratized child in the west. I would say the ramifications of such contorted systems of romantic relations may be attributable to space, time, and the highest of morbid pleasures within what the contemporary form of romance is predicated on in western humanity, novelty.

Within the inalienable freedom of choice, and yes, most do choose their own partners in the west, comes with it the burden of responsibility. If in craving some chocolate you embark on a nearby convenience store and thus buy a new brand of chocolate, you take the calories along with the novelty, and thus biological responsibility is granted to you as much as the freedom to choose from the range of chocolates has become your inalienable right. If you then consume the chocolate and suffer an allergic reaction as a consequence, one could parse the label of chocolate eaten to discover if allergies are thus absent, which would render the company legally responsible, if not, the responsibility still lies on you. If you were to buy a product that didn't turn out to be what it promised to be, the buyer was originally responsible, hence marketeers introduced refunds after a relatively short period of purchase. Buyers would thus become either more judicious, or more precautious, so marketeers galvanised the buyer's whimsical equivocations with the introduction of warranties, which in financial terms become expenses. Meanwhile, in the eagerness, or biological proclivity to acquire a partner as an asset expected to appreciate over time, a partner may gratuitously become a liability, and children that come from that liability create more liabilities and may suffer from such liability also, leading to a life of chaos for all involved. In as far as we see the caveats of being over-cautious in such affairs, one must clearly see that impetuosity is thus merely suboptimal. There are no warranties involved in romantic affairs as the subject does not develop emotions or feelings with the object, rendering the product replaceable. One cannot merely replace a partner with a better version of him or herself, hence we find the famous retort to romantic termination, 'But please, I can change!'

One will argue contraception freed the female, but I would argue, it has only freed her in so far as she *knows* how to use it, as emotions, feelings, and memories are biological processes that happen and consolidate relations over time.[138] Hence it is the male who weaponizes time and emotions to deceive the female, with disingenuously sincere promises, and when those fail, resort to utterly mendacious lies. So a philosophy approximating caution could be beneficial for the female, with the use of preclusive measures and a stern criteria, which in an environment leads men to raising such standards, but that would be germane to the philosophy of idealism which I would argue is anathema to the expansive sociological arena today. The most preclusive measure for the female seems to be, primatologically speaking, a *coy* strategy within the *prospective* phase, most especially seeing that technology is aiding and abetting males in deceiving the female through space and time. That goes for the essence of rule II, that *I must reproduce with the fittest I could possibly acquire as an organism; thus reproduce.*

~

'"Why do you look at the speck of sawdust in your brother's eye and pay no attention to the plank in your own eye? How can you say to your brother, 'Let me take the speck out of your eye,' when all the time there is a plank in your own eye? You hypocrite, first take the plank out of your own eye, and then you will see clearly to remove the speck from your brother's eye' (New International Version, 2000, Matthews. 7:3-6).

The desire for a better world throughout sociological circles in the west, has run in tandem with critical events of the last two hundred years, giving an innumerable population the world has not experienced thus far, unseen scientific and technological advancement, and what I believe to be overlooked by many, that out of the almost eight billion of the world, literacy rates sky-rocketed! Before dissecting the famous verses of the book of Matthews and thus giving five acts of deontology to approximate the idealist's claim, (outside of the initial two of Abraham Maslow's hierarchy of needs),[139] I do believe it essential to briefly address the political smoke of the left's stance on injustice and revolution, notwithstanding why someone believes such things, known well in many academic circles but scarce in much of the humanities.

As excited as Dumas was on setting his novel on those French and hostile politics, enlightenment philosophy was almost already on its rise throughout much of the west. For the French, not so many miles across the channel, was British abolitionist thinking, with prominent figures such as William Wilberforce and Granville Sharp founding the 'Providence of Freedom'. Now over a decade, the biggest colony in the world had officially declared independence, and this is by the start of the French revolution. Of the famous words that follow the first in the American reliquary:

'We hold these truths to be self-evident, that all men are created equal, that they are endowed by their Creator with certain unalienable Rights, that among these are Life, Liberty and the pursuit of Happiness' (Jefferson, 1997, p. 3).

Of the declaration's signers were liberals Stephen Hopkins, George Ross, and George Wythe, who opposed slavery and freed their slaves, sought justice for Indian tribes through peace treaties, and opposed repressive British policies, instantiating a philosophy of freedom and liberty throughout American politics which is still widely reverberated by patriots today.

Across the pond and of those subliminal modes of enlightenment philosophy, it was satire such as Candide, written by the French philosopher François-Marie Arouet under the pseudonym Voltaire, published thirty years before, that ludically questioned such unjust feudal affairs across political systems in the west. The satire

The Epistemological Quest

plays with the Leibnizian concept of optimism, which approximates epistemic modalities instead of alethic, disassociating it from necessary modalities as Candide moves away from his teachings of his mentor and the philosopher he highly extols, Dr. Pangloss, as he witnesses the plight of alternating experiences as he ventures the world in search of his dear Cunégonde (Voltaire, 1991).

Who Told You That Life Was Fair?

In the storming of the Bastille on July 14, 1789, the French revolution surely set course without the need of true machinations to overthrow the crown. It wasn't until 1792 that the Legislative Assembly, or the Girondins, were perceived as the moderate revolutionaries desired to make what they would have perceived to be feasible and interoperable concessions. The more radical of the revolutionaries were those who had gained the reputable positions in the chamber, the Jacobins, eventually conspiring the trial and execution of the king, Louis XVI and many of the allied Girondins. Partisanship was slowly bifurcating and eventually interspersed throughout French politics as those whose partisanship lent towards royalty had pledged their allegiance first to the rightful Bourbon successor, Louis XVI's son, who they would come to crown as Louis XVII. Unfortunately for the royalists, he unexpectedly died. This then would crown Louis XVI's brother in exile as Louis XVIII. Able to finally arrogate his rightful power, Napoleon took charge of the revolutionary proceedings, taking power officially in 1799 as First Consul, crowning himself Emperor by 1804. The irony came when Napoleon's administration endeavoured to mollify the church in an attempt to espouse a hereditary system where he would place his own family members, essentially creating his own aristocracy and placing family members on distinct European thrones. In an attempt to invade Russia in 1812, allied powers (Austria, Prussia, Britain, and Russia) successfully reprised in 1814, leading to Napoleon's exile on the island of Elba, in which the Bourbon king Louis XVIII returned to restore order onto the nation. In 1815, Napoleon, on making use of allied partisans interspersed within France, conspired his return, leading to the exile of Louis XVIII yet again as he successfully absconded to Belgium. In the famous battle of Waterloo, fought on the 18th of June, which lead to further exile of Napoleon but this time to the mid-Atlantic island of St Helena, rendering his brief spell of power as 'The Hundred Days'. The neo-monarchist belief that a hereditary system should be solely adopted to end the revolution is what was seen as counterintuitive to the entire revolution itself, as historians quarrel on the nature of Napoleon's perspective on such affairs, proving power to be much more of a mysterious phenomenon (Dwyer, 2010).[140]

In 1824, the Bourbon monarchy was restored for another fifteen years via a calmer succession of Charles X. France successfully evaded Spanish neighbours and volatile domestic affairs until aiding the Restoration of king Ferdinand VII to power. For the Spanish monarchy and for problems subsequently amassing, it was the *Salic Law* becoming the incendiary flame that would ignite the First Carlist War in Spain. It happened so that the monarchy had expected and quite rightly presumed that the brother of Ferdinand VII, Don Carlos, would succeed him on the demise of the queen in 1829. Be that as it may, within a mere few months, as a possibly strategic machination against acrimonious filial ties with his brother, Ferdinand married his young niece, who would later become a historically significant figure for Spanish politics, the Queen Regent, Maria Cristina. Ferdinand failed to produce heirs from his first three marriages as it stood, and so on the demise of his third wife, Ferdinand saw it wise to summon an abrogation of the *Savic law* by what became his infamous *Pragmatic Sanction*, a mere changing the rules of the fundamental legislative laws.

The Savic Law, promulgated by the first Bourbon king, Philip V (in 1713), held that only a son or fraternal mail heir was to succeed the throne if no son was born. But it was 1789 (the year of the inception of the French revolution), that king Charles IV had successfully abrogated the law with the approval of the constitutional legislative chamber, the *Cortes of Cadiz*, the first constitution of Spain (Parker, 1937). On Ferdinand invoking this abrogation to render the daughter to Maria Cristina, Isabella, heir to the throne, this polarised Spanish politics; one side allied with Ferdinand and the Queen regent, the *Progressives*, and their belligerents being the Carlist *Moderates*.

The proclivities of power throughout monarchies of Western Europe led the American constitution to the tenets of *Checks and Balances*, believing a successful separation of powers would be achieved through divided legislative, executive, and judicial chambers, thus approximating the philosophy of freedom and liberty throughout the colony. Meanwhile, with all that transpired in such a taxing revolution (even irrespective of Napoleon and the disquietude concerning his true means to his ends), in 1848, while the three-year Second Carlist War was still being fought by the Carlists and Liberals, a pamphlet was soon published, becoming one of the most influential pamphlets in political history, *The Communist Manifesto*.

Emanating oxides ensued from early industrial engineering, and as enlightenment philosophy became attractive recalcitrant philosophy, Marx and Engels saw fit the importune moment to write and publish a manifesto unfurling what could eventually liberate the so-called economic base of the *power struggle*. Disconcerted with feudal oppression that levelled monarchs, lords, ladies, knights, and vassals, precipitously down to the lower status of the common peasant, Marx and Engels availed themselves of the opportunity to disclose an industrialist philosophy of what they believed became the 'Alienation of Humanity' for an economic base, as the capitalist *mode of production*. This was germane to the polarisation of compatibilist and incompatibilist free will, in that freedom the proletariat worker was given by nature in the agency he would have of his own actions, thus diminish, as the worker becomes indoctrinated, confoundedly conceiving himself as an agent of his actions. For Marx and Engels it was a *Philosophy of Action*. The proletariat worker is thus controlled to produce the labour that benefits the bourgeoise (who own the *means of* production), rendering the proletariat worker more labour and less control of all means of production, akin to business competition among industrialists (Marx and Engels, 2015). In the classical Marxist perspective, the proletariat alienates himself from his agency and would merely become the cog of an industrialist wheel which fortifies the interminably functional *infrastructure*. Religion thus holds part of a *superstructure*, alienating the proletariat worker from his philosophy of action and thus continues to alienate himself from his own nature (body). A police force and a military institution are that which substantiate alienation successfully, through a coercive trepidation of both policed and military action. Art, or anything and everything meaningful becomes what ossifies a needed superstructure, being simply inexorable for the proletariat and hence a bourgeoise steers clear of the economic base, a proletariat class of factory workers and beyond. In so far as a narrative of 19th century political philosophy may meander towards its own pathway, the question invoked for 21st century political philosophy, is, 'how applicable, how true, or how obsolete can these views be seen through a 21st century prism?' Simply, 'why aren't we all Marxists?' In as much as the Communist Manifesto enrages the *true* egalitarian who consecrates her life to justice, we must approach Marx and Engels with assiduous reason, allowing emotions to subside briefly in the quest for truth before the revelations of 20th century history and why in the world those events would have *wilfully* transpired.

The Epistemological Quest

Before the acceptance of this untenable manifesto, it needs to be brought to our attention that eleven years after the significant publication of the *Communist Manifesto* was published, another more important and sophisticated science was published by a man (who has been a major theme in this book), from the humble and British county, Shropshire, both adulterating and discrediting the philosophy of both Marx and Engels forever. That book came to be *On the Origin of Species* by Charles Robert Darwin.

As some whimper and question, 'how would Charles Darwin pertain to their philosophy?' The answer is biology. As briefly addressed in the first part of this book as I sifted through a theory of sexual selection (which I hope isn't as misconstrued anathema to my sincerest intentions), tacitly shared between whoever you may be reading this book is a proficient understanding of evolution, and thus, behaviour. Evolution being, the necessitation of survival across all life and, well if any anomalous exceptions, for instance group selection, then the survival and replication of the gene. Tacit to this biological understanding will mean, if we endeavour to approach political philosophy carefully, that hierarchies are thus inevitable. For the egalitarian, the hierarchy may prove to be prejudicial to their doctrine, if one does not approach egalitarianism carefully. The harsh question often asked to the libertarian is, 'Who told you that life was fair?' which may seem caustic, but yet does not mean that we cannot endeavour to approximate equality or at least a civilisation approximating something just. According to the laws of biology, in epistemically biological modal possibility being that which may exist, we most always distribute ourselves into hierarchies.[141] So how must equality be addressed? The problem presents two possible solutions that entertained identity politics today. The first being equality of *outcome*, the second being equality of *opportunity*. The two solutions we are presented with are such complicated ideas and this must be tacitly agreed upon. It may seem evident to some that the equality of outcome may hold, meanwhile on enforcing disparate groups to hold a percentage of representability may, (and should) seem somewhat ludicrous. The ingenious would consequentially be absent from ingenuity, as a multitude adulterates the sciences. This is learned when addressing the serious caveat... what if a group constantly falls behind another group? So, what about the equality of opportunity? Before addressing the caveats of the equality of opportunity, we must continue to unfurl the political narrative to truly fathom fields that permeate 21st century humanities in the west.

A Political Hyperplasia

As Marx and Engels' prescient ideas were subsequently realised throughout the latter decades of the nineteenth century (as companies did monopolise), the recalcitrant doctrine of revolution grew in popularity. for those who misconstrue the rest of the twenty-first-century historical narrative and the subsequent events that led to such significant ramifications. What wasn't so prescient of both Marx and Engels, and in fact was extremely unprecedented, was how science and technology would become the driving tools that would eventually free the proletariat of their *Philosophy of Action*. Of the most important successes of science and technology in the west: literacy rates would increase along with education, people became more prosperous living longer lives, and the nature of economic sectors would change to the extent of *gamers* and *youtubers* becoming millionaires in less than 200 years!
But for the significant twist to the communist narrative of how we got here, what cannot be elided is what Vladimir Lenin had built, and what the Georgian, Joseph

The Epistemological Quest

Stalin, had availed himself of between the period of 1934 -1953, the Soviet-forced system of the Gulag.

1 in 4

It must be well understood between the years of 1934 and 1944, 12-14 million Russians passed through the gulag, while 10-13.5 million experienced some type of gulag detention in the post-war years between 1945 and 1954. This is of a population sporting around 110 million people at the time (meaning at least 22% of the population had at least passed through the gulag). It means, what Soviet specialists will fathom more than any, that the chance of somebody having passed through the gulag approximates 1 in 4, as the population increased to 110 million by the end of Stalin's gulag period. A statistic of 1 in 4 only becomes necessary once merely conceptualising the ramifications of life in the Gulag for those residing in a nation in such times.
Stalin's Great Purge of the 1930s killed hundreds of thousands as he targeted affluent peasants who achieved such wealth in the latter stages of the Russian empire and early stages of the Soviet Union, known principally as the Kulaks. The *Politburo*, the highest policy-making authority within the Soviet Union, successfully arrogated the authority to put passports, transportation and food under the gulag's responsibility, leading to even more irreparable administrative failure and chaos. Even during the time of the Second World war, close to 60,000 prisoners who had already served their terms were held captive by denied release, being coerced into remaining in their camps as wage laborers. The NKVD, the Soviet Union's police, also held an expansive estate of secret police who administered labour camps and colonies across such a genocidal period, and albeit it seems an assortment of mere names and numbers, they were the consequences of a Russian not coming to terms with the statistic of what 1 in 4 would mean for those who would eventually endure the gulag (Alexopoulos, 2005). Families were estranged, mothers separated from children, husbands from wives, employees from work, while millions became tortured and killed. Some may reasonably argue it attributable to the Marxist doctrine, while others may argue that Stalin's campaign to the Marxist doctrine is simply and fallaciously *post hoc*. The probability of you never seeing your family again was akin to the probability of you being detained, which was akin to the probability of you leaving once serving your term, akin to the probability of a prison guard yet inculpating you in another crime, akin to the probability of you being moved to an even worse off gulag so unreasonably far away.
As late as 1973 was when the published account of Aleksandr Solzhenitsyn transfixed the political mind of the world outside of the USSR; in what he named 'The Gulag Archipelago', encompassing what Solzhenitsyn collected in diaries, legal documents, interviews, reports, and statements, from all who experienced the gulags. Solzhenitsyn, deemed by the USSR as a dissident, was a soldier on the Russian front. Due to a broken pact between Hitler and Stalin, Nazi Germany attacked Russia unprepared, and Solzhenitsyn was one of the early bearers of that brunt coerced to fight on that very front line. Writing letters to his fellow compatriots as he was disconcerted by lack of preparation, the USSR were in control of his fate in which his life would quickly change for the worst. The Soviet prisoners returning to Russia were also unfairly imprisoned, which can be ascribed to two things for Stalin (1) the fact that no accords of the Geneva conventions were met in 1929, and (2) being that he now saw those prisoners as a threat to the class due to over exposure. The releasing of this compilation of gulag experiences led

intellectuals to vociferously question communism for the first time throughout the west.
Important to note is that the gulag became so nationally powerful, that the NKVD as an economic institution entailed vital sectors of the economy now depending on penal labour throughout Soviet Russia (akin to the sectors failing ignominiously), so administration would find surreptitious strategies to fill the camps, or not (Alexopoulos, 2005). One may recall the incipient stages of the catastrophes of communist genocide during the campaign of the Red Terror between 1918 and 1922, when Bolsheviks repressed political thought as anathema to the imperious Russian Empire and thus levied Russia with political executions via the Cheka; the Soviet's first secret police. As Solzhenitsyn vituperatively recorded in his compilation of events:

'A district Party conference was under way in Moscow Province. It was presided over by a new secretary of the District Party Committee, replacing one recently *arrested*. At the conclusion of the conference, a tribute to Comrade Stalin was called for. Of couture, everyone stood up (just as everyone had leaped to his feet during the conference at every mention of his name). The small hall echoed with "stormy applause, rising to an ovation". For three minutes, four minutes, five minutes, the "stormy applause, rising to an ovation," continued. But palms were getting sore and raised arms were already aching. And the older people were panting from exhaustion. It was becoming insufferably silly even to those who really adored Stalin. However, who would dare be the *first* to stop' (Solzhenitsyn, 2003, p. 27)?

Those familiar with the Orwellian novel, *Nineteen Eighty-Four*, will find this particular report from Solzhenitsyn evocative of one of the early events of the novel, in which the very same scene transpires in an attempt to lampoon totalitarianism, government surveillance, and propaganda (Orwell, 2003). The novel, by George Orwell, unveils the ramifications of authoritarianism and the miles taken from the inches of government privacy a government may arrogate in protecting the rights of its citizens. The novel has thus played a major role in the political philosophy of technology and power in communist regimes and authorities expanding outside of the threshold of its jurisdictions. In so far as a government will present itself as communist, the government will be authoritarian, and in so far as the government presents itself as socialist, it winds up being authoritarian. The question is, how? There are proclivities, some would often say biological,[142] that most necessarily should be understood in humans inhering evolutionarily, as other animals and life do so in the same manner. When an individual is thus given a modicum of power, by fortuity or intent, and when this particular individual lacks the necessary tools of introspection and a reasonable understanding of the ramifications of not truly fathoming the origin of this power, bad things invariably transpire akin to the very proclivity. The incessant issue molests the true notion of liberalism and occurs throughout civilisation almost daily, but the politically and morally prejudicial ramifications remain adulterated in as far as the issue remains untameable. Once the modicums of responsibilities (influences), or duties, are evaded by the inclined proclivity to offload responsibility, those very responsibilities are levied on to those beneath, and she who holds the power may be incognizant of this, as it was a **proclivity** under such faculties individuals experience that many constraints achieve throughout their lifetime, hence go unnoticed. The ontological offload could be said to replace responsibility with the ontological vigour, or security that guides the individual wielding that very power, manifesting itself into conceited self-pride, which most certainly plays itself out as the invariable catalyst of superiority. As ontologically insecure and incomplete an individual will be in her power, the more ontological offload the individual will have to offload and convert that offload into the pride and superiority justifying her ontological being, and so she may adorn and pamper her new ontological vigour to help fortify the new ontological figure, leading to the dismay of all around her.

Orwell presents such malevolent proclivities in what became one of the most important fables written, *Animal Farm*, published three years earlier than Nineteen Eighty Four (a few months after Adolf Hitler killed himself). The fable is merely predicated on the political events of Soviet Russia, as the Bolsheviks in their communist Red Terror campaign led to the tumult of power and consequential genocide. In the fable, the farmer is scurried out of his farm by his very own animals under the successful *coup d'état* the animals derived in unity. This is after an incendiary conference held by a pig-calling for the overthrow of humans. Throughout the fable, a gradual monopolisation of resources is experienced as power struggles, indoctrination, and conspiracy permeate the farm. Nascent hopes, promises, and dreams are continued to be reignited by those in control and even executions are ordered throughout. Over time, they notice Napoleon, the leading pig, adopts more and more human characteristics which was both originally and ironically antithetic to their initial constitution that led them to revolt against the farmer at the beginning of the fable.

The motif accentuated by Orwell must be carefully understood before buying what seems to be a highly attractive revolutionary idea of communist dogma. Subsequent to a successful revolution, Napoleon Bonaparte, a century before Stalin had played the indiscernible games of identity politics, in what eventually divided the Jacobins, also divided the three factions of socialism in Russia, i.e. the Bolsheviks, Mensheviks and Socialist Revolutionaries. Imminently succeeding the Russian revolution in 1917, we find towards the former years of the 1920s a struggle for power which seemed to instantiate itself in the vicissitudes of Stalin. It mirrored a same ontological offload that came of the animals in Mr. Jones' farm, which in reality resulted to millions of deaths and penal torture for millions under Stalin's regime. One must tread carefully in evading what may be deemed fallaciously *post hoc* in addressing these claims, as communism remains a complicated manner when eliding evolutionary biology. Be that as it may, one must summon the biological foundations in order to truly delineate the contours of how an incompatibilist free will can lead to such events and thus fathom the ontological underpinnings of superiority. We see the socialist dogma played out referentially in Venezuela today, in which scholars have claimed to be attributable to verbal and non-verbal discourse within a philosophy of soteriology, predicating itself upon a quasi-religion of reparatory salvation. The Venezuelan leaders, once Hugo Chavez and now Nicolas Maduro, were actors of such soteriological strategy who harnessed the games of identity politics, summoning the revolutionary leader Simon Bolivar to stress a resounding philosophy that 'I too will die and sacrifice myself for my 'patria' and the people' (Zuquete, 2008). Meanwhile, there has been economic turmoil that led to mass exodus from a land that was once the richest democracy in Latin America, and for those extraneous to the borders of Venezuela nothing but an advertent rendering confusion still plagues the mind, as does the same for North Korea, Xinjiang in China, and just as it did in the USSR as the world is starting to be informed of the same politically inhumane horrors.

Once truly ruminating on what became of the universal declaration of human rights after countless treatises, pacts and accords, the ruminator must do so successfully on the dogma of twentieth century world wars that many still alive today did experience. The ruminator must then fulfil his onus on how such a congested philosophy rendered us counting so assiduously generation by generation since the First World War. An archetypal motif can be found across dictators and those who

The Epistemological Quest

plagued the intemperate years of such a politically volatile century. The largest of instances that illustrate the archetypal claim is the polarisation of the same political event, that through the grace of literature unveiled the extremities of ontological replenishing and ramifications of such acrimonious thought and intention. The two stories are of the stories of Hans and Anne Frank. The former, the Nazi governor of Auschwitz, the latter, a precocious and charismatic teen, both ironically sharing the same name, but do not by any means the same experience until their homicide.

From One Frank to Another

Many accentuate the horrors of Hitler and Stalin when summoning the highest of demagogues, meanwhile it may also now be the onus of the lay ruminator to approach such thought more perspicaciously. The serious question I propose to the ruminator, is (in so far as I am do not believe in free will), what is the probability that you would have followed suit as either a prison guard in the Soviet gulag or a fellow Nazi, eventually espousing and enacting on their persevered beliefs? How can we disassemble the faculties of the ontological being that led to such horrors and have played out as the political motifs for many around the world in the same century which lay latently in many successors today? Is it merely the task of the criminologists and those of the legal sciences who tend to parse cases and the literature on jurisprudence and child psychology? Or is the responsibility in unfurling the proclivities that may lead to such events for all of the civilised lay people who expect one's neighbour to treat thy neighbour as they wish to be treated? I ascribe it to the ontological arrogations ensconced in the organism's proclivity to compete against the adversary, be that a stranger or as corroborated in the sciences, even siblings. The proclivity may arrogate the authority via the teleological endeavour to protect the ontological status of the organism in that an organism contributes more to a society than the other, notwithstanding how big that society may be. A dilapidation of such constructed security in the context of the social animal approximates nihilism and eventual suicide, and so an agglomeration of those necessary merits that harbour the organism manifest a coffer of those very ontic valuables that construct the ontological apparatus. It may be achieved in the arts, sciences, linguistics, culture, or anything the organism is able to successfully arrogate in order to construct such conducive security, as well illustrated through the countenance of Beecher-Stowe's dyspeptic character, Marie, who as a matter of fact revelled in constantly reminding her chattel they were inferior by the hand of god, played out in a psychopathology of her ontological replenishing, precisely meaning she was in fact endeavouring to convince herself.

The demagogic charisma for those unprepared, who as a matter of significant fact were not of the most susceptibly low faculties of intellect, managed to assimilate a narrative as importunely ripe for such ontological replenishing. One of the most exemplary foundations of the complicated Nazi problem lay in who I believe to be one of the perplexing figures of Nazi Germany, the Governor-General of Occupied Poland, Hans Frank. Most significant within the issue of identity politics, of the most core discussions of political philosophy, is of what should be necessitated, the *collective*, or the *individual?* As Phillipe Sands, the British and French barrister and professor of law, who achieved amazing work on researching his own familial background unveiling the subject of identity politics from a purely political and legal point of view, pertains the very perplexing issue which taxingly troubles an ever-interchanging world today:

The Epistemological Quest

'What did Frank want? Non-interference in the internal affairs of foreign states' was a fine idea supported by Frank to cover any criticism of Germany. So were independent judges, but only up to a point. He wanted strong government based on values that protected the vision of 'national community', which should prevail over all else. There would be no individual rights in the new Germany, so he announced a total opposition to the 'individualistic, liberalistic atomizing tendencies of the egoism of the individual ('Complete equality, absolute submission, absolute loss of individuality,' the writer Friedrich Reck recorded in his diary, citing Dostoevesky's *The possessed* as reflecting ideas of the kind expressed by Frank). Frank listed all the positive developments since 1933, including Hitler's new approach to criminal policy, one from which the world should learn. Innovations included 'eugenic prophylactics', the 'castration of dangerous moral criminals' and the 'preventive detention' of anyone who threatened the nation or 'national community' Those who should not have children would be sterilized (he described this as a 'natural process of elimination'), undesirables deported, new racial laws adopted to prevent 'the mixing of absolutely incompatible races'. To this international audience he made no explicit mention of the Jews or the gypsies, but those present knew of whom he spoke. He was silent too about the scourge of homosexuality, a subject addressed earlier in the year by the *Reich* penal code (which he helped draft), which criminalized all homosexual acts. The new Germany would be 'racially intact', he declared, allowing Germany to 'get rid of the criminal as a healthy body gets rid of the germs of disease' (Sands, 2016, p. 215).

The controversial execution of Hans Frank in the journey towards the infamous Nuremburg trials held Raphael Lemkin, the Polish lawyer, responsible for the coining of the term *genocide*, prejudicial to what Hersch Lauterpacht, the British international lawyer, had consecrated his whole work to, the rights of the individual as opposed to rights of the collective. Lauterpacht believed Lemkin to be too emotionally driven by the misfortune of his own family under Nazi Germany and saw *genocide* as a word that misarticulated and misrepresented such heinous crimes, hence rendering the grey area that the very despots had capitalised on. And therefore it was the global court of law that was the result of Lauterpacht's philosophy and ran parallel to the Versailles Treaty of 1919, necessitating individual responsibility and a tribunal that would merely exercise jurisdiction over individuals, without defendants cowering to the responsibility and authority of the state:

'Based in The Hague, the Permanent Court of International Justice opened its doors in 1922, aspiring to resolve disputes between states. Among the sources of international law it applied – the main ones were treaties and customary law – were 'general principles of law recognized by civilised nations'. These were to be found in national legal systems, so that the content of international law could draw on the better-established rules of national law. Lauterpacht recognized that this connection between national and international law offered a 'revolutionary' possibility of developing the rules so as to place more limits on the supposedly 'eternal and inalienable' powers of the state' (Sands, 2016, p. 110).

As stressed throughout the discourse of individual responsibility (or influences), once one acts upon the adoption of his inalienable (yet intraspecific) individual rights as a citizen of the world is it then that he must bear the individual responsibility of the acts against his fellow citizens (conspecifics).[143] What concerns the truly discerning ruminator is how easy one is poised to dispel of individual responsibility, failing to conceptualise the wonders of responsibility eventually being taken. The issue plagues both criminals and recidivists, inundated in a dangerous dogma of 'victim mentality' which controversially oscillates between the compatibilist/incompatibilist notion of free will. Notwithstanding the issues I endeavour to tackle in the later chapters of this book, I will audaciously propose that I believe individual responsibility (influence or supposed agency) to be the antidote to many problems one meddles with in the twenty-first century. Meanwhile, as Hans Frank wrestled with those demons under the holding cells at Nuremberg, in profound ponderance of the keys he so dexterously stroked in the height of such conceited superiority, we must firstly turn to the fate of the other Frank who bore the same surname, but bore the wrong race and creed, Anne Frank.

One of the most important documents written of Second World War literature, and I would certainly argue of the century, is Anne Frank's diary. Lugubriously published by a father in both posthumously bereft months and years, is when the diary came to be read by the thousands then to eventual millions. Encapsulated by a subtitle, are the experiences of *a young girl*. What I contend to be essential within the diary of Anne Frank (more than any other that could have possibly been found), is the paradoxical irony that Anne was a product of privilege and humble prosperity, as opposed to being a product of mere social privation, giving the twist to racism for those who endeavour to opine racism as ensconced in such superficial underpinnings, instead of the carefully assiduous approach it most perspicaciously deserves from the ruminator.

The infeasibility of the human mind to reify such occurrences in the metaphysics of space and time, lies in the literary misfortunes of the historian's hand. As we continue day to day and within the struggle for our own existence, we leave behind the facts that transpire far away or long before us and live moment by moment gracefully. The significant element of Anne Frank's diary that one should endeavour to understand is that, and quite reasonably, in as far as her epiphanies misguided her young mind, she did not believe she was going to die by any reasonable part of her *subjective* imagination. I do believe that her diary, which is often told inversely, is preferable to adulthood as opposed to adolescence (or both as even better). Anne Frank, who bore the privilege of being hidden due to her father's wealth and networks, fills her diary with often humorously daily gossip, thoughts, and disquietude endured by the often prehensile adolescents who endeavour to achieve a firm grasp of adulthood. But most importantly, she expresses her nascent hopes and dreams of what will become of her when the war is finished, as they huddle around a radio in hope of the allied powers to eventually announce heroic victory. Advanced beyond her years and writing over puberty, Frank writes of her indulgence in literature throughout, portraying her young and precocious character with told stories of outwitting her mother and the inclinations of the other boys to her inimitably innocent demeanour. In as far as her nascent desires, hopes, and dreams consoled her, and all the while psychopathological thoughts harnessed the minds of simple men, a young girl's eventualities were in a reality ensconced in what seemed to be an infeasible dearth of the upmost abject and wretched evil, as written on Monday evening on the 8th of November in 1943:

'At night, when I'm in bed, I see myself alone in a dungeon, without Mummy and Daddy. Sometimes I wander by the roadside, or our "Secret Annexe" is on fire, or they come and take us away at night. I see everything as if it is actually taking place, and this gives me the feeling that it may all happen to me very soon!' (Frank, 1993, pp. 114-115).

Anne Frank is only exemplary of the ramifications of evading the onus we have in approximating *logos* and harnessing an *eros* of human life and flourishing. Tacitly understood between many today, is the conception that one inherently holds a spirit more distasteful and ferocious that one will find inconceivable. The emergence of such young and premature fields of psychoanalysis targets this and is yet being more unfurled, as Carl Jung noted accurately, 'Suppressed and wounded instincts are the dangers threatening civilized man; uninhibited drives are the dangers threatening primitive man' (Jung et al, 1968, p. 266). Jung believed archetypes entailed of both logos (regularity) and eros (chaos and creativity), which could be a dichotomy of reason and pleasure, or knowledge and relatedness, represented in the sacred and symbolic agreement of marriage which archetypically traces back to a such mysterious and religious human substrate. Logos would be akin to the male *animus*, the male counterpart of the female *anima*, which can be discovered in both the sexes and can be reasonably analysed and dismantled in manifest and latent dream thoughts. Our biological proclivities are prejudicial to anything germane to human democracy and thus prejudicial to human flourishing, which had been the prelude of enlightenment philosophy of the last two hundred years. It's quintessential for the democratised citizen to conceptualise Anne Frank's story and what became of those in her Secret Annexe, while cognizant of all of the primitive and injurious instincts before we audaciously look at the speck of sawdust in our brother's eye. To use crystallography for an analogy, for a twenty first century prism, I contend there to be five facets of the human prism of agency I contend to be antidotal to the evil effacing many, which is what I also contend is the precursor of the sort of evil becoming what many alive today will be reminded as human suffering.

Anthony Beevor successfully articulates the *eros* through the calamities and political horrors of the Second World War in his book, *Ardennes: Hitler's Last Gamble*. General Dwight Eisenhower (who eventually became the 34th U.S president) had to meticulously cooperate with fellow American generals Omar Bradley and George S. Patton, while concertedly work with British field Marshall Bernard Montgomery in defeating the most dangerous political party a government has ever endured, attributable to the technological aid of unprecedented war vehicles and weapons (such as half-tracks, tanks, rifles, and grenades, i.e. new twentieth century war phenomena). Beevor conveys the sheer destruction of towns, cities, and the loss and self-torture of millions, predicated on military strategies and an inexorable stubbornness of the Nazi's to prevail. Effaced with the stringent deterrent of execution for those that would somehow find the temerity to defect, men walked on the ridges of hell as they were compelled to fight for a political landscape (Beevor, 2016). Mothers lost sons, brothers, and husbands, while men would lose so much more. Wars became the narrative of the twentieth century which would diminish and change shape as the horrors gave the twentieth century a good run for its money late into the 90s and into the twenty-first century. As the world population continued to increase, along with education, literacy, technology, and science, the world is in a matter of researched and substantiated fact a better place. Meanwhile for the ruminator today, it must be noted the ruminator is only human, and a dopaminergic system works its way on the ruminator just as any other as we are blind to our privileges in the struggle for existence; a simple and mere product of nature.

It was the German neuroscientist Wolfram Schultz who revealed the relationship between prediction, probability, risk, ambiguity/uncertainty, and reward that are played out in behavioural economics, as well as animal behaviour comprehensively. A practicable definition of risk can distinguish, most especially for the layperson who gambles, places bets, and invests, between the probabilities and expectation of reward that are dopaminergically scalable. Risk is thus predicated on the amount of information an individual has on the probability of such a thing transpiring, which for behavioural economics would be risk, if the individual has sufficient information on such a prediction. Conversely, ambiguity, or uncertainty would be risk predicated on insufficient information of a prediction which would render such a prediction uncertain. The scope of the salesman is to *present* himself as the former, while the onus of an investor seeking either mutual or index funds would be to find a broker; the former. Schultz concludes that as prediction is attributable to experienced reward, dopaminergic neurons activate when reward is subsequently better than the previous. A dopaminergic system hence resists the repetitious reward over time, while rats, pigeons, monkeys, and humans are also found to prefer 'sooner smaller rewards' over 'later rewards' (Schultz, 2010). The dopaminergic system encompasses two significant pathways, both the mesolimbic dopamine pathway and the mesocortical dopamine pathway. Neurological projections sent to a part of the brain; i.e. the nucleus accumbens, are from an ancient and necessary area close to the brain stem known as the ventral tegmental area (or the tegmentum). The tegmentum sending projections to limbic areas such as the amygdala and hippocampus engenders the 'mesolimbic dopamine pathway', while projections sent directly to the prefrontal cortex is known as the 'mesocortical dopamine pathway' (Sapolsky, 2017).

Let's look at this within the context of rudimentary physics in order to truly fathom what this could mean. Probability is founded on the likelihood of a certain event occurring within space and time, i.e. spacetime. The human organ, the brain, responds respectively, meaning dopaminergic receptors respond to reward within a proximity of its *prediction* coming to fruition, and hence activating, more specifically, the D1 receptors the less information there is initially given in the very *risk*. The more given information on risk would thus turn 'risk', or prediction for that matter, into *expectation*, and so expectation thus decreases the chances of dopaminergic receptors being activated akin to its expectancy, or judicious predictions. This is best articulated by Sapolsky when he states, 'What was an unexpected pleasure yesterday is what we feel entitled to today, and what won't be enough tomorrow' akin to insatiable predispositions as we are yet again harnessed by the power of space, time, physiology, and the environment (Sapolsky, 2017, p. 70).

Of Avarice and Ingratitude

Whether it's Siddhartha Guatama attempting to desensitise the very predispositions in Buddhist philosophy, or neurophysiological research attempting to understand the ancient tegmental region of the brain, one must be pragmatically considerate of the political ramifications. It's often retorted by the political left when challenged by the right that, 'In as much as I complain about the world's problems, I do so only because I desire and believe things could be better.' TEDx sensation and shame researcher Brené Brown in her bestseller *Daring Greatly* calls this problem **'scarcity: the never enough problem'**. According to Brown, *one* could never be:

- Good enough
- Perfect enough
- Thin enough
- Powerful enough
- Successful enough
- Smart enough
- Certain enough
- Safe enough

- Extraordinary enough (Brown, 2015, p. 25)

Scarcity renders insatiable appetites and becomes subservient to our wild imaginations in infeasible idealist philosophies, eliding all that is truly incumbent upon the ruminator, human history. In so far as we have learned that trepid Napoleonic and revolutionary narratives may serve well against tyranny, and the Marxist communist dogma served well against the monopolies of capitalism of the late nineteenth century, most of our twentieth century lives may find the communist dogma obsolescent, or a Napoleonic revolt a lot more infrequent than one may suspect. That's because, and I reverberate this case fervently, what may have been improvident, or at least unknown to both Marx and Engels, was a scientific and technological revolution that was going to improve the lives of humans in the twenty-first century significantly. Parties within the social sciences may revel in interspersing the horrors of capitalism and the pusillanimous spirit of capitalist avarice, nevertheless a pervasive antidote of responsibility could render immunity against such young and forceful phenomena. Brené Brown's scarcity problem emanates from the contours of the individual and is thus pervaded towards loose dicta we often hear, such as, 'the reason why there is *scarcity* in the world is because the rich have too much money'.[144]

Once more still pertinent to Brené Brown's scarcity problem is the case of Bjørn Lomborg, the Danish president of the research institute 'think tank', who in 2006 edited '*How to Spend $80 Billion to Make the World a Better Place*'. Between the 24th and the 28th May, 2004, 38 economists were amassed in Copenhagen, Denmark, to both practically and theoretically debate the world's most pressing challenges in what became known as the Copenhagen consensus. Consisting of eight distinguished economists, each economist came prepared with a draft on how to economically *solve the world's problems*, ranging from issues dealing with hunger and clean drinking water to disease and climate change. On presenting such results, 20 more preeminent researchers were summoned to challenge those papers; inculcating eight economists (some of which with Nobel laureates in the field), playing as an expert panel. The necessitation and summoning of economists, as opposed to other specialists, was to distil the issue of partiality and to prune both politics and journalism that has overcomplicated such issues. The consensus understood that specialists, or those who had more *epistemological purchase* than others, may be partial in leaning to their own field as the specialist may be emotive to their own work and experiences. Using benefit-cost analyses, experts agreed the method to be indispensably organised with having some issues taking precedence over others, through *challenge* and *opportunity*, ranking each proposal commensurate to how much should be spent in their value. These were the resulting proposals:

Challenges: Very Good

1. Communicable diseases ~ **Opportunity:** Control of HIV/Aids
2. Malnutrition and hunger ~ **Opportunity:** Providing micronutrients
3. Subsidies and trade barriers ~ **Opportunity:** Trade liberalization
4. Communicable diseases ~ **Opportunity:** Control of malaria

Challenges: Good

5. Malnutrition and hunger ~ **Opportunity:** Development of new agricultural technologies
6. Sanitation and water ~ **Opportunity:** Small-scale water technology for livelihoods
7. Sanitation and water ~ **Opportunity:** Community-managed water supply and sanitation
8. Sanitation and water ~ **Opportunity:** Research on water productivity in food production
9. Governance and corruption ~ **Opportunity:** Lowering the cost of starting a new business

Challenges: Fair

10. Migration ~ **Opportunity:** Lowering barriers to migration for skilled workers
11. Malnutrition and hunger ~ **Opportunity:** Improving infant and child nutrition
12. Communicable diseases ~ **Opportunity:** Scaled-up basic health services
13. Malnutrition and hunger ~ **Opportunity:** Reducing the prevalence of Low Birth Weight

Challenges: Bad

14. Migration ~ **Opportunity:** Guest-worker programs for the unskilled
15. Climate change ~ **Opportunity:** Optimal carbon tax
16. Climate change ~ **Opportunity:** The Kyoto Protocol
17. Climate Change ~ **Opportunity:** Value-at-risk carbon tax
    (Lomborg, 2006, pp. 166-167)

The results of the Copenhagen consensus challenge the pervasive myths of the world's pressing issues as contrary to what one is pressed to prioritise. It seems the issues that should be prioritised are a steady hold of communicable diseases such as HIV/Aids and malaria, an effective provision of micronutrients to prevent malnutrition and hunger, and for trade barriers to be practically liberated. The issue of global warming obtrudes to those extraneous to the expertise of the world's pressing issues and thus become the whistle-blowers on carbon emission percentages and resort to the puppetry of figures such as Greta Thunberg to support the very avatars of their ideological scarcity problem they fall victim to. Climate change is a complicated issue in need of more research, giving the oscillating theories on carbon taxing, the debility of the 2005 *Kyoto Protocol*, and the timespan that can control for such issues. It seems that a dearth of realism has led richer countries to support and fight diseases of the developed world such as cardiovascular disease, rather than defeating scourges such as HIV/AIDS and malaria, which was also claimed to be indicative of our implicit inclination to value life *less* in poor countries than in wealthier countries.[145] Bill Gates in his *How to Avoid A Climate Disaster* advocates for Green Premiums[146] as *the solutions we have and the breakthroughs we need* for climate change. Pivoting from 51 billion tonnes of gas emitted yearly, Gates shows how 31% of emissions are ascribed to making things (cement, steel, and plastic), 27% plugging in (electricity), 19% growing things (plants and animals), 16% getting around (planes, trucks, and cargo ships), and 7% keeping warm and cool (heating, cooling, and refrigeration) (Gates, 2021). With the use of Green Premiums and a careful eye at how we can get to 0 emissions, Gates highlights necessary actions we can take to fight climate change.

The Pentagonal Prism of *Agency*[147]

How can one effectively remove the plank from their own eye in order to *see clearly* that there is a speck of sawdust in their brother's? How can one atone for such sanctimony once inculpating the other? After thorough deliberation and much philosophical thought, it seems that, for one outside of the perimeters of Abraham Maslow's pyramid, there is *agency* for the twenty-first century human who can attempt to ameliorate things for themselves, families, communities, country, and fundamentally humanity. It's been articulated succinctly by Yuval Noah Harari, that when a rat dies, or any other animal for that manner, all of its information dies with it. Its deoxynucleic information has died, and hopefully for the rat, the ribonucleic information had copied itself sufficiently through sexual selection and a sheer struggle for existence. There have been many impressive animals who have adapted to and flourished in the earth's atmosphere, and that information can be best sought paleontologically by a look through the Triassic, Jurassic, and the Cretaceous periods of the Mesozoic era. As per space and time, it seems that the most impressive organ evolved out of over 4.6 billion years of our earth's history, is what lies capped in our skulls, the human, and more specifically, the *Homo sapiens* brain. And so as Harari points out, once the survival machine of the gene dies, it's information does not die with her, as her ideas are recorded through her dexterity harnessed by a brain to write, rendering a start-from-where-she-left-off practice. Over two hundred years ago, over 90% of the world were illiterate and also victims of extreme poverty. Our dexterity is slowly and computationally bequeathed to the keyboard for future generations and so our responsibilities are evidently changing. Our morbid pleasures, procrastination, and insatiable appetite for novelty, choice and consumerism are the issues we face with everything besieging us from television adverts, billboards, and the general social topics of conversation. But through all of which that distracts the human mind, I would contend that I have found nothing that supersedes the craft of parsing literature. Hence, it's why I say the first facet of the prism of agency is reading. So let's envisage a small community in which through all imperfections and strictures of each facet one faces, one endeavours to approximate a prism of the five facets of agency to be an effectual member of their own community.

I have divided the pentagonal prism into three compartments I believe to be quintessential to this theory: *literature, ontological security, and welfare.*

The Epistemological Quest

Literature

By 1450, the German goldsmith and inventor, Johannes Gutenberg, would pioneer the proliferation of literature with his *printing press*. By the 21st century, reading became an implicit act of gratitude with the access of an eclectic genre of books, as by the third decade, over 86% of the human population could both read and write. It would allude to the appreciation of the survivorship machine that has taken the time to express idea, weltanschauung, or philosophy throughout spacetime, notwithstanding how useful it may be for the community to perspicaciously understand the subject, and inadvertently curtail errors from being recommitted. It must be said that the true miracle of reading the symbols of humans that pre-exist us for thousands of years is impressive in human evolution. Only by briefly parsing through Jungian psychology one can understand that a disaffected and resentful countenance to religious documentation for the irreligious proves unnecessary, as a philosophy of such documentation is ensconced in your local movie theatres.

It is akin to what may be seen between the Kantian conception of *Weltanschauung*, or worldview, consciousness, the unconscious, and altered states of consciousness. As recalled by Gert Malan in Kant's *Critique of Pure Reason*, Kant believed that we perceive things in this world not as they are in themselves, which would be *noumena*, but as they appear, *phenomena*, and only that our senses act as the filters for our consciousness in understanding the world. Our conscious faculties act in subjectivity, poised for introspection and is thus what society holds the citizen accountable for in their subjectivity of the world. The unconscious, which Malan presents throughout the Freudian perspective of the *id, ego* and *superego*, holds that our ego is in constant oscillation and mediation between the stringent superego, a case of moral consciousness, and the id, the case of instinctual sexual or aggressive behaviours harnessed by the ego and eventually the superego. Our altered states of consciousness are experienced in such a variegated spectrum of mental functioning that range from 'sensations, perceptions, cognition, memory, sense of self, identity, body, environment (time and space), other people emotions, attention, perception, inner speech, arousal and volition', which Malan claims within consciousness are variegated between such aboundingly multiple states such as 'dreaming, sleeping, hypnagogic state (drowsiness before falling asleep), hypnopompic state (semi-consciousness before waking) regression, meditation, trance, sensory deprivation, dissociative states, reverie, daydreaming, internal scanning, stupor, coma, stored memory, expanded consciousness, hallucinations and states induced by psychoactive substances' (Malan, 2016, p. 3).

The Epistemological Quest

Significance of consciousness, the unconscious, and altered states of consciousness, notwithstanding religious distinctions and pertinent to our reading responsibilities, instantiates itself in our collective responsibilities to be engaged in a better *Weltanschauung*. The issue within the twenty-first century presents itself as what Malan has mapped out contiguously, which he calls 'a social multiverse'. A social multiverse would entail a:

1. Traditional cultural and religious social world
2. Technological modern social world (workplace, social media, entertainment)
3. Globalised modern social world (workplace, travel, and relatives abroad)

Attributable to such scientific and technological prowess, and attributable to extant deleteriously traditional, cultural, and religious beliefs, we seem to be dissociating with a healthy traditionally cultural and religious social world. In as far as parts of this world may be often branded as anachronistic, this world is deeply grounded in our human history (for better or worse), and quite significantly, mythology. As Malan well articulates, mythology would be the study, interpretation and the total corpus of myths, while myths would be the narratives explaining how societies came to function as they do. Given the rise of democracy and literacy, our *agency* in understanding and ameliorating our society is to harbour the traditionally 'cultural and religious' social world that bequeathed to us the technological and modern 'globalised' social world and bequeath to the succeeding generations an even better world, all through the agency in world literature.

Writing

Whether writing dreams, novels, poems, songs, or a memoir, merely involved would be the organisation of thought. We know linguistically, excepting anomalous circumstances, we speak with circumlocution instinctually, and do so both garrulously and freely. And so, in so far as debate may prove effective, the cul-de-sac of *eristics* may often be met. One cannot write and be sure of what's written unless those very ideas are somewhat organised, and so a deeper process of organisation are both the cognizant and incognizant occurrences within the practice. Those with an inclination towards the arts may do the same, in so far as it remains practically expressive, but it seems writing proves most comprehensive.

Ontology

To fruitfully capitalise on our bipedalism would be to avail oneself of the locomotion ancestors endured millions of years to be as motive as we are through spacetime. The human child does so unabatedly, as she tours the playground in such ludic activity. On understanding we're primates who spend much time in boxes, sedentary lives are part and parcel of the twenty-first century cost of our *articles of peace*. It's often argued we do not live *naturally* attributable to foraging-hunter-gathering ancestors. This is misleading. If it exists, it's natural, as natural as the molecules that compose plastic, it seems, we must get used to it. Heart-related illnesses that have plagued humanity were congruous with an unprecedented market for choice and consumerism one had no reasonable philosophy to withstand, and so the susceptible have suffered as a consequence. The access to our ludic capacity outlives us generationally, whether it be a competing vicenarian MMA fighter to an octogenarian taking a walk around her garden.

A War of Nutrition

I often call this the war against, or the antidote to capitalist choice and consumerism. The laws that endeavour to protect the buyer will improve over centuries. Passing from the agricultural revolution around 10,000 years ago when the agriculturalist's diet was contingent upon the energy of our closest star, to moving to a time in which technology can measure our nutritional intake, would seem bewildering from the eye of the ancestral agriculturalist. There is a pervasive movement of nutritional information involving experts disseminating facts and debunking myths akin to our nutritional illiteracy. The new twenty-first century is held in scientists who have engineered concentrated nutritional information for the layperson to avail of in cellular applications tracking our nutritional intake. Whether it's a comprehensive manipulation of your *being* or briefly observing individuated calory deficits, information is there, offering the benefit of being free of charge and at the cost of taking agent responsibility.

Welfare

Financial Management, Investment, and Space and Time

What normally springs to mind when one elicits the term *investing* is most usually something financial. But what should be to nobody's surprise, we are almost always investing. We could define investing simply as, foregoing small costs for future benefits, and not only are small sacrifices endogenous to our species, but so happens from a young age. This could be as simple as a teenager walking ten minutes through her neighbourhood on the chance that her friend is at home and available to hang out. Book-swapping your favourite novel on the chance your friend's novel is also a worthy read. Saving a bit of your ice cream for your little brother on the chance he'll remember the small act of kindness, and so on. On conversion into adulthood, sacrifices, or costs, become vast, and as victims of the very nature compelling us to compete and thus flourish, similar pleasures are thus sought by multitudes, so if one wants to make profitable sacrifices to make life a little easier for themselves, it might be a good idea for one to organise what one invests, which is a feature of the human species one can never evade, also akin to the philosophy of space and time. Those with higher faculties of intellect are thus stimulated by investments that one could say are slower to burn, and conversely those with lower faculties stimulated by speedy returns. The fun and games that lie within investment is that uncontrollable variables occur, thus shrewdness lies upon the act of making what one sees as the best decision in a specific context or circumstance, which is the very locus of space and time, thus one will never see a concrete set rubric for investing in places like the stock market. To give a simple instance, imagine I invest **all** of my time training to become a basketball player and due to unforeseen circumstances, I lose a leg. Normally, and conducive to my ontological horror, I spend from nine months to a year in utter despair. With an athletic disposition, I take up training once again and now train to become a worthy Paralympic basketball superstar, relative to the *space* that can now be covered within a specific span of *time*.

In terms of capitalist costs, it seems in financial terms one looks past the value of such commodities. The commodity one will never give up easily is *time*, thus we tend to whine when feeling overworked or summoned to gratuitous meetings in places such as those nine-to-fives. But one must not be blamed for a lack of shrewd ability to understand the value of such commodities, as humans thus far have never been induced with the numerous luxuries we see on the capitalist markets today (those of high value, utility, and adornment). Competent marketeers continue to clamp down on our caprices as we get peckish on the way home from our workplace, breaststroke in the pool of novelty, and look twice at anything red or displaying the word 'sale'. We communicate faster, wash quicker, travel farther, and indulge more frequently than ever before, so it seems the **only** true vaccine to the pleasures of capitalism is agency an illusion of personal responsibility. We must deliberate on the gratitude for the luxurious commodities of twenty-first century trade that make such luxuries accessible in the form of capitalism, as capitalism continues to ameliorate across the world.[148] So then what must be done in the face of such luxuries, easy, **quantify**. We've all heard the saying 'numbers don't lie'. It's true, numbers don't lie, but they can deceive, and so it's on the lay statisticion to provide a comprehensive picture of the issue as competently as possible, and this is achieved by asking the many feasible questions that numbers can account for, which multiply sociologically. But thanks to the evolution of fields of economics and finance, the numbers game is already compartmentalised for everybody's use. Useful for the lowest of earners to the highest brokerage firms, banks, and conglomerate companies, is what one knows as the *balance sheet*.

A balance sheet is principally used for an investor (which happens to be everybody), to *balance* both income and expenses, assets, and liabilities. A look at one's cashflow tells the story of what lies superfluously for one to invest. American businessmen such as Robert Kiyosaki has been famous for the *Rich Dad Poor Dad* brand in simple finance books and games on helping investors to generate enough passive income to leave what he calls the 'rat race', which is simply to acquire enoug1h assets to pay more than your income does monthly in areas such as real estate, while understanding exactly how one's cashflow should look like.

For the anti-capitalist who winces at the disparities of those high earners are the manifest advantages and disadvantages for both parties. High earners have more cashflow, yes, however, it's only commensurate to how much they earn, as individuals hold the propensity to amass liabilities (becoming the disadvantage). Of low earners, it takes a less amount of money to generate passive income that may free them from their day job, while they simply have a less amount of cashflow to reinvest. But both high earners and low earners are equally sensitive to bankruptcy, which would be amassing more liabilities than assets, while losing all of one's cash! So it's important to keep one's balance. Thus they call it a *balance* sheet.

So, is it a good idea to quantify? One must quantify their expenses. What children cost are also quantified and play more than a financial asset, but a life asset one may naturally invest in. The question is almost always invoked with the luxury, 'can I afford it?', and unfortunately the question seems to a lot more complicated than it originally appears, as such a question instantiates itself in the becoming of financial literacy. Many dispel of their cashflow to alcohol, other drugs and unnecessary vices, and in turn deem bills, fines, and sporadic moments of payment *expensive*. So it seems given the advantage of a lower necessary passive income that would free the lower earner from her day job, the disadvantage of a decreased cashflow can only be met with the act of personal responsibility. That is watching her liabilities and expenses against her assets and income. Any vice or poor financial habit would ruin her balance sheet. But yet again vices such as drugs and poor financial habits are both purely sociological and psychological, in so far as an individual manages to retain her *free will* from such injurious philosophies.

Thus I call it a war with capitalist luxuries! Some wars you will not win, such as the requisite of carrying a mobile phone which is needed for the world of work and communication. Meanwhile your investment into a worthy mobile phone gives you internet access, an mp3 player, an alarm, email, camera, video recorder, map, and a cornucopia of many of the tacitly amazing features that mobile phones bring with them. Now, in terms of the compatibilism between free will and determinism, can we ever be held responsible for messages we do not see and calls we do not answer?

~

In our free trade markets, assets mustn't always be liquid, in as much as they start out otherwise. When J.K Rowling was first writing her Harry Potter novel before it was published, it was speculatively liquid as there would have been a plethora of unwritten novels in the world ready to be published. It's just that hers prevailed! The same would apply to ideas such as Facebook, Amazon, and so on. In this way music, arts, entertainment and even networking platforms such as influencing managed to monetise more than even sophisticated academia and art, which falls to many of a critic's frustrations. The sort that monetises of unsophisticated means has its ephemeral tendency in meretricious displays by individuals living decadently and driving themselves amiss of financial flourishing. It may never be necessary to covet your neighbour, as it's always far too complicated to compare. As the founder of logotherapy, Victor E. Frankl, prefaced in his famous book:

'For success, like happiness cannot be pursued; it must ensue, and it only does so as the unintended side-effect of one's dedication to a cause greater than oneself or as the by-product of one's surrender to a person other than oneself' (Frankl, 2008, p. 12 pre).

The misconception of financial literacy believes investors to become economists, financiers, brokers, bankers, and/or even mathematicians. But if this were the case, then such experts would be indubitably rich. Financial literacy would, thus, be more akin to a simple philosophy of an individual earning an income. Irrespectively, financial illiteracy *appears to be* financial irresponsibility and may thus hold one accountable for expenses, and consequentially liabilities. The essence of financial management is absolutely necessary, and for many will understand that an evasion of financial responsibility leading to parenthood is merely a suboptimal recipe for parent planning. Every adult has experienced childhood, and so most adults have been part of a family in some shape or form, and so one will be aware of families estranged in the world because of what is thus bequeathed from generation to generation, rendering financial hostility. More precisely, the issue isn't the amount being bequeathed to the next generation, but a literacy which could serve the individual practically, and there is countless literature that unfurl the myths of spending and the financial propensities instantiated in the *poor person's philosophy*, being unfortunately and incessantly bequeathed. A humility to the financial field could render the learner financial gains and with those financial gains be as altruistic and independent as she chooses to be. Financial illiteracy leaves the individual necessitous and resentful, under the misconception that she fulfils her deterministic fate in the capitalist struggle. If she can read and write and chooses not to avail of herself the opportunity to research her opportunities to gain more financial stability and consequentially independence, she takes responsibility, in so far as she should take responsibility to seek help if there are large hindrances to her daily life.
Investment also instantiates itself upon the philosophies of space and time. There are the yolo (you only live once) philosophies, which one may brand as analogous to impetuosity and short-sightedness, and, conversely, those who think of 'long term'. Take an example of gaining over $1,000,000 in stock ownership and retirement in what Eric Tyson terms *social security choice* via placements of just over $200 a month over a 45-year period on $50,000 a year in the US:

'Working folks are paying a lot into the current U.S Social Security system. For people who work for an employer, the employer withholds Social Security taxes of 7.65 percent (out of their ow coffers) for a total of 15.3 percent of the worker's compensation. Of that amount, 2.9 percent is earmarked for Medicare – the federal government health insurance program for people age 65 and older. The remaining 12.4 percent goes toward Social Security. So, for a worker being paid $50,000 per year, that amounts to $6,200 that gets sent to the federal government and used for the Social Security system…

So, returning to our example of the worker earning $50,000 per ear, he and his employer are paying about $5,022 into the Social Security retirement/pension program. Imagine if just half of the money (Say $2,500 per year in today's dollars) could be invested in a highly diversified, low cost U.S stock market index fund – such as Vanguard's Total Stock Market Index fund, which invests in thousands of U.S. publicly traded companies of all sizes…

Over a 45-year working career, this worker's stock fund invested Social Security money would have grown to more than one million dollars - $1,377,870 assuming an average annual rate of return of 9 percent. Of that, just $112,500 is the amount paid in by the worker, and the vast remainder, more than $1,265,300, represents the investment returns' (Tyson, 2021, pp. 58-59)!

Without earning $50,000 dollars a year and half the amount of Social Security they would abound to make over a million dollars, anyone in employment can open IRA (individual retirement accounts), and upon contributing their limit, put more money in *annuities*. The essence here is a perspective of investing conducive to space and time and the *decisions* that present themselves upon the individual (or the Prefrontal Cortex for decision-making). For the individual to wince at the thought of their post sexagenarian and retirement responsibilities is to think about the agency of amassing enough capital to pay for your child's education.
Take the example of the Affordable Care Act that sees taxpayers earning over a total taxable income of $200,000 as a single return or $250,000 on a joint return to pay 3.8 percent extra on tax (through gains or a 'modified adjusted gross income'). This adjusted tax is commensurate to how much one earns. Taxpayers subject to this 3.8 percent tax increase can find ways to reduce taxes by either opting for tax-free money markets and bonds, investing in tax-friendly stock funds, or investing in small business and real estate as the wealthy rollover gains in properties, all whilst watching tax laws (Tyson, 2021).
The issue of the wealthiest branded avaricious, and, in turn subject to tax laws, may be vindicated by many means. The incommensurate distributions of wealth that see the wealthy blamed and targeted is by no means the solution. It seems to level itself out in the communist arrangement when anti-capitalist and pro-egalitarianism predicates itself on a world we simply do not live in as we endure a struggle for existence. Socialist and communist states thus suffered hyperinflation in which bootlegging and racketeering emerge. Black markets then set themselves out akin to the tenets of capitalism. Our ontological faculties (founded upon what's contributed to the species), most certainly for males in reproducing with exceptions playing no part of the rule, thus returns.
Instead of shifting responsibility and calling on the wealthiest, using unfair contrasts of luxury and privation as reason, what must be necessitated is how those who suffer from financial privation can be freed from their unfortunate circumstances. It presents itself as another *post hoc* fallacy, that one person's wealth is another person's poverty, put into a rather too simplistic paradigm of benefit and cost. If this

is so, how can extreme poverty have been reduced to less than 10 percent in the world from more than 10 percent enduring extreme poverty over two hundred years ago?

Space, Time, Normative Ethics and the Five Facets of the Pentagonal Prism of Agency

Unfurling consequentialist theories against the quality of action must be carefully doctored and explained. Ethically, the consequentialist believes *consequence* for action in society is fundamental and necessitated good. Necessitating good of the consequence undermines morality of action in which the deontologist whimpers, who calls for the necessitation of action to be of *categorical imperative* in ends free of ulterior motives (Kant, 2021; Pinker, 2012). In as far as deontological ethics and consequences may prove effective, the virtue ethicist believes the character of the subject should hold *traits* of virtue, as opposed to *traits* of vice, which is conducive to the outcome as best. For instance, we could reveal the complexity of moral ethics when looking at an example we make in our daily lives. If I make a promise to you, irrespective of the outcome, we could say that the good could be held ensconced in the *action* that intends to be good. If we can't trust each other with small and simple promises, how can we cooperate as a species? Meanwhile, if the consequences of such an action could be highly risky, deleterious, or rewarding to the promisor, promisee, both, or many, then we could raise the moral good of the promisor to make such a promise in all her severity, while the consequences are hardly necessitated. But here ethics attributable to deontological laws of morality render responsibility ubiquitous! Let's say deontologically obligatory laws may be useful for such trivialities, in which consequences are not of great importance. Take the instance of child play.

Those trivial deontological positions are duly instantiated in what we perceive as honour societies greatly founded on honour, notwithstanding the consequences of what may be asked, as the moral good is in servitude and even character of the giver of service to the subject being honoured.[149] For virtue ethics, the character of the individual who expresses such actions can be ascribed to her traits. Let's look at this again from the context of space and time, what we are in a constant struggle and endeavour to adapt to in one's environment. We thus confer with ourselves whether the action is morally *better* over a longer time in larger space than of its ephemerality, thus rendering the consequence of an action measurably and morally *better* than its ephemeral moral good. A character who possesses traits that opt for the moral good are profitable against a struggle and adaptation of space and time, as that is what a character would be, hence mollifying the virtue ethicist. Meanwhile, of the more complicated predicaments, we cannot know the future and thus must measure total cost and benefit, which unfurl the teleology of 'the' or 'an' individual against a group a subject is part of against the group that the subject is extraneous to (implicating the *Us vs Them* in moral ethics). Thus the consequences of a languidly *good* deontological act ramify into what could be perceived counterproductive to the flourishing of humanity as a whole. Professor Jordan Peterson was shrewd in pointing this out in his 2021 book, *Beyond Order: 12 More Rules for Life*, in which Peterson calls attention to Freud's catalogue of phenomena germane to repression. They include defence mechanisms, reaction formation, displacement, identification, rationalisation, intellectualisation, sublimation, and projection; strategies of repression that often leads to mental illnesses. Peterson tells of an elision of Freud's mediation between dishonesty and psychopathology; that of *sins of omission*, as the assumption that things experienced are things understood. The *sins of omission* will play a pivotal role in the rest of this book:

'People generally believe that actively something bad (that is the sin of commission) is, on average, worse than passively not doing something good (that is the sin of omission)' (Peterson, 2021, p. 97).

Let's simplify the *act of omission* with an instance of voluntary manslaughter before I return to this issue in the discourse on racism:

A pilot is flying her plane, on landing the plane, and after many years of experience landing planes in which it appears second nature to her, she is also pondering on her co-pilot's unfortunately bad day. On pondering on her co-pilots bad day, she recklessly replaces a procedure with a good act of empathically pouring the co-pilot a glass of gin and tonic in which they both rejoice, consequentially killing everyone on the plane. We know of her act to seem morally good, but the consequence of such an action was catastrophically bad.

A prism of agency would be an individual approximating these five facets, some daily, others weekly. With it being merely infeasible to follow all five facets every day of the week, month, year and our lives, it would be an initial effort in removing the plank from one's eye before attempting to remove the sawdust from your apparently irresponsible brother's, as he necessitates the speck of sawdust in his apparently irresponsible sister's eye; as Jordan Peterson's sixth rule stipulated best, 'put your house in order before you criticise the world.'
If one effectively subscribes herself to the work of literature, ontology and welfare, she gages in the onus of competing with herself effectively concerning the largest of issues in the world. It would be indubitably supercilious to believe that one is more informed about the world's most pressing issues than those of the likes of academics in the Copenhagen consensus. As learned from Lomborg, it might be better for an individual to first work on themselves, and on being slightly more informed and financially stable, he can choose to redirect his offerings to helping communicable diseases, malnutrition, and hunger, instead of financing many a confounded cause.

~

As I would like to resound in this part of the book, one must trace the trajectory of evil, or at least dismantle the apparatus of evil that fortifies the ontological being to be held responsible for some of the most inhumane acts within civilisation. The danger of such philosophical discernment is falling into the valley of *causal* fallacies. Be that as it may, I do only find that *ontological security* can compensate for anything that runs misconstrued to such a theory, and thus should be summoned in the light of those evildoers. I do believe those five facets of a prism of responsibility to hold true, but I must augment to such an understanding of literature. What those approximating a philosophy of realism may agree with, is that I do believe it is best not to evade the literature concomitant with the most wretched and cruel occurrences transpired in human history. The onus is on the twenty-first century literate citizen to apprehend the composition of such evil being and to both be able to detect such comportment and thus hold the person accountable, which would effectively be to save him from himself. It's a biological argument, giving the genetic miracles of our ontological status, we necessitate our scarcities and imperfections, as a constant struggle for existence instantiates itself in sexual selection and the tendency for nature to select for the most profitable characteristics in a vast complicated gene pool.

Fyodor Dostoevsky, the nineteenth century Russian novelist, depicted this ontological apparatus in the notorious demeanour of his most complicated character, Rodion Raskolnikov in *Crime and Punishment*. On eventually allowing his crime to come to a successful fruition, Raskolnikov holds out on his responsibility to be punished through his self-conceit and self-pity. Throughout the novel, Dostoevsky is able to both disassemble and reassemble the ontological apparatus of Raskolnikov's evil as he lives life concerning relations with others, encompassing true friendship, family, opportunities, and ephemeral misfortunes many endure. He is successful in being able to manifest his *will* into the very resentment that many endure through the offload and mere evasion of the taxing responsibilities of life (Dostoevsky, 1991). Today there are a plethora of organisations and groups to choose from that help in fuelling the motif of resentment. Such resentment is played out incognizant of why Lauterpacht became one of the pioneers of human rights we may so often remind people of today. The game of identity politics (what most should certainly construe and which I express very carefully), is played which there is and will always be someone to blame for given misfortunes. The only antidote for such occurrences is a sense of agency and illusion of responsibility, which includes rendering those accountable as time is a powerful force that to be utilised.

Group responsibility, as we have learned in jurisprudence and the legal sciences for millennia, understand this most, so a game of identifying oneself with a group in order to blame another group is not only injurious to reason, but dissembling! If a group of five criminals manages to rob a bank and three victims are murdered in the process, whosoever shall adjudicate understands the proclivities of responsibility not taken, which comes in the face of blame. Somehow it will be one who incited the other to commit the act, the other who incited the other, *ad infinitum*, until someone can be successfully blamed. Be that as it may, in such adjudication, the whole group will not share responsibility of the crime, each will be adjudicated relative to the responsibility taken for their act and mitigating circumstances that may exonerate them from such commission.[150] The evolution of such legal science is antithetic to the anachronistic and now obsolete idea of acting through impetuosity, or the *id*, and holding all criminals accountable (hence *innocent* before *guilty*).

What starts out as a progressive group challenging disparate social wellbeing turns into an arena in which individuals abuse others in shirking responsibility. Thus you find such groups are not only no longer worth following but have converted themselves into armies that aim to harm those who do not espouse and recognise their personal misfortunes. Irresponsibility, rather than irresponsible people (as irresponsibility is garnered from individual to individual and some certainly more than others), leads to political parties adopting horrific and counterproductive policies. One of those policies is 'affirmative action'. If a group has been oppressed politically for hundreds or thousands of years, is it possible that there isn't any deleterious collateral damage to their ethos? Is it possible that such environmental influences render them irresponsible also, which in turn renders their social disparity? Affirmative action would only be throwing a man more fish and possibly a rod, without him both knowing how to use it or even what it's used for.

And so one of the most injurious philosophies of all, in as far as human philosophy is concerned, is the victim. Justice has troubled philosophy for time immemorial in that the issue finds itself irreparably woven within reason, which finds itself woven within more reason. The politician, whether cognizant or incognizant, may use that of *telos* to gain purchase for her party without ever having to entirely become accountable for the ramifications of such narratives. The essence of the victim, and I make this an indubitable claim and distance this from reasonable politics intentionally, is that every living organism under scientifically classified *life* is a victim. From the viruses that have tortured the earth recently, both literally and politically, to the big-headed primates we are, we are in a non-consensual *struggle for existence*. We are in a constant state of eating, sleeping, drinking, blinking, itching, coughing, sneezing, walking, breathing, and all through a process of homeostasis while simultaneously doing so against the forces like gravity. As I have hopefully resounded through the core this book, we have also a relationship I deem non-consensual with space and time, as we know is the same thing in essence. That means, we have, as products of nature, another non-consensual relationship with competition. We are thus hardwired and bioengineered to compete, and I have stressed this enough through the philosophy of ontological being. Our hardwired competitiveness is instantiated in the noun *protection*, which could be easily replaced with *defend*, as to protect oneself is to defend oneself against the forces of nature.[151] The state for every member of our species is to protect oneself.[152] My retort would be that suicide would still fall within the realm of protection, which would be to protect oneself from oneself, realising that the self is not that which exists. A jellyfish or a centipede has evolved instincts to sting at its propitious moment to survive, while a rose will have its prickle, and the human his myriad ways of doing so. So, the most deleterious philosophy for the human to appropriate as exclusive to the next member of its species is the philosophy of the victim. That's because where an individual will search for the subsets that constitute the victim, remaining merely incognizant of his biology, is what he will certainly find. The very subsets that constitute the victim are the very tools used to hijack his franchise and radicalise the masses, as we see perpetuated throughout the world today, and may continue to do so unabatedly.

A more comprehensive understanding of history will edify the victim, along with a pure biological conceptualisation of her nature in merely coming to terms with the idea that we are ephemerally part of a world under the illusion of space and time metaphysically scalable by such intuitions. In terms of epistemological circles, and in so far as an individual may consider themselves religious (who will know more than anyone they have a *reasonable* right to have), the onus of the individual to propel science is certainly warranted. For the citizen to engage in the theological philosophy of science and reason is required as those religious philosophers were the very pioneers of the science we capitalise on today to live longer and more prosperous lives. Many endure the psychological trauma from the scholastic experience in denarian years to irrationally eschew the responsibilities of returning to science. They may shroud their ignorance of science with the epistemological cloak by branding those profusely meddling with science as atheists and believe there are no philosophically facile pathways to understand it, which returns to the prism of human agency in superseding oneself and an endeavour to venture the abode of humble responsibility.

Stupidity

Our ontological proclivities, which I advertently resound to make my point both fervently and necessarily, instantiates itself in what best evades 'stupidity', which we may term 'uneducated', 'unlearned', 'uncultured', and so on. It also presents itself in the relationship one has with her relative phenomena, which we linguistically accentuate in the verb, *to know*. For instance, let's look at the maths problem:

$$2 + 2 = \_$$

The Epistemological Quest

This math problem sits well between ontological perimeters of all who will read this equation and does not doctor with one's ontological apparatus in finding the solution to what two with two will allow the reader to intuit, which is obviously four. Let's look at this problem:

$$\zeta(s) := 1 + 1\,2s + 1\,3s + \cdots = \infty\, n=1\,1\,ns\,.$$

How does one's ontological apparatus sit when seeing this problem. It should not doctor with one's ontological apparatus as it is what is known to be part of, by few, as the *Riemann Hypothesis*, dealing with the nature of a prime number arithmetically compared to atoms, as prime numbers are divisible only by 1 and themselves. In fact, the Riemann Hypothesis is a solution so complicated that an Indian physicist in the city of Hyderabad, India, claimed to have solved this problem. Irrespective of approbation by preeminent figures deeming his work laudable, authorities still await the solution to the Riemann Hypothesis, allowing a one million USD prize pending for whosoever shall solve it (Sadam, 2021).

But the ontological apparatus within the locus of the former equation, moving slowly towards the latter, is what triggers those ontological nerves, and so one will do away with *knowledge* by accentuating its futility in the larger context of human experience. As democratic entities in our taxonomical species, and in the case for political philosophy, many will reach a speedy and tacit conclusion that it's not necessary for an individual to *know* anything nearly approximating the Riemann Hypothesis. Meanwhile, one will certainly vacillate of one erring on the former equation. But within the scope of rationality one does not err as often as you should expect, which I claim to be an intricate problem ensconced in either the art of manipulation, or what others have called misology.

Of Misology: *Shame, Shame, Shame!*

'The danger of becoming misologists, he replied, which is one of the very worst things that can happen to us. For as there are misanthropists or haters of men, there are also misologists or haters of ideas, and both spring from the same cause, which is ignorance of the world. Misanthropy arises from the too great confidence of inexperience;- you trust a man think him altogether true and good and faithful, and then in a little while he turns out to be false and knavish, an then another and another, and when this has happened several times to a man, especially within the circle of his own most trusted friends, as he deems them, and he has often quarrelled with them, he at last hates all men, and believes that no one has any good in him at all' (Plato, 2012, p. 88).

It was Socrates' trial and death that the art of misology prevailed, promulgating his apology to those present as misology became the art that shrouded his defence. But where would it play itself out and what is of its relationship with misanthropy in the case for science and reason today? Given my strenuous reverberation of the ontological apparatus, it seems misology and misanthropy can be well instantiated in a benefit-cost for the ontological subject, which may also play out through the experience of *eristics*. It will mean anything that approximates a precarious ontological apparatus must be dispelled of while anything deleterious to the ontological subject also endures such effect (what we may also call *shame*). We observe this in the philosophy of honour or sense in authoritarianism. What a subject may have the proclivity to enhance is anything which may safeguard one's position (that's notwithstanding the dearth of reason within civility), hence we find both the insurgency and antipathy of Napoleonic revolution and the disquietude of Marx and Engels.

The monarchs, lords, ladies and knights/vassals were willing to forego reason to safeguard an intraspecific ontological position commensurate to as much as what could be garnered. This is akin to the case of misology enduring a *struggle for existence* in nature. Repetitious throughout authoritarian regimes, it was imputed to what we define as power, anything adducing such ontological and intraspecific security within a particular **environment**. The sense of misology is thus extended once the subject is extraneous to the particular environment and extenuated when harbouring an ontological status. You will find him disconcerted when enduring ontological asymmetries from his environment and equanimous when within, as the Nigerian poet Niyi Osundare best expresses in his poem *Not My Business*: 'So long they don't take the yam from my savouring mouth' (Mardlife, 2018).

This is played out in the subtlest of contexts as asymmetries are thus innumerable. They are both unscalable and immeasurable for lucid political discourse. As noted by the invocation of John Maynard Smith, evolutionarily stable strategies must be played out through chaos before it finds its order, yet evolutionary games are most always being played, and quite frankly, misology will be a tool eschewed if injurious to the precariousness of security (Maynard Smith, 1982).[153]

What has been heretofore deemed immeasurable and unscalable is what the advent of science and reason prevailed. Science, or that which scales and measures, is to be found in the responsibility of the twenty-first-century democratised citizen who honours her inalienable rights. The bearer of her irresponsibility becomes her children and the children after those, and so what becomes incorrigible irresponsibility approximates a said *infinite regress* beyond our generational responsibility in bequeathing reason beyond future civilities.

The misologist thus becomes the sufferer of 'just'. As mentioned earlier within the philosophy of *telos*, the adverb, 'just', within our instinctual linguistics and threshold for communication, is merely utilised to expedite a necessary process of intraspecific communication. and so is used in such manner in the English language. But teleology may quickly be espoused as anathema to human reason, and so must be cautiously evaded in formal philosophical discourse.

A balance of the necessary evils in biological propensities must be assiduously measured in misology and reason. It manifests itself in other inclinations that become resentment and/or a nihilism of humanity; which we may call misanthropy, as scaled in the polarised extremes of the radical feminist and misogynistic Incel. Ontological predisposition engenders the by-product of hierarchy and must be significantly considered when approximating an alacritous political philosophy. What will be deemed as psychological is what politics may impetuously eschew as a dearth of evidence, or commission, so the onus on the subject to ruminate justly on the subject will allow for a better understanding of such a complicated problem. The problem is what becomes the epistemological problem of a society today with a vast population of literate and democratised subjects entering the realm of a philosophy of knowledge. And so, as stressed earlier, our epistemological consensus is merely and contractually induced rather than knowledge which is abduced or deduced. It's within induction of such phenomena which is what will necessitate deduction for that which we can trust throughout the species, which we tend to call a process of civilisation.

Turning to the Socratic position of knowledge for an epistemology that approximates *the good*, I find it more than congenial to end on this section on the famous *allegory of the cave*.[154] Plato divides worlds into both the *visible* and *intelligible*. The visible world begins with the sun. The sun gives its visibility to objects of sense which we have from the power of 'observation'. The intelligible world begins with *the good*, which is the source of reality and truth, giving intelligibility to objects of thought and thus the power of knowledge to the mind. Both forms of knowledge could be seen as a straight line separated into unequal parts (with both lines folding into themselves in the same ratio representing the distinction between what's visible and intelligible).

The faculty of knowledge then manifests itself into knowledge (*epistēme*), and opinion (*doxa*). The former sense concerns itself with the inferior aspect of epistemology discovered by deductive reasoning, which is how one comes to *know* a thing. When more contradictory information presents itself, one must resort to dialectic reasoning (*noesis*; to seek truth), and this is achieved through the dissection of forms. The latter sense, opinion (*doxa*), concerns itself with belief as its principal aspect in the visible world and illusion, thus inferior opinion. Belief could be said to correlate with physical things we perceive and illusion to correlate with both shadows and images.

Socrates believed that those who approximated intelligence (*noesis*) were the philosophers that must govern as they approximated what was *the good*; the intelligible world. The analogy of the cave I believe is best represented by what was previously referred to as *the epistemological cake*. It seems the prisoners who have glared at shadows and images are those that meddle profusely in the most superficial layer of the cake (concerned with humanities and politics alike), and the part we call natural philosophy, parents of the scientific revolution, is that which holds most epistemological purchase. This would be the prisoner who was freed and became inured with staring directly at the sun once freed from the cave (Plato, 2007).

The scientific revolution distanced itself from the visibility of the sun as it manifested for the new philosopher of today a larger truth, or larger intelligence (*noēsis*). The ingenuity of Socrates in understanding the sun to be pivotal for what's epistemologically *known* was not only prescient but cosmologically accurate with Darwinian evolution in the retinal development of the eye. This is extended to anthropic features such as circadian rhythms, melanated skin, and eyebrows evolved to protect perspired sweat from reaching the eye. It is thus germane to the Copernican notion of the sun being central to our solar system, discovered millennia after Socrates professed his notion of *epistēme*.

I then must confer that it may seem futile to approach any form of knowledge amiss of a teleologically scientific substrate. Or approach such knowledge amiss of attempted philosophical explanations.

Our democratic evolution in the west, in which Winston Churchill believed was the best of the worse choices, instantiated the false sense of *doxa* to what we claimed to *know* in the twenty-first century. This is often recursively defended in the entitlement we remind each other of with the *right to hold an opinion*. But it seems this enfranchised *right of opinion* becomes diminished the more the severity of the circumstance is raised.[155] Everyday political skirmishes by the populace hold little severity, and so it seems epistemological purchase is appropriated where necessary. Where exceptions or anomalies may not seem to be injurious when viewed myopically, individuals do compose society and so such poor philosophy both adulterates society and condones such poor philosophy, in turn pervading larger parts of society. So the inquisitive must ask herself better questions of what morality and ethics truly are and inquire into what she thinks she knows through a philosophic and scientific prison. Quintessential of the cave analogy is the ideas that prisoners who see those *doxastic* images and shadows hold such views with epistemological conviction, and so are most arguably poised to kill or die for what they believe to *know*.

Socrates' gives a precise analogy of those various ways of *knowing*:

'Suppose the following to be the state of affairs on board a ship or ships. The captain is larger and stronger than any of the crew, but a bit deaf and short-sighted, and similarly limited in seamanship. The crew are all quarrelling with each other about how to navigate the ship, each thinking he ought to be at the helm; they have never learned the art of navigation and cannot say that anyone ever taught it them, or that they spent any time studying it: indeed they say it can't be taught and are ready to murder anyone who says it can. They spend all their time milling round the captain and doing all they can to get him to give them the helm. If one faction is more successful than another, their rivals may kill them and throw them overboard, lay out the honest captain with drugs or drink or in some other way, take control of the ship, help themselves to what's on board, and turn the voyage into the sort of drunken pleasure-cruise you would expect. Finally, they reserve their admiration for the man who knows how to lend a hand in controlling the captain by force or fraud: they praise his seamanship and navigation and knowledge of the sea and condemn everyone else as useless. They have no idea that the true navigator must study the seasons of the year, the sky, the stars, the winds and all the other subjects appropriate to his profession if he is to be really fit to control a ship; and they think that it's quite impossible to acquire the professional skill needed for such control (whether or not they want it exercised) and that there's no such thing as an art of navigation. With all this going on aboard aren't the sailors on any such ship bound to regard the true navigator as a word-spinner and a star-gazer, of no use to them at all' (Plato, 2007, p. 210).

Socrates' seems to discredit in democracy what we would seem to value the most in the west, liberty. As many are cajoled into the narratives of freedom as it appears antonymous to colonialism, oppression, and slavery, it seems the narratives of liberalism gain purchase. From American independence to Simon Bolivar, the French Revolution, and decolonisation in sub-Saharan Africa through to the latter stages of the twentieth century, it seems anything evocative of restraint on such freedom is *de facto* bad in essence. Meanwhile the philosophical standpoint, and from what I hope will be successfully reverberated, many may be unwilling to take responsibility along with such freedom also taken.

Socrates unfurls the ramifications of liberalism using the dichotomy of pain and pleasure. The top would be the most pleasure and the bottom the most pain. Those experiencing the middle when comparing their pleasure with the pain of what's previously experienced believe the middle to be top, and thus oscillate between the middle and bottom. A simple instance of this instantiates itself in ephemerally morbid pleasures such as copulation, drugs, and insatiable appetites. Democracy then also puts the son at equal standing as the father, the student at equal standing as the teacher, and he being led at equal standing as the leader. The result is that discipline subsides, the father panders to the *pleasures* of the son in his fear of abandonment, the teacher panders to the *pleasures* of the student, and the leader to those being led.

In terms of the illustration given, achievement of pleasure as the top is never fulfilled, attributable to the lack of discipline and the insatiable appetite of such ephemeral pleasures:

'They bend over their tables, like sheep with heads bent over their pasture and eyes on the ground, they stuff themselves and copulate, and in their greed for more they kick and butt each other with hooves and horns of steel, and kill each other because they are not satisfied, as they cannot be while they fill with unrealities a part of themselves which is itself unreal and insatiable' (Plato, 2007, p. 327).

Experience of the higher senses of pleasure are that which concern themselves with the higher faculties of intellect, thus approximating the Socratic *epistēme*. The desire for gain and our ambition followed by a guidance of knowledge and reason achieves this, and the pleasure that wisdom indicates, which is truth, will be the highest of all pleasures.

Part III

A Brief Discourse on Racism

'Knowledge of the Past is the Key to the Future: Upside Down Jesus and the Politics of Survival' (Colescott, 1987)

The Epistemological Quest

Before we explore this brief discourse on racism, let's go back to the three forms of knowledge that helped preface this book: *doxa*, *episteme*, and *gnosis*. As one will gather, discourse on racism seethes between all three constituents of knowledge, and racism, for the most part, is gnostic. I argue in this part of the book, that, if racism persists as *gnosis* or *doxa*, it does so unreliably (for those not experiencing it) and thus becomes implausibly denied. To keep racism unreliable, Hollywood and Penguin may sell and publish both movies and books that help fortify the evermore unreliable doxa. Be that as it may, it's the advent of neuroimaging measures such as magnetic resonance imaging (MRI) as recent as 1977, and functional magnetic resonance imaging (fMRI) of the brain in 1990, that allowed both the neuro and cognitive sciences to home in on both a reliable and epistemic authority over knowledge. Since the mid-twentieth century we've seen the mathematical sciences of statistics and probability theory play as incipient forms, and that's notwithstanding causal fallacies being few and far between. This part is founded on exactly that. The part that says, 'well, how do you know it's racist?'

The Epistemological Quest

Chapter 6

The Negro

~

In the little time preceding the 3rd August 1492, a Spanish slave trader, Martin Alonso Pinzón, was poised to avail of a lucrative opportunity to fulfil nascent hopes of adding to his estate. A wealthy trader from the small town, Palos de la Frontera, a town in the province of Huelva towards the southwest of Spain, Martin Alonso had imminently returned from a trip to Rome after selling a shipload of sardines that he successfully dispensed of. He had heard from the Vatican library about the ambitious legends relating to undiscovered lands in other corners of the seas.
Martin Alonso and his younger brother, Vicente Yañez Pinzón, were part of a family known in Palos as the Pinzón brothers, occupying a family seat in the centre of Palos as the wealthiest people in town, as they held ample influence in even the political affairs of the town. It was Martin Alonso who was well acquainted with a Franciscan, Father Antonio de Marchena, also acquainted with another inimitable Italian man on a speculative mission that may have been of interest to Martin Alonso. Martin Alonso and his sailors were nonetheless indebted to the crown's administrators for a past-misdemeanour, and thus it was through Queen Isabel's command they were compelled to give over two of their caravels to an Italian for the period of approximately one calendar year. The Italian, who went by Christopher Columbus, with the help of Antonio de Marchena, now had his two million maravedis needed to fund his trajectory, with also sufficient sailors to set sail and the final support from the crown (after both a tedious and fought-after Commission).
The Talavera Commission, brought together by the Spanish councillor of Queen Isabel, Hernando de Talavera, saw Columbus's ideas as not only infeasible, but inauspicious. With a lack of trust for the sailor's competence and cartographical skills, Columbus, along with his idiosyncrasies, adduced to the confutation of the commission's approval of his expedition. It had been the high spirits of the last Commission with Queen Isabel and her magistrates, and others involved in the Commission that had been conducive to Columbus's ultimate imploration as the Moors were finally driven from the fortress, La Alhambra, Granada. In as far as Columbus was initially disappointed with the 'paltry' 1.14 million maravedis from the Royal finance (precisely Luis de Santángel), it would become the Pinzón brothers who would play a financially larger role in the significant part of this part of history. Of the three caravels, Martin Alonso would captain the *Pinta*, Vicente Yañez, the *Niña*, and Columbus the *Santa Maria*. There were three latent and then manifest ideas that played out the secure substrate for the expedition. It was, as experienced traders, Martin and Vicente motivated in acquiring slaves, Queen Isabel in gold, and for Columbus, in a legacy as a worthy discoverer (Brinkbäumer and Höges, 2006).
However, and considering the technological excellence of the Spanish and Portuguese caravel that was newly able to sail an extremely vast Saharan coast, it would be another similar story of three ships that left Portsmouth after 61 years in 1553. Captained by a Sir Thomas Windham, the *Lion*, the *Moon*, and the *Primrose* set sail with both a Portuguese captain, Anthony Anes Pinteado, and *Henry VIII*'s intractable *Acts of Supremacy* between 1534-1558. Their quest was to envelop themselves in the profitable gold of the Ghanian coast. They came back with stories of successful trade of Malaguetta pepper, along with the tropical fevers and a

The Epistemological Quest

decimated crew but with demonstrable potential for British inculcation. But it was the voyage in 1562 led by John Hawkins, a ship owner from Plymouth, who sailed to Sierra Leone plundering at will and selling such plunder on the island of Hispaniola that became the first successful mission of a British slave trade. This would mean that from the founding of the Royal African Company over a century later in 1672, the English share of the Atlantic slave trade increased from 33-74%, successfully purloining the cruel business of the trading of African humans (Olusoga, 2017).

~

The story of the term 'race', most especially within the realm of political philosophy, I say should and somehow starts here, and even so much long before that an apparatus of an indestructible and socially constructed paradigm. The biologically selectable variables of strength, seasonal distribution, and bioactivity of both Ultraviolet A and B prove attributable to the dispersal of hominins around 1.9 million to 80 thousand years ago, as human pigmentation involved the number of dispersing hominins in UVR rich environments against methods of sun protection in seasonal patterns, intensity, and wavelength mixtures of UVR (Jablonski and Chaplin, 2010). It's quintessential for any ruminant to comprehensively fathom this matter of fact before entering the realm of identity politics in what appears advantageous or conversely disadvantageous, as it most certainly leads to a domino effect and to deleterious ramifications and augmentation to the injury of reason. Before propelling ideas onto a complicated political philosophy of race in both a sixteenth then seventeenth century (when a post-Columbus transatlantic slave trade would truly set forth), it began around the contours of fifteenth century Portuguese exploration, in which names such as Nuno Tristão, Antão Gonçalves, and Afonso Gonçalves Baldaia play of importance. It was the term 'negro' first by the Portuguese then subsequent Spanish explorers, thus onto English verbiage as 'black', which truly commenced those *ontic* modes of inscriptions in sociology.
The early usage of the word 'black'[156] by those explorers on articulating sub-Saharan Africans would use the term as, to the phenomenological eye, all that one subjectively perceives, a just *a posteriori* expression of sense-based judgement. 'Black' is merely sensical as distinguishing the subject as another type or form of oneself after its predicate. The European ontic distinction of oneself and those alike. This would characterise itself for a long time as synthetic to the European explorer, settler, or trader, conducive to a mere dearth of understanding, which would come as late as the nineteenth century, and would be to misconstrue humanity in its most severe and reasonable sense. The mistake conducive to ignorance was in politicising a term which would prove to be epistemically confounded, all the way to *civil rights* in the United States of America to the approximation of the latter years of the twentieth century. In a twenty-first century, we may confoundedly continue to use the term 'black' with the same mistake as those earlier explorers, in so far as we use the term to correctly identify its social construct, of such an erroneous and impetuously expressed pastime.

The Separation of Races

As per our biology, we understand in rudimentary and formal scientific classification, that, since Carolus Linnaeus' work of a mid-eighteenth century, we primarily classify living organisms into *life, domain, kingdom, phylum, class, order, family, genus,* and *species*. Cladistics continue on in distinguishing organisms into its distinct clades, while phylogeny looks at the evolutionary history of specific organisms or species and ontogeny observes the present state of the organism or embryo for the sake of well-founded nomenclature. What we understand of humanity today is that we are scientifically classified as a species, that of *Homo sapiens*. We enfranchise ourselves in brandishing the term human/*Homo* as we are classified as the only humans to exist (as *Neanderthals* were the last to perish around 33 thousand years ago). Around 45, 000 thousand years ago our species inhabited Australia, simultaneously leading to the extinction of megafauna weighing more than 60 kilograms. The same was of our species inhabiting the Americas around 16,000 years ago, again leading to the curious distinction of megafauna of more than exactly 60 kilograms. Around *the cognitive revolution*,[157] around 70,000 years ago, did fictive language evolve and *Homo sapiens* slowly move out of Africa and spread. This was only after 130,000 years of when we first began to evolve in the eastern plains of Africa (Harari, 2015).

The geomorphology of land and ocean activity on the Bering strait was once terrestrially sufficient in allowing those *Homo sapiens* to cross and thus inhabit American land due to a long and expansive ice age (which could have feasibly led to the allopatric speciation of the species if left for hundreds of thousands or millions of years). An ice age so cold that moisture evaporated from oceans, soaked up by clouds that did not rain but snow. As snow fell in this age (akin to such consistent low temperatures), it agglomerated and lead to the varying altitudes of land and ocean. The sustaining accumulation of snow was the path allowing herds of those early travelling *Homo sapiens* to cross *Beringia* and continue on in inhabiting the Americas, while after the ice age (when the climate had risen in temperature), snow would evaporate and raise sea levels once again, the body of water that came to be the Bering strait.[158]

On October 10, 1492, Columbus endeavoured to preclude the invariable mutiny of his crew through (a) incessant reassurance, (b) constant beseeching of his crew to cooperate, (c) threats of the queen's wrath on return, and (d) piratical cajoling of the wealth rewarded upon discovery. Columbus not only had to display pre-eminence in meticulous estimations of the speed he was sailing but was also compelled to work against the current and drift of his vessel. As Brinkbäumer and Höges suggested, 'Everything depended on accurate estimations. Dead reckoning might be more an art than a science; it would seem Columbus was an artist' (Brinkbäumer and Höges, 2006, p. 127).

Columbus' *Santa Maria* and those of *La Pinta* and *La Niña* would arrive at a small island on October 12, 1492, known today as San Salvador, or as the Indians had known, Guanahani, an island south of the Bahaman archipelago.[159]

A little earlier (between 1434 and 1447), before Constantinople would fall to the Ottomans in 1453, after over a thousand years of Roman imperial rule and the birth of Christopher Columbus in 1451, it was Prince Henry the Navigator's rule that would be the first to explore the western coast of Africa and conclusively lead to judicial regulation of mercantilism from Europe to Africa with relatively harsh deterrents. It was incessant skirmishes and a premeditated endeavour to kill, capture, and enslave that would be the cause of many a death for both the Portuguese and the Berbers, then an arduous time would become their fate once reaching the more numerous sub-Saharan African tribes of Cape Verde. Once reaching Guinea and the Cape, the Portuguese, in such pertinacious endeavour, lost their most valuable knights and commanders; Gonçalo da Sintra, Nuno Tristão and the Danish knight Valarte. The death of Valarte had (ironically for some) been conducive to a reconciliatory approach in attempting to trade and dispel of such hostility, as Prince Henry had become aware of the intricacies of skirmishes across the coasts of West Africa in achieving the goals he had first set out. Compensation and the tutelage of widows and children stressed Henry's administration, as

between 1453 and 1500 six privileges had been inserted in a codex, granting 'to widows and descendants of knights fallen in Africa either a yearly monetary pension or rights to the part of the revenues collected earlier by the deceased from rural estates and royal towns' (Tymowski, 2014, p. 13). It was when the Europeans would first begin to use the term 'Negro', a time we know as the *Age of Discovery*.
The clear separation of what would slowly adulterate scientific philosophy, or political philosophy, whichever side of the coin you choose to necessitate the claim, made the separation of races a clearly feasible method in articulating what could be phenomenologically perceived as race. One white, one black, one India, and so on... Politicisation of such an impetuous stratification of human races would be feasible until, principally, the European enlightenment would compel the ruminator to sagaciously discern a predicated idea and see it through to the end through necessary syntheses. The issue would become of time and space which will trouble the twenty-first century mind. As the rise of technology transpired throughout the technological revolution and through the nineteenth century, so did the rise of what we confoundedly described as interraciality. That would be those of suspected races reproducing together and creating what was also confoundedly described as the *mixed-raced* child. In such psychopathological hostility and thus a belligerence with the true and veracious nature of race, it seems a pure scientific observation and precise definition had been contorted and until today, run amiss.[160]
Our proclivity or need to drive home a group, a group in which we can identify, or concertedly conquer, is what I observe to be merely attributable to identity politics. It was the same mistake Hersh Lauterpacht had sifted out of Raphael Lemkin's disquietude with *genocide*. Hence pertaining to the biology of blackness and what being 'black' would mean is, simply, nothing. As a species, in a vast and large gene pool we refer to as *life*, 'race' would be a sub-spectrum of our species. Characteristics would differ such a height, pigmentation, bone structure, susceptibility to one such thing and the other, running congruously with the biology of adaptational differences in an environment for as we know, is naturally selected. Melanin, one of those characteristics that would protect the skin from our solar star (akin to your climatological disadvantages within Milankovitch cycles). It would mean as a species, race would merely be a spectrum of **differences** (variability), not anything approximating classifications, and as the technological revolution has exploded over the twentieth and twenty-first century, an unabated conflation of the species has capitalised on the contraction of space and time and have thus reproduced, and reproduces *interracially*, engendering unidentifiable organisms we still almost fatuously fail to *identify*. This issue of misarticulation may not be as innocent as one would suspect, in so far as the philistine will evade the prism and the facets of agency I already mentioned previously.

That Being Said

A former professor of mine, principally in the latter years of achieving my honours degree, compelled me, along with his students today, to think evermore critically around the issues of race. In his book, *The politics of Islamophobia*, David Tyrer accentuates how a misconstrued inclination of treating race as a merely phenotypic entity is a rather obsolescent confutation of the issue of the discourse in racism today. Instead, he reverberates how race has been socially constructed through what he best articulates as *ontic inscriptions*. Irrespective of what you may be, what boundaries are emulated and superseded for a race to lose its encumbering stereotypes, the ontic inscription of the human eye is what manifests itself in such a phenomenology, and thus psychology, and what I would finish as the psychopathology of racism, attributable to a premature and debile criteria for diagnosis. Tyrer lays out a plethora of cases substantiating his claim, in which such examples where profiling instantiates its *ontic inscriptions* leads to the manifestation of a socially constructed and unjust cause of what he terms as racism. One of the famous cases in British politics in which *ontic inscription* is played out, was in the shooting of a Brazilian man, Jean Charles de Menezes, by officers of the London Metropolitan Police Service in 2005. On the 22nd July 2005, the 27 year old was wrongly suspected as a fugitive involved in a failed bombing attempt the previous day. Shot 11 times in the head as officers boarded the London Underground in what became a physical confrontation, the outburst to political theory of racial profiling indignantly transpired (Tyrer, 2013).

Yet again, particular to scientists, some have and may incessantly claim that as Islam is a religion, one cannot therefore be racist to an individual Muslim subject, as a Muslim is not a race. In so far as it is true in its literal scientific sense, it is also untrue. For instance, one could also argue that as countries do not bear any natural borders in the world, countries neither exist. Once somebody, or a group of people, act upon something intraspecifically, or social constructed to the extent that politics are thus meddled with, the social construct enters the realm of political philosophy. The Muslim, as a mere social construct, can be envisaged and organised in the human mind as what is best defined as a race. Attire, physiognomy, language, cultural tendencies, and demeanour are all added to a social construct which will be loosely related to reality in the form of what is healthily received as stereotypes. Augmented to the Muslim subject is then the most heinous sense of the barbarism that occurs within the 'Muslim *race*', merely extrapolated from the most miniscule on to the whole label of the social construct (Tyrer, 2013). It is achieved by the profitable tool of fear, which is what should be tacitly understood is a sensitively biological emotion attributable to the most injurious effects of misinformation, that which may be called xenophobia. The emotion cannot be harboured in as far as the fearer plans to shroud such fear once approximated with the racial subject, and so what thus appears is *action*. One could clearly impute the death of Jean Charles de Menezes to such political philosophies, through an *action* of confounding *fear*.

This harks back to the prism of agency from what was mentioned earlier as *Literature*. As subjects who must endure the responsibility of literacy, the most ominous dangers efface us in what manifests itself as misinformation. Misinformation comes in the form of a languid approach to information that at first presents itself as privy, then approximates the more conspiratorial realm which may become political parties predicated on a latent sense of fear, always cajoling the most languid subjects. The neurobiology of *Them vs Us* is best mapped by Robert Sapolsky, showing the biological proclivities we are most always amenable to (Sapolsky, 2017). Hence awareness of such crucially scientific information must be risen to evade such politics, or at least to approach fear in its assiduously deserved manner.

Africa

To make another significant case of *ontic inscriptions* exemplary within this epistemological quest, we may look at a geographical instance, or as I will call it, *general knowledge*, which also sits between the lines of identity politics in the west. At least the western European is thus recognised within the demarcations of his own nation, as linguistically, we tacitly, or epistemically fathom the French to be of his nation France, or the Chinese man to be the subject of his own demarcations also, the Japanese man to be of accord with his culture and history, the Australasian to be

The Epistemological Quest

of the demarcations of his own enterprise, or the Brazilian, Columbian, Mexican or man of the United States to be of his own historically liberated land. In as far as the *general knowledge*, or geographical distinctions dilapidate from continent to continent in the edified man, the identity of the African, most particularly the native sub-Saharan African, is unknown.[161] I merely attribute this to an *ontic inscription* and the instantiation of the linguistically philosophical term, '*black*'.

Firstly let's keep this both endogenous to a European and rest of world problem and exceptions of the American continent. The black man becomes his race. He is not inquired into through curiosity or inquiry as the information of his origin is implicitly superfluous. Incognizant of what has transpired, the subject has already been inscribed with the ontology of his race which bears all the social construct he is. Shrouded and instantiated by the canopy of ignorance that separates the Senegalese from the Liberian, the Angolan from the Zambian, or the Ethiopian from the Somali is such *ontic inscription* of his *black*ness. The German may be known as the German, the Indian, the Indian, the Thai, the Thai, the Russian, the Russian, in as much as the Kuwaiti the Arab, but the Nigerian, the *black* man.

Identity politics of all in which the black person entails, good or bad, man or woman, we call stereotype. Adduced to the stereotypes comes a philosophy imputed to such intricate political history concerning the racialisation of groups through social constructs and thus oppression through the actualisation of those very ontological and political social constructs. Turning back to the USA we see the issue of identity politics incessantly re-emerge at its highest given the accolades of individual Americans since 1776, in industries such as science, art, and technology ever more laudably so. Be that as it may, the issue is rightfully more sensitive and complicated within what is often presented belligerently in discourse.

Most recently in the USA, a wave of 'black' conservatism has rightfully begun to emanate in US politics which is successfully challenging Democrat's policies and the issue within *African American* communities. Larry Elder, a Republican American conservative and also radio host, may be lauded as a pioneer in the modern conservative movement democratising black Republican voices in the US with an endeavour to challenge those very issues within the 'black' community. Elder achieved both fame and notoriety after his interview on Dave Ruben's *Ruben Report* broadcast on Youtube earning over 1.7 million views. Elder dismantles Ruben purporting the existence of systemic racism as being a major problem black people face in the USA. Ruben goes on to claim that a white cop is more likely to shoot a black perpetrator as opposed to a white one. Larry Elder informatively unfurls statistics that prove otherwise.[162]

Returning to the issue of *ontic inscription*, Elder claims the use of the term 'African American', as he claims was initially coined by the black Democrat Jessie Jackson calling for African Americans to be engaged with an *African homeland*, is merely farcical. Elder points out that Americans with immigrants from countries such as Romania, Greece or Italy do not bear a hyphenated identity while their ascendents had immigrated the USA after his own, hence being African American would be farcically absurd. It yet ensconces itself in an obscure *ontic inscription* of race, as did my professor David Tyrer also claim of the term 'second' and 'third' generation in the United Kingdom in which the subject is identified as secondary or tertiary to their native identity, whilst being naturalised in their very nation. Elder believes the leading issue for black people in the USA is the family, as he prefaces in Candace Owens' book, *Blackout*:

'We know the statistics – that children who grow up without a father are five times more likely to live in poverty and commit crime; nine times more likely to drop out of schools, and twenty times more likely to end up in prison' (Elder, 2020, p. xv).

Candace Owens, who emerged primarily off Youtube as an acolyte of Larry Elder, espoused her conservative views on the black issue, rose to fame as a political commentator with partisanship leaning towards administrations emanating the Republican party. An advocate of free speech and critical discourse, Owen holds Democrat politicians accountable for the injurious narrative feeding a psychopathological *victim mentality* to black voters in the USA. As the 46th President of the United States, Joe Biden, had said in an interview on 'The Breakfast Club', irritating Owens, 'But I'll tell you what, if you have a problem figuring out whether you're for me or Trump, then you ***ain't*** black' (Herndon and Glueck, 2021). Owens is yet again a disciple of responsibility as the solution for black America today, as she writes:

'I have given consideration to the idea that recognizing our equality might make some black people uncomfortable, because with no one to blame but ourselves for failures, the weight of our own irresponsibility may seem too heavy a burden to bear. It is much easier to go through life with a white supremacist boogeyman' (Owens, 2020, p. 104).

An act from politicians who instantiate the *victim mentality* on voters is such an injuriously and surreptitious act, and as a sociological crime, I would render heinous. The victim mentality strategy has been and continues to be availed of in radicalisation, in extreme cases leading to the deaths of innocent lives, and is availed of in grooming, leading to principally underage girls being victims of abject sexual crimes. The implicit reason why the victim mentality strategy will always work so well is, which I have already briefly mentioned, because we *are* all victims to the struggle for existence. Anyone who truly understands the rudiments of biology will be cognizant of how this transpires, as there is no sense of homeostasis for the human mind conducive to the weapon of time and space. The danger of those susceptible to victimhood proves merely vitriolic when not doctoring the discoursers words cautiously, akin to the teleology of one's purpose to find oneself in a less precarious position in the political world holding no accountability for such action that appertain to avarice and consequent power. The sordid machinations used to espouse these strategies on the populace are duly significant in journalism as there is manifest disinclination of voters to turn to manifestos. A poll from 2017 indicating this revealed of 1,370 British people aged 18 and over, 67% either do not read manifestos or know what they are. As the BMG Research Poll found:

- 10% always read manifestos
- 23% sometimes read manifestos
- 28% don't read manifestos but use websites/newspaper comparisons as a guide
- 10% have never read a manifesto but would be interested in doing so
- 19% have never read a manifesto and would not be interested in doing so
- 10% do not know what a party manifesto is (BMG Research Poll, 2017)

It is also attributable to an unrelenting permeation of Marxist and feminist dogmas in the highest institutions of thought, quite specifically, universities. In as far as Marxism, feminism, and the Black Lives Matter movements are proving effectual, what always misfortunately transpires within such social politics is that either the movements are hijacked by the most irresponsible; criminals, or the susceptible cede such responsibility to their internal victimhood, in turn adulterating the entire purpose of the movement. Responsibility is thus seen to abdicate to our innate forces that render us victims to the struggle for existence. As seen in more detail throughout the other parts of this book, it's paramount in fathoming the complexity of life and the hierarchies engendered by natural selection and thus sexual selection. Hierarchies work on themselves both constantly and unwittingly and have done so for megannums. The Austrian doctor and pioneer of *Individual Psychology*, Alfred Adler, spoke about the proclivities of such thought from a biological perspective most extensively in his book, *Understanding Human Nature*. On responsibility, Adler points out what he finds incumbent upon such nature:

'Whatever he makes, he betrays one thing, and that is that he wishes to be excused of further responsibility. In his manner he has an apparent justification and avoids all criticism of himself. He himself is never to blame. The reason he has never accomplished what he desired to do is always someone else's fault' (Adler, 2010, pp. 10-11).

Adler goes on to accentuate how the individual seeks to adapt to such an environment from pure teleological endeavour. Innate proclivities to such goals and purpose manifest itself in adaptation and are hardwired in endocrine systems and a temperamental psychology. Every individual of our species Adler claims entails of an *Inferiority Complex*, and why? Well, because of our inferiority to the vast plethora of animals that outperform the human in their nature. Other animals are known to jump higher, run faster, bite fiercer, scratch deeper, *are* stronger, bigger, warmer, and even digest meat quicker, rendering the man without his tools as inferior to the system of survivorship! Hence, we should duly comprehend where such victimhood may descend upon our species, as ontological security in an evolution of such social primates will be predicated on how much one contributes to the band of hunter gatherers, and in contemporary terms, the group, community, town, city, country; civilisation, and then species as a whole; the world, sifting itself into consequent hierarchies of contribution.

Teleological Dimorphism

The ontological contribution is anathema to a vain motif expressed in our species as no other species hitherto experienced. We may thus value those contributing more to the group than she who doesn't. But sexual selection coerces the male via bioengineering predispositions to select for the worthy contributor with the invaluable reward of life; more *options* to mate! Hence the teleological competition may morph between the male and the female, germane to such compelling rewards. If the female finds that she contributes much more and thus rises in value, due to the misfortune of life, her act does not prove profitable in the case of reward, it may be merely financial, or yet akin to laudability itself. So, here we find *telos* thus morphs. The reward for the male is what is merely ensconced in his *telos* and thus approximates the compatibilist claim as his *will* (intention) is further reduced. Thus this case of sexual dimorphism is only germane to a determination of what we say is a teleological will, thus his will cannot be free in this drive.
As the individual is rendered defenceless without tools, a standing of ontological value suffices to dictate to her the very options she has in her environment. Such free will is appropriated attributable to compulsory education, germane to a *philosophy of action* as one becomes appraised of their labour, what we thus term *society*. Our sociological ontology becomes enlightened, and incandescent, while appearing to escape biology, and in between lies philosophy, which succours the

social and biological world from belying each other. An inferiority complex is thus endured, which distances her from her tools, and likewise a superior status is relished on the approximation of her tools that we hold of an intraspecific value.

To return to the epistemological foundations of what I have written and what we must come to tacitly understand, I wish to return to Candace Owens and the re-emergence of *black* conservatism, which I deem quite healthy for the scale of politics. In as far as Owens unfurls both responsibility and a victim mentality that darkens the case in the United States, I do not believe racism to be a merely political issue. So, let me start with 'equality' to lay my claim. Equality can only be of either opportunity or outcome. In as far as the equality of opportunity is all that can be offered, and the equality of outcome a corollary of something akin to the USSR turmoil, there still remains the issue. The issue may be that the equality of outcome doesn't seem representative of eclectic *groups* within society, and so liberals seem stuck. Why? Because the issue of equality of outcome (which I reverberate is all that can be offered), isn't a political issue which advocates for the equality of opportunity, but both a psychological and a philosophical issue for *individuals*, as the right-wing are repetitiously invoking examples of *individuals* within oppressed minorities that somehow break the fetters of their cumbersome determinism and shape their free-willed 'destiny'. I will continue using racism to demonstrate this and show how it continues to be both a philosophical and psychological issue, rather than political.

An Epistemology of Racism, Plausible, and Implausible Deniability: 'But I Didn't Even Do Anything'

Returning to normative ethics, looking precisely at consequentialism, in which we elide the significance of deontological *action* and necessitate *consequence*, we unfurl the issue of racism at hand. The consequence will be, the ramifications of action or non-action that leads to the moral bad, irrespective of the moral good of the actor in his action or non-action. Thus we return to *sins of omission*. The onus one has as a twenty-first century literate voter to live more concertedly and fathom the consequence of not doing so, has proven time after time injurious to any group or individual. Sins of omission are never entirely complete inaction, which is merely imputed to a proclivity of time and space. Unlike all life extraneous to our species, we brandish the prowess of reason. In reasoning, we disincline ourselves to those proclivities and through the edification of the civilised and democratised voter, inscribe every individual with a socially constructed and inalienable right, professing the right of one individual to be *equal* to that of another. The more antiquated and obsolescent forms of racism may be said to necessitate an *act of commission*; i.e. racial slurs, attacks, and so on. Today we find it imbued in the *act of omission*, which many may see as akin to ignorance, which in the legal sciences exonerates the individual through *plausible deniability*. But in the case for reason, how far should such deniability be plausible? We **reasonably** hold individuals responsible for deniability which in such cases become implausible. Let's venture this with two simple scenarios, one legal, and the other illegal:

Scenario 1: *You forget your wedding anniversary. Your wife is very upset and spends the rest of her day in peevish response to all interactions.*

Scenario 2: *You are driving and don't see the traffic lights are red.*

In scenario 1, we would treat his wife's response as a rational. His inaction (not 'remembering' in this case), would be a *sin of omission* and thus recorded as a possible indicator for his wife reckonings. Brené Brown brands this type of marital activity disengagement. His wife stipulated certain requisites that deem him responsible for his *actions* akin to her own need for approval in the agreement they made when they decided to get married.

In scenario 2, *attention* is necessitated for one to hold her licence and be attentive of her surroundings, and thus a denial of seeing the traffic light as red is rather implausible for law enforcement. The *act* of not knowing thus becomes implausible. Responsibility is a mandate in the context of driving a dangerous vehicle and so we see deniability, or ignorance in human behaviour implausible when responsibility soars.

Thus implausible deniability is also compatible with responsibility which doctors with the cause of will in the becoming and maintenance of a globally civilised world.[163] We use implausible deniability in the travelling world when in airports, planes, customs clearances, immigration, etc. I recall a time when a fellow traveller in an airport asked me if New York was a city or state on filling a small form to enter the USA. Customs in the USA are particularly and evidently notorious for their interviewing techniques on entering the nation.

So to what extend should and would deniability be plausible when dealing with the epistemological domain of racism? How much should people *know* about race? It seems when dealing with deniability, a part of our will is truncated as a mandate to *act* upon something for the comprehensive benefit of the whole (utilitarianism). In scenario 1 we see the cost of having to remember his wedding anniversary is a requisite for the benefit of keeping his wife content, for whatever that may be worth. In scenario 2 we see that the cost of being attentive and driving safely should be outweighed by the benefit of the continuation to drive at all. This form of implausible deniability as a mandate for *action* has been hijacked by many extreme leftists to the extent of criminals and the resentful enforcing such mandates in Black Lives Matter and Antifa militant mobs, which had even led to a case of a white man being drawn from his vehicle and beaten unconscious (Brown, 2020). I would still argue that a gradual instantiation of responsibility (agency) in individuals would be antidotal in achieving a deracialised world one desires, using reason and both literary and spoken discourse to achieve the auspicious goal.

Epistemological Deduction, Induction, Abduction, and Defence Mechanisms

As I profess racism to be a psychological issue through such *sins of omission* within implausible deniability, I see racism being also an issue appertaining to philosophy. As is the case with *Homo sapiens* as an animal, there are various ways in which we come to *know* a thing, one of them is though what must be deduced. To draw the distinction explicitly between what we induce, abduce, and deduce, I shall offer a lucid example that perplexes the accolades of critical race theory.

*The subject, a black man, enters the subway. On entering the subway he chooses to sit down on an available seat. The available seat happens to be between two people, two white British people. After a bit of shuffling which we impute to the suboptimal proximity of strangers, one of the white British people, a white British lady, gets up and walks to another carriage as what it **seems** is finding another seat.*

Now, for an epistemological perspective, how do we *know* that his is an explicit act of racism? Starting from the solipsist claim for such an approach, we can never be sure of what transpires extraneous to the subjects mind akin to existence, but in

order for an epistemically cooperative world, let's say we act is if we do with endeavour. Now the person could have stood up and left for myriad reasons. Maybe she didn't like to feel congested sitting next to someone, maybe she had a family member on the other carriage etc. This would lead to what the subject has come to know as simple abduction, as the black man does not contain enough information to entirely know the reasons of the white British lady changing her seat.
But now let's reduce both the black man and the white British lady to space and time. First, we add time. The British lady who stood up is characteristic of a lady in her late septuagenarian years. This we will agree has been reasonably deduced, as one may deem it impossible her being anything younger than fifty years old. Now let's add space. Space would be the location. The subject boarded *knowing* he was in a predominantly white British district along with the surrounding areas of the county, with a history of racial hostility. Now the subject understands these areas as he was born in the late seventies in one of these specific areas. All of his phenomenological experience assimilated is thus considered. His space, meaning where he is located in the area he boarded the subway, in so far as the possibility of him ever standing in the exact locus of where he was standing in the area, on the subway, is thus deduced.
Her shuffling next to him cannot be entirely known, so is thus abduced. Her momentary wincing before leaving is thus begins to be induced. The agglomeration of such knowledge is how he comes to turn from abductive reasoning to inductive reasoning in his experience of racism.
But one can choose to be rather apprehensive about his claim. What is the motivation of such apprehensiveness? It may be ascribed to the discomfort of hearing of such gratuitous cruelty as a fellow white British person, merely standing in as a *defence mechanism*. If my mother puts clothes out to dry early in the morning and we leave for work coming back later in the evening, having rained all day and then stopping approximately half an hour before arriving home my mother suddenly asks me to see if the clothes are dry. I test the first, second, and third item of clothing, a t-shirt, trousers, and a jumper, and observe that the clothes are soaking wet! I return indoors and tell my mother the clothes are wet. I have made a mere induction regarding the rest of the thirty-odd items of clothing. What would incite my mother to compel me to observe the rest of the items of clothing against my simple induction? The answer would be *defence mechanisms*.
If any of the other items of clothing being dry would exonerate me from a heinous crime as evidence in a large case, as the **defendant**, my defence mechanism compels me to observe the other items of clothing. We see it's a clear defence mechanism, and rightly so. This appertains to our epistemology of racism and the many phenomenological experiences of many enduring such predicaments. This is misfortunately used to undermine the shame endured through racist experience and is also capitalised on to adduce a victimhood philosophy, rendering the politics of racism throughout the west.
A victim will claim she deduces racism, while it is merely induced, and those who are shamed by racism in adapting to such environments are inscribed with said racial inferiority, manifested in gestures, stares, eye-contact, tone of voice, and the general plethora of such *sins of omission*. All racism experienced is therefore mere induction, characterised by the suppositions of ontological inscription.
As racism can only be explained as an intangible phenomenon, it must exist in a society. A society is composed of individuals as individuals compose a society that approximates the psychological sciences. The ramifications, or consequences of such literary irresponsibility[164] manifests itself in the psyche of an individual as a biological proclivity,[165] in which those biological proclivities embed themselves in the *id*. The sins committed, or omitted, are assimilated by the subject who thus

experiences these very sins and must discover for himself what such sins of his extraneous individuals are attributable to, which he finds and inductively concludes is his race.

Privilege

On commencing privilege from the Adlerian perspective of inferiority, we know that no one can be privileged, in as far as everyone has privilege in the world. The bear is privileged with his warmth, the cheetah and the lion with her speed, the ox with his strength. Privilege thus extends intraspecifically in that everyone will find such privilege from one individual to the next, and so on. We create sports from privilege within criteria of privilege, and reason has evolved alongside what the thresholds of privilege shall be. It seems that privilege inhered is loathed while privilege attained postnatally lauded.

If one sufficiently concerns herself with privilege she does not have, her inferiority is thus born, and breathes as a reified entity. If one concerns herself with her own privilege, quite possibly, the struggle for existence isn't so futile.[166] This becomes the work of clinicians.

Teleologically, we are in a constant battle to gain more and more privilege in life, and it is only from such *telos* we find ourselves effaced with the privilege we do not have rather than that which we hold. A competitive demeanour is thus lauded and considered healthy, without excess or boast. But within racism, and through reason, it must be carefully dissected to find the unjust privilege that transpires postnatally, in what we find and call social constructionism.

As we know race to be socially constructed, we know it to be sociologically reified through the composition of such individual psych. As I have hopefully convinced the reader of the agency one holds of what he may not do, as opposed to what he may, our approximation of racism is manifested. Individuals are thus incognizant of such proclivities, and it happens that once the individual is an actor of proclivities assimilated by the racial subject, racism thus becomes *social* and is therefore an intersubjective reality. Racism suffered by those of sub-Saharan African heritage is attributable to intricate networks of recent history appertaining to a principal transatlantic slave trade, unlike other forms of slavery in which clear racial divides could not be feasibly achieved, rendering the work of racism expansive. By this I mean with explicit racial division for the work of ontic inscription to work expansively.

State v. *Is*

Often conferred is the state of which a perpetrator is deemed racist. For a dearth of knowledge appertaining to other languages I shall use Spanish as a brief example. In the Spanish language, what would be the infinitive *to be* is divided into two uses, *estar*, which would be a state of being, e.g., merely *being* hungry, and its counterpart *ser*, meaning being without subject to change, e.g., one's sex, (either having the XX or XY chromosome). The division of being is *essentially* political, which may turn to the legal sciences as forms of slander, libel, and general defamation as one may be subject to ostracization and even violence through acrimonious mischaracterisation. Hence the mechanism for defence within the human domain of epistemology is *identifiable* and hence we see such a mechanism exercised. And so, Occam's razor reasoning is therefore utilised as a defence mechanism within the politicisation of races, suggesting the most innocuous experience may take

precedence. As per inductive reasoning, such experiences lead to the political left and right which encircles the philosophical trajectory of normative ethics in which it is either action, consequence or character that's queen of all ethics. The politics of racism is thus an epistemic war of the unknown; a war of knowledge.
One may therefore *be* racist, as to exist as a racist entity, biologically. In so far as xenophobia is embedded in our DNA as the opposition to anything extraneous of the organism's own ontological being, rendering the subject racist, it may not be actually so. This can be ascribed to social constructionism, or intraspecific realities, and so we slowly turn to Piaget's notion of what's *accommodated* and *assimilated* during our significant operational stages. The sensorimotor, preoperational, concretely operational, and formal operational stages of an individual's neurological plasticity is thus significant, as we find populous individuals of the species *not racist*. Through sins of omission and commission, and the representation of the ontically inscribed, we find such information both accommodated and assimilated in the human psyche, in order to place oneself in the adaptation of one's environment with a teleological goal to flourish in both survivorship and reproduction.

Id vs Superego

We find that within societies (which I must reiterate is composed of individuals), acts of commission take place. It seems evident to those most particularly in the west that within civilised spaces acts of commission are suppressed psychologically. To discover racism within a **society**, it is best to seek **individuals** within such a society who encounter themselves with the *id*, either through impairment or vice. Hence when one finds individuals of such a society inebriated, the probability of racial slurs heighten significantly as acts of passionate sins of commission. Or if an individual in the said society suffers miserably from mental illnesses, the probability of racial slurs heighten when encountering the racial subject also. Individuals who have lost control of the *id* are merely indicative of how hard a *superego* works for other individuals of the same racial inscriptions in harnessing control of the *id*. Hence those very few individuals who do lose control of the *id* in such societies are the apertures of what makes a society *racist*, in as far as individuals in that said society may not be racist at all.
We may reason that a society is inductively racist, hence the term 'racist' is used by many in order to articulate the very nature of a society. One will argue it unfair to brand a society racist conducive to the error of generalisation. Be that as it may, you will find we generalise most all of the time, as we simply cannot foresee the future and cannot be in all places at the same time, ascribed to the laws of general relativity.

Paranoia and Probability

The ramifications of such insidiously surreptitious forms of racism lead to those in the discourse of racism branding certain racist experiences as paranoia. When treating such accusations assiduously, and thus sagaciously, paranoia cannot be the precise term used. Paranoia would be defined as an *irrational* conclusion to such a thing. With what the racial victim should induce, as opposed to abduce or infer, is what makes such a conclusion *rational* as opposed to irrational. It's all a question of probability and the gravity of that thing transpiring. To give a clear instance, if one is anonymously overcharged for a given product when purchasing the product online (without a name which could adduce to racialisation of the customer), a churlish reaction could be deemed paranoia. Conversely, if one is overcharged in an

The Epistemological Quest

offline purchase in a 'racist society', after already experiencing such numerous prejudices before the particular event in that specific environment, the probability could be said to naturally increase. If it is a 3 in 4 chance, a 1 in 2 chance, or a 1 in 4, the probability does not matter. Why? Because the adducing of all *sins of commission* and *omission* assimilated by the racial subject increase the probability of such events being induced.

Miss Ophelia

What must be contended quite carefully will be an adducing to the entire claim I profess. An individual, in as far as she is shrewd and discerning in her nature, witty, decorous and of the higher faculties of intellect, cannot escape from her *sins of commission and omission*. When an individual assimilates sufficient experience in psychological or psychopathological racism, the trauma of the ontological inscription is alert as it has most probably been both assimilated and accommodated in childhood, rendering her detection in such acts expert. The perpetrator, inexperienced in such *sins of commission* and *omission*, and I reiterate, *inexperienced*, makes two **inevitable** errors. The first error is she overly *commits*, which could be characterised as superfluous decorum, for instance. The second error is insufficiently acts, which could be characterised by what isn't being said, asked, gesticulated, and so on, which manifest itself in what we figuratively define as **distance.** This was ingeniously portrayed by Harriet Beecher Stowe within Topsy's and Miss Ophelia's relationship in *Uncle Tom's Cabin*.

Miss Ophelia was given the responsibility of one of Augustine St. Clare's new slaves, of raising a young slave girl on her own, Topsy, cleansing her of impropriety, churlishness, and an inclination to steal and commit other indecorous acts. Topsy was a product of cruel and harsh treatment from her former master while confused of ontology and a superfluous feeling of burdening all company. Miss Ophelia was to raise her with decorum and teach the way of proper being, of hope into a fine and respectable little lady. After unsuccessful and hapless endeavour, trying time and time again to free Topsy of impudence and bad habit, Ophelia implored Augustine St Clare to convince him Topsy was incorrigible. Topsy confided in Augustine St Clare's child, Evangeline, telling Eva it's hard to change or love Miss Ophelia *knowing* she'd rather kiss a toad. As Augustine St Clare replied to Ophelia's disquietude:

'But I believe that all the trying in the world to benefit a child, and all the substantial favours you can do them, will never excite one emotion of gratitude, while that feeling of repugnance remains in the heart; it's a queer kind of a fact, but so it is' (Stowe, 1995, p. 262).

Of the Atomisation of Victims

As I have laid out a few exemplary causes of that experienced in an environment within compositions of individuals that may deem whosoever a perpetrator, we must also turn to the victim! We may understand the racial subject to induce such racism, but the question is, how well? It seems, and I will further contend this audaciously, that those victims of higher intellectual faculties will approximate inductive reasoning in such circumstances, and those with lower faculties, abductive. Let's return to the instance of one overcharged in an offline transaction. We would say the racial subject of *higher* faculties has more substance to induce racism, ascribed to substantial experiences within space and time of a particular

society (environment), agglomerating into what would thus be 'reasonably' induced.
Conversely, those of *lower* faculties may **not** nearly have the substance to induce such a claim, which may be ascribed to an amenability to the indoctrination of victim-bound agendas, thus abduces!
In as much as it is necessary to atomise the racist society into their respective and individual psyches, we must also do so with the victim. To say that all instances of what a victim may *induce* is racist, is illogical! In logical terms, possibility fluctuates between the victim, perpetrator, inductive, and the abductive reasoning in said environments.
Induction to abduction turns to inference, or mere generalisation of something unrelated. It is surprising how one is incognizant of the proclivities for us to fall into fallacies, especially when encumbered by a teleology of such agendas, thus one must approach a situation carefully, especially when dealing with the surreptitious nature of those insidious *sins of omission*.

~

In 2017, Bloomsbury Publishing aided the British journalist and author, Reni Eddo-Lodge, in publishing her controversially titled book, *Why I'm No Longer Talking to White People About Race*. Eddo-Lodge writes brilliantly of the elided British history and the scourge black people both had and have faced in Britain (principally across the twentieth and into the twenty-first century), invocating significant events and figures, invoking the likes of Charles Wooton, Dr Harold Moody, Stephen Lawrence, and many more. Corroborating the initial apprehensibility of the use of politics as a tool to solve racism, Eddo-Lodge continues her book with colossal claims about the nature of racism, as she invokes the most implausible concepts such as *intersectionality*, instantiated in such wild and aimless propositions, even about possible solutions of racism. In no part of her book does she consider the meta-narrative of science, history, or a possible determinism (Eddo-Lodge, 2018).[167]
The issue with intersectionality further complicates the matter, as we must deal with the sexes as a mere biological phenomenon, compared to races being a social construct, thus we must approach phenomena carefully, as we should fathom the benefits of diving into the proclivities of the victim through *identity politics*.
The consequences of political pessimism in Britain are not only injuriously counterproductive but are also hostile. To elide the places in which the system has evolved in the United Kingdom is not only untrue, but is unfair, hence I find myself wincing at such political commentary. It is often said by the right in British politics, that, 'if you don't like the country, you should leave', which is rather unrighteous. I would say, 'if you don't like the country, find an opportunity to live in other countries for a while, then return'. I say this as someone who has lived in ten cities for a period of at least three months and significantly longer across three continents of the world. Even travelling across nations in eastern Europe not so far from the UK, I say one of the best places to live as a *black* person in the world, if not the best, is the United Kingdom. The by-product of the dearth of reason in sub-Saharan African countries is telling, that's if you are an advocate of returning to the so-called-motherland of Africa.
Be that as it may, it does not exonerate or absolve the horrors of racism across the United Kingdom *black* people so painfully endure, all I suggest is a more carefully assiduous approach for an issue we tacitly know to be so complicated. And so a look at the philosophy of racism is quintessential in meeting at a locus for such an issue, an issue presented as long ago as the early 1930s.
Carter Godwin Woodson, a figure duly canonised for his role in unfurling the first philosophies of the *negro* (principally from experience in the USA), in his prominent

The Epistemological Quest

work, *The Miseducation of the Negro*, gave due purchase to the Adlerian perspective. He believed the inferiority of the *negro* is instilled in him in almost every part of education, and in almost every book the *negro* studied. Compare Woodson's quote to former President Barack Obama's written at the end of the 20th century:

'When you determine what a man shall think you do not have to concern yourself about what he will do. If you make a man feel that he is inferior, you do not have to compel him to accept an inferior status, for he will seek it himself. If you make a man that he is justly a outcast, you do not have to order him to the back door. He will go without being told; and if there is no back door, his very nature will demand one' (Woodson, 2010, p. 48).

...

'Just think about what a real education for these children would involve. It would start by giving a child an understanding of *himself, his* world, *his* culture, *his* community. That's the starting point of any educational process. That's what makes a child hungry to learn – the promise of being part of something, of mastering his environment. But for the black child, everything's turned upside down. From day one, what's he learning about? Someone else's history. Someone else's culture. Not only that, this culture he's supposed to learn is the same culture that's systematically rejected him, denied his humanity' (Obama, 2016, p. 258).

Woodson claimed philosophers long conceded man having two educations, one being that which is given to him, and the other, which he gives himself. Lauding Woodson for what some may have described ingenuity, is what I endeavour to extend as we continue our journey on this epistemological quest.

The Epistemology of the Negro

Within black epistemological circles knowledge and discourse oscillated between the two educations Woodson suggests. It seems, for a rational and conducive response to such perplex political philosophy and history, the educated black man intends to alleviate himself from his education given to him. For those who have never thought of the epistemological nebulae of principally sub-Saharan African history with what it has and does contribute to the species, picture the age of discovery and the transatlantic slave trade completely extirpating what sub— Saharan Africa could have contributed through the minorities who may have read, written, or *known* of such significant mysteries. Hence figures have endeavoured to resuscitate such information, such as *100 Amazing Facts About the Negro: With Complete Proof* by Joel Augustus Rogers (Rogers, 2014).
The epistemological nebulae of sub-Saharan African history in such complexes that a 'social' and politicised construct of race engendered, has rendered a plethora of individuals fabricating science and history in order to compensate for such misfortunes. A radical example would be a figure such as Dr Umar Johnson who opposes interracial relationships, perceiving other races to be adulterating the purity of the *negro*, which he professes as a psychologist with an evidently vacuous understanding of genomics. This form of epistemological nebulae rendering the inferiority complex was also conducive to an event transpiring in the summer of 2020, in which Nick Cannon, the multi-talented, actor, TV show host, musician, and comedian (in his own right), was harshly reprimanded after what was deemed to be anti-Semitic comments he made on his podcast and YouTube show, *Cannon's Class* (Limbong, 2020). Cannon involved himself in alluding to other races somewhat

coveting what 'melanated' people have in some part of history, ascribing it to the misfortunes the black race have endured. This is merely evocative of such ontological compensation that's needed for the complex to live and flourish in the species, which is merely a corollary of the misfortunes of racism for figures such as Nick Cannon. In the USA, this could be easily ascribed to the infeasibility of economic reparations for a fabricated and social construct that politicised such fabrication and enslaved such people, which has come to have contorted psychological ramifications for *black* people, affecting an epistemology of the negro.

To fathom the cruelty of reified slavery in the USA while truncating novelist aspects (which could also be subject to hyperbole which as a matter of fact can never really be exhausted giving its proximity to the twenty-fist century and lack of ability to psychologically deracialise), one must look at the story of Frederick Washington Bailey. Known historically as Frederick Douglass, he was compelled to change his name throughout his life. As Frederick Douglass recalled of Mr Auld, his master, when he was only a child, Mrs Auld naturally found the onus to teach him the ABC and thus learn how to read. As Mr Auld said as he reprimanded his wife:

"'If you give a nigger an inch, he will take an ell. A nigger should know nothing but to obey his master – to do as he is told to do. Learning would *spoil* the best nigger in the world. Now," said he, " if you teach that nigger (speaking of myself) how to read, there would be no keeping him. It would forever unfit him to be a slave. He would at once become unmanageable, and of no value to his master. As to himself it could do him no good, but a great deal of harm. It would make him discontented and unhappy." These words sank deep into my heart, stirred up sentiments within that lay slumbering, and called into existence an entirely new train of thought' (Douglass, 1995, p. 20).

That would become the catalyst of Douglass' lifelong endeavour, to first learn how to read and write, and escape slavery altogether. Douglass gave instances of the cruelty endured by other slaves, lacerated by cowskin whipping and hard labour, as the slave's unfortunate fate was in endured in perpetuity. Douglass' perspicuity and aperture of literacy (which other slaves did not have), came to allow Douglass to understand the strategies of slaveholders. Douglass used this to play the epistemological games all the way to freedom, as literacy and charismatic intelligence is what would become conducive to his fame, articulating his very narrative throughout recent nineteenth-century slavery.

Returning to the two concessions Woodson gave for the black man to be educated, it seems resentment and/or disaffection instantiates itself on the *negro* for that which is given to him. It thus rendered the *negro* more susceptible to conspiracy as it is a political science that is either, irreparable, or at least difficult to repair. This is merely indubitable for the black man that it is the white man that who prospered in history. One can hark back along the history of both the greatest warriors and thinkers of all time. To give a few examples to drive home the point, of thinkers, we see Socrates, Plato, and Aristotle. Warriors such as Alexander the Great, Julius Caesar, Marcus Aurelius and Constantine. Explorers such as Pliny, Ptolemy, Henry the Navigator, and Christopher Columbus. Scientists and inventors such as Euclid, Copernicus, Isaac Newton, Benjamin Franklin, Carolus Linnaeus, Charles Babbage, Thomas Edison, Nikola Tesla, the Wright brothers, Albert Einstein, Edwin Hubble, Werner Heisenberg, Alexander Fleming, Alan Turing, James Watson, Francis Crick (and now Rosalind Franklin), Neil Armstrong, Buzz Aldrin, and Stephen Hawking. Finally, artists and poets such as Geoffrey Chaucer, Shakespeare, Michelangelo, Leonardo Da Vinci, Vincent Van Gogh, and Pablo Picasso.[168] The examples one could give of males of European ancestry is rather inexhaustive.

The Epistemological Quest

Only a fool would say this has no bearing on the complex of the *negro*, in so far as such Europeans were rightly so such thinkers, warriors, explorers, scientists, inventors, and artists. If reason may alleviate the black man from such historical (or as I believe 'scientific') misfortune, it must be understood that those thinkers, warriors, explorers, scientists, and inventors were so minuscule of a minority, that most of their work and legacy was met with contempt by the masses, as is often the case. And so, as we reasonably should, have done and still do, atomise continents, countries and societies into their respected individuals. Thus one cannot hold a society, nation, or continent responsible for large numbers of individuals in history for legal issues. Not only is it infeasible, but unreasonable. The closest we can harness accountability is within an institution, holding a social organisation accountable for a by-product of historical misfortune through academic reparations, as the only kind of economic reparations that can be feasibly given is the *equality of opportunity*, rendering *outcome* the responsibility of one sole institution, the family (Narayan, 2021).[169]

As an individual is only responsible for their own achievements, in as much as the individual lauds other individuals who helped them achieve such success, it is the *causal agency* of the individual moving forward.

The Illusion of Responsibility: Past, Present, Future, Free will, and the Baby Mama - Music, More Music, Sports, and Even More Sports

An illusion of responsibility. As a necessary theme in this book is not solely a noun but, for many, a philosophy. As *misfortunate* as it may be, once an adult, in as far as legal sciences appertains to biological changes, all falls under individual responsibility. The cantankerous and rebellious understand this and abscond from the familial institution once anathema to authority. A destitute woman in the occidental world with three children and an absent father may resent the world for her misfortunate condition. In as far as her sexual intercourse with the very *irresponsible* father was consensual, her situation becomes of... responsibility. She will be cognizant of those religious philosophies that compel one to abstain from intercourse until marriage, but yet exercises *freedom* to repudiate. She may fall into despair and thus blame the political institutions for why single mothers aren't protected by the state. What she must understand is that everybody makes mistakes. She must realise this is her past and cannot change this as she can only take *responsibility* for the present and future. Her demands for the type of partner may change and this isn't a moral phenomenon but biological, of the probabilities of benefit, cost and her mating strategies.

Be that as it may, as we do not experience a multi-conditional world she must only work with her condition. She must fathom responsibility for herself and thus bequeath such responsibility to her children. This would be to take responsibility for the present and future. In so far as you believe that free will and causal determinism are **incompatible**, you will eschew a philosophy that alleviates responsibility for the actions that you *willed* in the past. And in so far as you may cohere with a philosophy of alleviated responsibility for the sexual, you will find legal systems not to do so for violence.

It seems that many will inexorably focus on exceptions to the rule and accentuate single mothers as great ones in order to alleviate her responsibility and drive home the *independent woman* narrative. A great mother is a great mother, and responsibility, a different case. If a woman rides a dirt bike without a helmet on the most perilous dirt bike-tracks and comes out unscathed every time, akin to the danger of the consequence, her exception certainly wouldn't compel authorities to advise the disuse of helmets![170]

As we have learned through the Piagetian prism, a philosophy is only inhered through mere assimilation and accommodation from the child. Such processes of assimilation becomes the philosophy in which the child thus accommodates. It must be argued, from the Adlerian sense, this process of accommodation is analogous to the child adapting to his environment and in as much as it may not be admitted, a case of sexual selection is thus instantiated. I write this carefully after much consideration.

In an environment, akin to the male child's pre-frontal cortex, for what he is able to accommodate profitably, the principal philosophy the child assimilates winning the female in such ontic subsets of sexual selection is what the child will approximate. It falls successfully within the bearing of space and time. The child who has assimilated within the perimeters of a specific environment as opposed to a child who assimilates from environments multifariously, will embody such ontic subsets he perceives being profitable to attain female attention and thus selection. Hence we brand those ignorant or incognizant of larger spaces and time short-sighted, parochial, insular, provincial, myopic, blinkered, and so on. The boy in assimilating such philosophy wins the female, which is thus germane to her faculties of intellect, and what she warrants profitable for her selection. The female with possible-higher faculties of intellect (germane to *variability* within the species), which may be inherent and/or conducive to experiences of larger spaces and time, may feel like the black sheep of her environment, implacable to those males who have the provincial[171] bearing of space and time, and thus what she finds profitable will be distinguished. My focus is of the provincial types for what the female requires and within groups of such societies who become intertwined due to similarities that consolidate to form subculture.

The female, essential for what she deems profitable, which may be engendered by the female or male but conclusively accepted by her, is what the male approximates in his assimilation and thus further accommodation. It first manifests itself in what the male contributes to his subculture before what he contributes to the world, which is merely indicative of operational stages through to late denarian years. What the male contributes I shall call once again ontological security, which variegates from male to male, group to group, and society to society, in so far as they become distinguished. For instance, an imam evidently would not hold the ontological security in a predominantly Roman Catholic country such as Lichtenstein than he would of a Muslim country such as Saudi Arabia, if all his peers are Roman catholic.

For what racially, as some believe to be a social construct, manifested itself into a majority of shared communities for what became deemed utilitarian in black communities in the west, and still successfully does so, became music, and for the physically precocious, sports. How the black man consecrates himself to what is deemed profitable by the female (in the said environment) for all that he has assimilated and accommodated is unparalleled. This then becomes his philosophy as no individual has existed independent of other individuals that form the society they both accommodate and assimilate from. This thus becomes *archetypal*, as a philosophy is inhered and assimilated by the society in which they are in, and attributable to a decreasing technological compensation of space and time when travelling from present to our past groups becoming more *racially* segregated, hence encapsulating such philosophies.

A major encapsulation has been a philosophy of music that has dominated the west for a black society. Ontologically, and akin to a history of colonial and political oppression, what music has offered the black male is unspeakably unparalleled.

The Epistemological Quest

Coinciding with an expanse in capitalism and its variegated opportunities, the inferior black man has thus been able to become of the highest contributors of society with access to its riches, transcending and superseding sports and music into the world of business. The exponential increases in wealth and thus access to the female through sexual selection has made this offer to the inferior black male attractive, along with the pleasure of the aesthetics and its significance for the arts. Hopefully the claim thus far is merely evident, so let's extend on such philosophy. With such an attractive offer comes gratuitous consequences for the *equality of opportunity* in such *black* determinism. As Socratic writings and as Plato professed, a man can only consecrate himself to one field to truly master it, especially if placed across operational stages for a child to precociously accommodate such assimilation and master a trade. The attractive offer in either sports or music I argue is a permutation of 'offer' into 'inclination', and for what becomes of one's inclination simultaneously becomes one's disinclination of doing; *sins of omission*.
Conclusive to such inclination, which plays a part of what is either induced or instantiated in common sense reasoning, the black male has flourished in both music and sports. So much so that the eminence of music and sports have evolved to remarkable levels ensconced with more philosophy and meaning than ever before and continue to do so. I must also add, which perplexes the situation even further, played out in capitalism, subjectivity of music means the sophistication of art may continue to go undetected in its evolution. What's monetised is what a population is willing to benefit from and is most always ensconced in novelty, which is unfortunate for the old-time musicians. Hence it's why we see new musicians in both denarian and vicenarian years emerge in lustrums. So for the black man in music, a long and robust career is rather unlikely. Meanwhile, variables in sports are much more controlled which is why sports may prove slightly more beneficial for his short career. And so it's such philosophy that seems to explain for startling numbers such as the fact that from a population of over twelve percent of black people in the United States, almost three-quarters of the men's NBA are observed to be black.
This can be expressed as profitable inclination. But in unfurling such inclination, one must also unfurl such disinclination, which may approximate the completion of a larger narrative. In as far as the black population in the west, which also entails of symmetries across the Atlantic Ocean in western Europe, may flourish in both music and sports, a disinclination has been archetypally manifested. Before I further adumbrate such disinclination, I must add that those of European descent across the west, politically referred to as *white*, are evidently larger in population. That being said, one finds variegated groups of white people with their own inclinations and disinclinations concomitant with their distinguished groups and thus distinguished philosophies. The black disinclination, against the inclination of the white group that prospers so well, is information, for what we more importantly define as knowledge. Many will object to this claim with due impetuosity, but there is good reason to suspect knowledge being more of a philosophy than a mere and insipid noun.
After my referral to all those *white* and canonised explorers, inventors, thinkers, philosophers, heroes, and so on, of Woodson's claim, this most certainly augments to the inclination of conspiracy in the black man's weltanschauung. I claim it also becomes augmented to the disinclination of knowledge, but why?... Because a sociological knowledge that approximates the most serious of matters requires cooperation, not mere information. As explored in earlier parts of this book, let's look at this further..
Many alive today wouldn't have witnessed the Nagasaki and Hiroshima nuclear bombings in 1945, just as those slowly approximating their denarian years in the next few years wouldn't have existed or been sufficiently aware of the 9/11 attacks by al-Qaeda. In as much as people didn't exist who are alive today, billions wouldn't

have witnessed such events with their 'own eyes' or had known anyone directly related to the events. So we work on probabilities, possibility, benefit, and cost. If one had reason to suspect organisations fabricating the event, one would have to first inquire into the benefit of anything untrue against the cost, or ramifications of such untruthfulness. Thus probability and possibility are entered. Reliability, meaning trust, would come from the source of such information, whether it be someone you know or something appertaining to periodicals. Against information, probability and possibility will be thus calculated and one may be incognizant of such calculations in so far as they purport cognizance of such reckonings.

Contiguous Circles and Spheres

To turn back to the black man, it seems that trust, or reliability, is that which is instantiated to and adulterates the epistemological issue in cooperative information, meaning his apprehensibility. When institutions of which he was educated condoned slavery of his ontological kind, and thus canonise a deluge of white males of which purported and officiated science such as phrenology, one would reason with such apprehensibility. The disinclination is thus politicised against that of the white male, and the white male of today is divided into so many distinguished groups that politics suffer from both languid and ill-thought politicisation. One is able to divide those white groups into their variegations, with inclinations and disinclinations that are both similar and dissimilar to other groups.

I will map this out to further elucidate on this point. When envisaged two-dimensionally, society appears as a large circle. Within this circle you find compositions of other large circles related by 'social constructs' such as race and class. Those circles can run contiguously. Other circles within races, religions, and classes of different sizes run contiguously with other circles, and so on. Within these circles you will find the very inclinations and disinclinations due to the commonalities of whatever they may be. Complicating things further is the addition of technology, and so it is better to now envisage society with an extra dimension. You may find individuals who are of certain spheres, who may find a liking to an aspect of such inclinations highly celebrated in other spheres far away.

An instance would be wealthy rural young white males who resonated with hip-hop when it was first pervasive in the 1980s across the west. Their clothes and taste of music would have been the same for spheres far away. Meanwhile black males in the spheres of the artists who resonated with clothes, music, poverty, police brutality, single-parent families and so on, would be rather sceptical of the white male's induction.

Another instance is of a black child adopted by white parents in a white neighbourhood. In her adolescent years she may capitalise on technology to connect to other spheres of those sharing distinguished commonalities, she may even wish she had a part of those other spheres racially. Without understanding what she doesn't share in common with the racial spheres she longs for, she may become resentful, disaffected, and recalcitrant to authority. It may be the very worst of those aspects of inclinations in the spheres she longs for that her parents saved her from.

It appears one finds many commonalities across many spheres that are principally pervaded by specific races, and thus we find stereotypes we deem to be racist if offensive, but the question is perplexed by a nebulous understanding of what is or is not offensive. My rule of thumb is that if a stereotype seems to be over fifty percent, it is no longer a stereotype, but a general truth or a problem. Sixty-four percent of black or African American children living in single-parent families is not a stereotype but a problem to be solved in the variegated spheres of pervasive black or African Americans, attributable to inclinations and disinclinations of distinguished philosophies which are evidently injurious to spheres within the largest sphere of society.

Affluent spheres pervaded by *whites* can have such inclinations cherry-picked, and in order to do so, one must have the patience to dissect such spheres. This has been and is being done by many, as we find individuals interspersed flourishing in finance, science, creativity, and much more around the world.

The Epistemological Quest

Chapter 7

On Intolerance and Intention

It was May 18th, 2013, when the fifth edition of the *Diagnostic and Statistical Manual of Mental Disorders* was updated since the first was published in 1952 by the largest psychiatric organisation in the world, the American Psychiatric Association. The date of the first edition is telling of how young the fields of psychiatry and psychology were, significant in being able to understand the nebulous case concerning the psyche.
Through accommodation and assimilation of phenomena we learn operational stages to biologically develop. Our concern should focus on the assimilation of what has been sociologically constructed, becoming what has intersubjectively, or intraspecifically become human realities we see as races. Moving onto a claim some say is more sensitive, or what virtue-signalling types may confound as untrue for themselves, I have developed a quasi-yardstick for the way in which the psyche can be measured of its uncomfortable sensitivity. This is through *racial preferences* in mating.
As explicated earlier, we know the incipient understanding of race to be merely biological, in that there is one race and individuals fall into distinguished areas of the spectrum entailing of different features of similarities and differences through *variation*, rendering them the same within a large and vast gene pool; *Homo sapiens*. We must also understand scientific classification considers such similarities and differences, whether it be organs, cells, the ability to reproduce healthily and so on.
Assimilation and accommodation of phenomena in operational stages conduces an individual to survive *instinctually*, which we may appertain to the *id*. We understand instinctual impetuosity is not a feature of *Homo sapiens* that led humans to cooperate in such large and flexible numbers, and Socratic reasoning shows us that when contradicting information presents itself, the intelligent of our kind refers to reasoning, first deductive and subsequently dialectic.
As the population exponentially increased and networking expanded so quickly in the last two hundred years, society became that which had unprecedently existed before, and so migration, immigration and emigration transpired expanding the intricacies of living and cooperating in the largest society that has ever existed.
Some individuals of that very society were and are thus exposed to much more heterogeneous features within the large spectrum of race, and many homogenous. In as far as features may vary, racial preferences are most always rendered attributable to what's assimilated and accommodated in operational stages and are rarely remedied in an ephemeral lifespan we endure in spacetime.[172] This thus renders a process of racialisation, which is merely akin to the survivorship and reproduction of the *id*. Individuals will entail of their own unique phenomenological experiences within a society in which an individual could be haphazardly exposed and experiences distinguished features, and another individual not so much. It then happens the individual who passes her operational stages in *multiracial* societies increases her chances of heterogenous predilections, as with an individual passing her operational stages in *uniracial* homogenous. It complicates itself further with the technological revolution that became inculcated in both the assimilation and accommodation of phenomena as the child operates what she comes to understand, which is then also akin to her proclivities, moving back to the Sapolskian perspective that it may be what her ancestors were involved with millennia ago that answers for such proclivities.

The Epistemological Quest

I believe racial preferences to be akin to *survivorship* for the female and even evolutionary psychology may explain for some part in this. For what's accommodated and assimilated, we are constantly distributing ourselves into hierarchies. For sexual selection sees the female to seek the best female increasing the chances of her own survival. For the male, he endeavours to attain the female who *represents* his position in the dominance hierarchy. We know the tropics to have been the most difficult to survive compared to temperate regions in the world, along with the ability to *culturally diffuse* germane to an *x-y* axis, geographic isolation, a reaching hold of the domesticable crops and animals, and access to such technologies that make life easier. It means probability factors kick in. In those tropical regions of the world, melanin is an indicator in many **environments** that the probability of that individual (out of entire populations) being of the economically elite, *logically* plummets.

Mating choices and preferences is a useful topic to unfurl the egalitarian shroud of what appear to me to be a case of virtue-signalling. It turns the discourse from idealism to realism, in which realism has proved efficient in the approximation of solutions. In this case it is merely ethics. Irrespective of language barriers, I do purport that giving an individual a smart and intelligent partner that is personally tailored to her preferences, cherry picked from the species, is not all that is needed in her tailoring. The added criterion of race I do believe must be added, in so far as she may endeavour to deny this. To understand the reality of this issue, let's break this down ethically.

When an individual exercises her freedom of choice she is ethically pure. This would be germane to an act which then pertains to her deontological and pure action. As mentioned previously, this is akin to the precedence of her *actions*. If her character (her virtue) takes precedence, then she ensures she edifies her character through any apparatus of a *deconstructing* of race, in order that her preferences are diminished to the extent that race would **not** be a criterion for preference. This is merely unfeasible, so her virtue must be elided from any part of this being purely virtuous. It seems consequentialist ethics appertain more than others in such scenarios, as the consequences of segregation in even multiracial societies draw a distinguished ethical picture, and as mentioned before, individuals segregate either out of a necessitation of one's own racial kind, or racialised individuals falling extraneous to an individual's racial criterion for mating. This leads to a *consequence* of segregation.

The *DSM* attempted to catalogue symptoms of the 'Intolerant Personality Disorder' (IPD). By developing the following taxonomy, taxa of an IPD diagnosis would encompass an amalgamation of criteria similar to ten other personality disorders. Symptoms of an IDP individual would:

a) hold a rigid set of beliefs that assert the intrinsic superiority due to race, religion, culture, or gender of the person's own group (this could include self-perceptions that indicate a belief that his or her own group is somehow unique, special, more privileged, or more deserving than is another).

b) lack empathy for one or more particular populations, such as, but not limited to, Latinos, African Americans, gays, lesbians, or women (conversely, the person may believe that only persons of his or her own group can truly understand him or her).

c) exhibit interpersonal behaviour that ranges from covert or overt antagonism and hostility to exploitation toward one or more specific or targeted populations.

d) seek to overtly or covertly block, deny, impede, or cancel the social, organizational, psychological, or financial advancement of someone of a group believed to be inferior.
e) use power or other means to inhibit or prevent free expression of contrary or intolerable ideas.
f) have a sense of entitlement based on membership in a privileged group and believe that others should recognize his or her superiority without commensurate achievements or valid credentials.
g) manifest a pervasive pattern of disregard for the human rights of members of particular populations.
h) show a lack of remorse as indicated by being callous or indifferent to having hurt, restricted, mistreated, or maligned members of selective populations. (Guindon et al., 2003)

This poor amalgamation of a diagnosis is demonstrable of two issues that concern our discourse on racism. The first of a languid political discourse resorting to an epistemic surface of knowledge. The second of how relative it is in its various contexts. Who or what, for example, decides what *actions* or *thoughts* convey a sense of entitlement?

Evolutionary psychologists Darren Burke and Danielle Sulikowski attempted to organise such truths with the initial premise of an own-race attraction in their paper, *Is There an Own-Race Preference in Attractiveness*. Interestingly enough, as they experimented across African, Asian, and Caucasian males and females, results of the overall experiments led to African female faces perceived as the least attractive. This is attributable to the *sin of omission* rather than the *sin of commission*, as preferences were elicited as opposed to what the least attractive faces objectively were. Burke and Sulikowski experimented how participants perceived faces as masculine and feminine, which they ascribed to why African females were perceived as least attractive, as their faces appeared to be more masculine than both their Asian and Caucasian counterparts. Meanwhile for females, and irrespective of the participant's own race, females rated European faces as most attractive (Burke et al., 2013). Burke and Sulikowski admit there wasn't any straightforward explanation for why the results turned out as they did, they also claimed it may have appertained to an affluence when certain faces were displayed. Be that as it may, their study contributes to the unspoken issues in psychological thought.

Henry Hays

In one of the mysteries of the USA's penitentiary system, Anthony Ray Hinton, a black man wrongfully inculpated for a heinous crime he did not commit, tells of his incarceration and time on death row of over thirty years. Unfurling racism within the American legal institutions across the southern states and the latter stages of the twentieth century, Hinton awaited the day he would once again see his mother and see her as a free man. Sadly, his mother passed while on death row. But Hinton's story wasn't merely predicated on the sorrow and melancholy one would expect amid any misfortune of providence but was centred on a level of compassion one would seldomly have the circumstance of experiencing through causes. So for the compatibilist freewill thinker, this story proves intriguing.

Sat in a five-by-seven cell, no air conditioning, and resorted to a foetal position to sleep, Hinton would endure losing fifty-four fellow prisoners to the cruel system of the death penalty in Alabama. But there was something profoundly perspicacious and *good* in what Hinton told of his compassion.

It was one of those days Hinton was sat in his cell that he heard a man crying from a nearby cell. Hinton recalls the wailing one would often hear in his corridor, along with the silent hostility between prison mates as each prisoner would evade other prisoners from having the chance to adulterate what would be a slim chance of being exonerated in court appeals. The ramifications of such negligence was certain death. Hinton had heard the same wailing from the same cell for days on end and finally poised himself to say something. On inquiring on his sorrow, he found a man, Henry Hays, recently discovered his mother had passed away. Hinton's own love for his mother allowed him to empathise with the experience of a man losing his mother. Henry and Hinton proceeded to talk and later became friends. They shared dreams from the extraneous world to death row and the simple things that would make them happy. It was on this topic Hinton mentioned his desire for some southern fried chicken that Henry inquired a little further, surprisingly finding Hinton to be black.

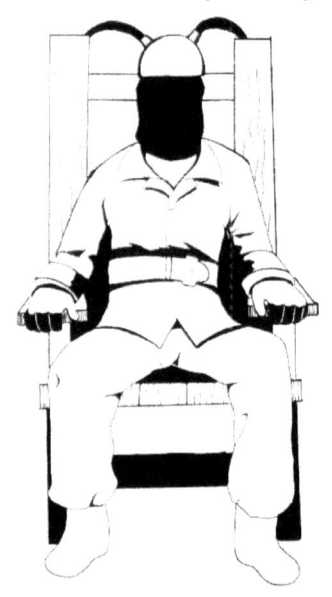

Henry Hays, a former Klansman, was sentenced to death row for a notorious lynching in 1981. Michael Donald, a nineteen-year-old black man, fell victim to Hays as he was violently hung from a tree in Mobile, Alabama. Hays' father was a notorious leader in the Ku Klux Klan.

It was a visitation day with his father in which Hays had planned to introduce to his father Hinton, referring to Hinton as his 'best friend'. As Hinton reached out his hand Hay's father rejected. His father made his act of distaste rather clear. Be that as it may, Hinton spoke of the courage it took of Hays to discomfort his father. As Hays prepared for his death on June 6, 1997, becoming the 392nd inmate executed in the United States since 1976 (and the 16th person executed in Alabama, and the 132nd person executed by electrocution),[173] Hays was granted one final meal and one final speech, in which he uttered:

'The very people who I was taught to hate, were the very people I came to love' (Hinton, 2018).

For Hays, what *will* did he have that led all the way up to his execution?

The Snitch, One Cannot Have Their Cake and Eat It!

On returning to contiguous spheres and the philosophies thus instantiated in their variegating shapes and sizes, the philosophy of the snitch has become an epistemological caveat for information outputs in the west. We may understand this as merely conducive to the relationship race has had in the US with authorities, in which authorities policed *blacks* in housing, marriage, and communal rights. And so

The Epistemological Quest

in order to dissect the philosophy of the snitch I must drive home the point again, which holds the politically scientific substrate that a dearth of trust is attributable to a lack of cooperation, notwithstanding that dearth being counterproductive to moving forward as a nation would endeavour to do so.
When the first census was taken in 1790 in the US, Blacks sported a population of around 760,000. In 1860, approximating the inception of what would become a four-year long American civil war, the black population increased to 4.4 million, but against the entire population, blacks had dropped to 14% from being 19% of the entire US population. The majority of the 14% were slaves, as 488,000 were 'freemen', and by 1900 (passing the abolition of slavery in 1965), the population doubled to 8.8 million blacks.
As blacks became refugees in a large and perplexing country, attributable to the hostility of the South, blacks were compelled to find a way to abscond as 90% of the black population were stranded. Black refugees with an endeavour to ameliorate their condition were forced to travel north, and by 1950, blacks sported a population of 15 million, and by 1980, 27. Representing more than 12% of the population today, the black population moved from 30 to over 40 million, the same *proportion* as 1900 (Bennett et al, 1993). The *numbers* and *proportions* are significantly important in contextualising a narrative for the United States and political science and philosophy today.
In 1882, the number of lynchings of whites in the US was at 64 while blacks at 49, decreasing from a total of 113 down to 0 in the years 1952, 1953, and 1954. The worst years for lynching of blacks came in the 1890s as over 100 lynchings were in 1891, 1892, 1893, 1894, and 1895, reaching its peak in 1892, as 161 blacks were lynched compared to 69 whites that year. This must be contextualised with the proportion of blacks being at around 12% of the population by 1900! Numbers of lynching for blacks consistently dropped to less than 10 lynchings a year from 1936 (Tuskegee University, 2021).
In the US, racial politics unfurling an honest narrative are already starting to be done, in an effort to compel individuals to exercise *responsibility* and thus impart a wisdom of accountability. This stems from what Owens believes are myths when one is found to take the fear of being lynched by such groups like the Ku Klux Klan seriously, in order for one to adduce myths to a victim mentality, profitable as it alleviates one of effectively competing and contributing to their society. This Owens sees is maniacal. Notwithstanding the diminishing numbers of the KKK, there are still approximately 3,000 Klan members in the US, as well as associates and supporters. The KKK, subdivided into groups, have groups with comparatively more members than others. Irrespective of having small numbers, groups such as the North Carolina-based Loyal White Knights (LWK) are possibly the most active Klan group in the US, with a particularly expansive geographical reach. As of 2015, between 150-200 members drew attention through an interspersion of fliers, as the group predominate the southern and eastern regions of the US (Anti-Defamation League, 2016).
A belief of superiority still lurks nascently behind closed doors in the west and may be conducive to the by-product of the political left in so far as they are of the political right. irrespectively, this form of extremism still exists.

Of Realism and Idealism

I augment with the case of implausible deniability what many *blacks* experience in commiserating with their European counterparts; the mandate that one should *ignore* such ignorance and turn such focus on those defined as *not racist*. To use an example of such implausible deniability, let's go to a city in England.

A black man in his late 20s is walking down a street in what seems to be a fairly popular and vibrant area he has never visited before. On waiting to cross the street, a car drives past consisting of three men (who also appear to be in their 20s) and shout 'Nigga' to the black man waiting to cross the road. Surrounded by whites and feeling utterly mortified as the car picks up its pace once the racial slur is both successfully released and received, what appears to be an uncomfortable white couple (who appear to be in their 40s), magnanimously console the black man by telling him, 'Just ignore them, they are idiots!'

In such situations, it seems the humanitarian medium is what is kindly suggested by those who aim to alleviate the victim. In as far as the white couple console the black man, ignoring the men is an implausibly deniable act, notwithstanding one's contrasting opinions. The black man is responsible here in so far as he can be by making necessary adjustments to perform one of two acts. One, ensure this type of mortification doesn't reoccur, and two, ensure that such racist occurrences do not escalate into anything worse for himself.[174] Such an instance gives him a gage of time and space and the possibilities of what he may be effaced with in such scenarios.
If the black man experiences another act of the receipt of racial slurs then call him fortunate, as it only adduces to the bearings he endeavours to make reckonings in such an environment. As touched on briefly, such experiences only give him the reckoning of the psyche of the society and so he must take necessary precautions.
A nocturnal plan of *going for a quiet few drinks* seems to fall into his biological responsibility. On the inebriation of such individuals in this new society, those individuals merely approximate the *id* and so approximate those classically tribal instincts of violence, territorialism, jealousy, and out-group belligerence. This is only stoked even more when females of their *white* in-group may socialise with the black man.
It seems he has three options if he is the only black man of this new city. The first is to render research into this area and find cosmopolitan places to enjoy his night with his white friends. The second is to inquire into his concerns about going to a safe and cosmopolitan area, simply asking his friends. The third is to avoid going out by finding other activities to do. Notwithstanding one's opinions about what he *should* be allowed to do, what he *decides* to do falls into his scope of biological responsibility.[175]
Thus we find segregation to be of use in such predicaments as the black man is **not alone**.[176]

~

Amid the inception of Reagan's campaign did such occurrences coincide in the US as of 1981. Management and Training Corporation (MTC) was founded in the same year as Reagan's appointment, which was a large corporation whose interests simply remain in the management of private prisons, operating and serving thousands of offenders who enter their prisons. But the privatisation of prisons that seek profit in a capitalist nation state will have evident ramifications for political scientists today. A nation that less than two decades ago (from when the MTC was founded) entailed of a disenfranchised race, and in 1983 found the Corrections Corporation of America (now Corecivic), owned and operated private prisons while the Wackenhut Corporation (now GEO group) came a year later, which concerns itself with both real estate and mental health facilities across the US (Watson, 2015).

As immigration laws fluctuated and quite evidently continue to do so, agreement needed to be bound in contract with the US Immigration and Customs Enforcement (ICE), ensuring occupancy rates were between 80-100%. They compelled lockup quotas, as the founder of GEO Group, George Zoley, implicitly told investors in 2008:

' "The federal market is being driven for the most part, as we've been discussing, by the need for criminal alien detention beds. That's being consistently funded' " (Tory and Blankenbuehler, 2017).

With immigrant detention being the fastest-growing form of incarceration in the US, costing taxpayers $2 billion a year, some still lay claim to lockup quotas attributable to advocates of Reagan's 'war on drugs' and 'tough on crime' laws during his campaign. Reports found two-thirds of private prisons including lock-up quotas in their contracts.

As of 2017, while blacks represented 12% of the US population, 33% of the sentenced prison population were black, and while whites represented 64% of adults, they comprised 30% of the sentenced prison population. Conversely, Hispanics represented 16% of the population and accounted for 23% of all inmates. Meanwhile, between 2007 and 2017, the black-white gap seemed to decline as there was a 20% decrease in the number of black inmates, outpacing a 13% decrease in the number of whites who were sentenced. By the end of 2017, blacks composed 475,900 of the 1,439,808 inmates (Gramlich, 2019).

The dissuasion to cooperate with a nation (or the originally white-harnessed United States), is evident when one both necessitates the time and occurrences of such situations, manifesting itself into resentment and disaffection in a lucrative endeavour to fulfil lockup quotas. An optimism of blacks to be in a state of pure gratitude and vigour for their franchise across the 60s, along with whites purged of both racism and disquietude of the black franchise, is beyond false. There are those still alive when the black franchise was ceded, and so the epistemological fragment of an expectation to cooperate is still generationally nebulous for many affected by a caustic and recent history.

For the US, such issues did not subside or ameliorate and only exacerbated itself across the demise of the Reagan campaign on the inception of Bill Clinton's campaign in the 90s.

The state of Washington in 1993 that was the first state to pass the 'no-nonsense Three Strikes policy that Clinton saw fit to impose on the nation. The 'Three Strikes, You're Out' policy encompassed the goal to cut down on recidivism merely turning three separate sentences into one life sentence. Be that as it may, research by the National Research Council revealed between 1980 and 2010 the 222% increase in the rate of incarceration in state prisons in the US as a function of *changes in policy* rather than of *crime rates*. It moved from a prison population of 330,000 in 1972 to 2.2 million in 2016, while approximately 4.6 million people were under probation or parole supervision (Lacourse Jr, 1997).

It seems that as blacks predominated the proportion of incarceration statistics since the 1970s, Clinton's administration endeavoured to tackle the issue with the old philosophy of deterrence, a product of virtue ethics in the most complicated fragments of ethical philosophy, as even Daniel Nagin, a leading deterrence scholar concluded:

"'[t]he evidence in support of the deterrent effect of the certainty of punishment is far more consistent and convincing than for the severity of punishment" and that "the effect of certainty rather than severity of punishment reflect[s] a response to the certainty of apprehension."' (Mauer, 2018).

Mauer goes on to give an implicit analogy of why such deterrence laws fail, as he believes Daniel Nagin's claim to make intuitive sense:

'Consider a person who is thinking about stealing a car or burglarizing a local business. If he is thinking rationally, he will take into account a variety of factors when considering how to commit the crime, including time of day, ease of entry, presence of security personnel or technology, or his ability to leave the crime scene. He does this to avoid being caught in the act because being arrested and prosecuted will impose significant burdens on him. Additionally, because he is not planning on being apprehended, he is unlikely to be thinking about how much time he might spend in prison and whether his sentence will be three, five, or seven years.
Notably, this example looks at the behaviour of a rational person, which rarely fits the picture of a substantial portion of those who actually commit a crime. Many are teenagers seeking peer approval for their illegal behaviour, individuals under the influence of alcohol or drugs at the time of the offense, or are motivated by economic challenges. Many of these individuals are not even thinking about the risk of being caught, let alone know how much prison time they may face' (Mauer, 2018).

The issue for an increase in incarceration numbers for blacks in the US prove to invoke a deeper realm of knowledge when flirting with deterrence, as even Clinton came to admit himself. Pertaining to the issue of both crime and punishment, and both the good and the bad, we must turn back to philosophy and into the realm of what one will call *free will*.
Aristotle believed for those to successfully study the nature of virtue, one must discerningly distinguish the voluntary with the involuntary as virtue concerns itself with both *passions* and *actions*. A thing that approximates the involuntary and so happen by force may be exculpated, while others of the involuntary take place by reason of ignorance (extenuating or mitigating circumstances). The voluntary in such actions become of bestowed praise or blame. It's of these extents Aristotle may find virtue, of what may be praiseworthy or exculpatory between the voluntary and hence the involuntary.
But to eschew the extensive philosophical discourse that is useful for the political arena, and focusing on the mere act within *free will* and responsibility, Aristotle wrote:

'Self-indulgence is more like a voluntary state than cowardice. For the former is actuated by pleasure, the latter by pain, of which the one is to be chosen and the other to be avoided; and pain upsets and destroys the nature of the person who feels it, while pleasure does nothing of the sort. Therefore self-indulgence is more voluntary' (Aristotle, 2009, p. 58).

In as far as action may be ascribed to cowardice or fear to approximate the involuntary, therefore approximating the exculpatory nature of actions, we see self-indulgence not to be so. Why? Because the clear distinction between self-indulgence and cowardice is pleasure and pain. I would also instantiate time and space into the act of free will. Those induced by the short-sighted pleasure forego the long term pleasure of gain for the family, community, and society. Naturally, cowardice and fear is what Aristotle believes keeps one away from such actions in so far as it compels him to act, rendering action or inaction voluntary. But research by the neuroscientist Patrick Haggard et al. (2002) show *will* or *intention* to be a mixture of **intentional binding**, creating an illusion we experience as the stream of consciousness.

~

For *blacks* in the west, to live free of responsibility in a Leviathan is predicated on benefit, cost, freedom, or the lack of it. The inclination for political phenomena to free itself from the fetters of racism also coincide with the classically libertarian approach, attributable to the *free slave* of abolitionist sentiment. But we have seen a rise in black conservatism that has begun to decry the counterproductive philosophies for the direction it heads. The perplexing nature of what's found in the dichotomies of human history are both the cruel and caustic events of the past, along with a technological revolution that ameliorated human existence. We made concessions in society that adolescents must forgo much of their freedom in order to become useful contributors of our species once reaching vicenarian years. And thus those benefits include; the fridge that preserves his food, the washing machine that washes his clothes, and the dryer that dries them, and our tribalism brings about the race that invented it. But through a cooperative civilisation such amenities were achieved and thus laws must be instantiated that incite cooperation and the superfluous freedom that grants it marvellous.

So, as Aristotle implicitly believed, virtue of the *society* is the highest of all virtues, predicated on the necessary continence for humans to flourish that freedom duly forgoes. To better this cooperative system, one must take the responsibility of his voluntary action (or agency) as opposed to shrouding his actions involuntary, which seems to be the inclination of liberals to do so with such freedoms hitherto flirted with.

Of the Male and Female

As the number of lynchings through US history were predominantly males across the late nineteenth century and onwards, lynched by white males or institutions operated by white males, the epistemology that frees itself from conspiracy theories thus continues to pollute upon itself. Hierarchies that render themselves within our capitalist framework, which do so through ontology, has only led to the pollution of epistemology of that of the black male, as a male's ontological position increases his value within the struggle of sexual selection. As mentioned in the first part of this book, akin to *ontic* subsets of the working power of sexual selection, those ontic powers are quintessential for the teleological endeavour of his goal and the by-product of his struggle for the goal that results in happiness, or being content with his work. Interracial marriages were permuted in acrimonious discourse to being duly instantiated in US policy, with the effect of ensuring black males were out of the ontic race within sexual selection of the *white* female:

'Section 4189 of the same Code declares that 'if any white person and any negro, or the descendent of any negro to the third generation, inclusive, though one ancestor of each generation was a white person, intermarry or live in adultery or fornication with each other, each of them must, on conviction, be imprisoned in the penitentiary or sentenced to hard labor for the county for not less than two nor more than seven years' (Pace v. Alabama, 1883).[177]

The popular case that rose from infamy to fame entailed of two individuals indicted under this code:

The Epistemological Quest

'In November, 1881, the plaintiff in error, Tony Pace, a negro man, and Mary J. Cox, a white woman, were indicted under section 4189, in a circuit court of Alabama, for living together in a state of adultery or fornication, and were tried, convicted, and sentenced, each to two years' imprisonment in the state penitentiary. On appeal to the supreme court of the state the judgement was affirmed, and he brought the case here on writ of error, insisting that the act under which he was indicted and convicted is in conflict with the concluding clause of the first section of the fourteenth amendment of the constitution, which declares that no state shall 'deny to any person the equal protection of the laws.' *J.R Tompkins*, for plaintiff in error' (Pace v. Alabama, 1883).[178]

The Dog, Thou Shall Be Outshined

Within the last two hundred years we learn from American history both the intricacies and the teleology of the racist purpose. If multitudes of white Americans knew the negro to be inferior then one would only assume such knowledge to be left in their axioms, just as one does with a dog, cat, or any other kind of pet. But the evasion of literacy and black codes that followed the abolition of slavery in 1865 instantiated the amalgam of hatred and scorn.

Like species who are either or in between the eusocial, prosocial, and social spectrum of ethology, as a species our livelihood and purpose in life are quite clearly predicated on what we contribute to each other, not what can only be individually profitable. An individual branded as arrogant is only branded in such ways for what is valuable intraspecifically, as opposed to what may be valued extraneous to his species. We have evolved and been hardwired to such evolution, whether it be the hippocampus storing a memory of social embarrassment (in order for the act not to be repeated), the ventral-medial prefrontal cortex making an empathic decision to donate money to charity, or the dopaminergic hit from the mesolimbic dopamine pathway's receipt of a compliment on good performance in sport. What we contribute to society is hardwired in our mental health and teleological purpose to flourish. This we simply call human ontology.

Segregation laws that came subsequent to 1865 on into the twentieth century were problematic not because Americans were segregated, as we see many segregate today and continue to do so in a civilised fashion, but because ontology of the blacks was thwarted and oppressed at all costs. To point this out carefully, between sub-Saharan African and European sexual dimorphism, the contribution of black males to society would raise an ontological status that would increase chances of the white American females inevitably ceding to, which the white male would be most certainly aware of . He would not need a reminder of this, as every day is an ontological struggle and thus such teleological information is inhered, intuited, and induced.

To tarnish the black man's ontology would be the only necessary form of counteraction and thus the primitive tendencies of the black man portrayed within the entertainment spheres pervaded with such strategy.

My point here is not to incessantly reverberate a summary of black American history, which others have done laudable work in doing so, but to raise a question and yet return to the philosophic solution to such a colossal problem. In so far as economic reparations became infeasible, and politics would remain unjust for a taxing length of time, the problem is ensconced with the ethological starting point of what one contributes to the species, and so what renders this increase in contribution is only the adoption of individual responsibility (agency) as antidotal to black suffering. One can refer back to the prism of human agency for a recapitulation of how I believe this can be achieved effectually.

The instantiation of sexual selection and the in-group/out-group phenomenon has not disappeared within the ontological foundations of racism and many who go without being democratised continue to endure human experiences of this sort, whether it be the 180 to 220 million Dalits in India, or those of sub-Saharan African descent across the west. As the trial judge had stated in the infamous *Loving vs Virgina* case in 1959:[179]

'Almighty God created the races white, black, yellow, malay and red, and he placed them on sperate continents. And but for the interference with his arrangement there would be no cause for such marriages. The fact that he separated the races shows that he did not intend for the races to mix' (Loving v. Virginia, 1967).[180]

The state of Virginia in section 20 to 54 of the Virginia code also provided:

'Intermarriage prohibited; meaning of term 'white persons'. It shall hererafter be unlawful for any white person in this State to marry any save a white person, or a person with no other admixture of blood than white and American Indian. For the purpose of this chapter, the term 'white person' shall apply only to such person as has no trace of whatever of any blood other than Caucasian; but persons who have one-sixteenth or less of the blood of the American Indian and have no other non-Caucasian blood shall be deemed to be white persons. All laws heretofore passed and how in effect regarding the intermarriage of white and colored persons shall apply to marriages prohibited by this chapter' (Loving v. Virginia, 1967).

Such laws only allude to the idea that if a person believes racism to be an obsolescent issue in the US today, is simply naïve. Naïve not because of malice, but of inexperience or omission if mature. Such naivety only conduces an idealist philosophy and lacks the necessary vehemence of the cruelty and reality of sexual selection from a purely ethological prism in how racism can be instantiated, affecting males and females in different ways. As one of Maya Angelou's famous quotes suggested:

The black women in the south who raises sons, grandsons and nephews had her heartstrings tied to a hanging noose. For this reason, Southern Blacks until the present generation could be counted among America's arch conservatives (Angelou, 2009).

The amalgam of black history in both the US and UK, reiterated in agglomerated form on education and politics, has only been recipe to both tumultuary and misunderstanding in black British history. As always, it's optimal first to dissect issues with that which may relevantly appertain and subsequently to those that differ, and after successful dissection then turn to viable solutions.

What must be carefully conceived before deliberating racism in the US (as for nations such as France), which slowly eschews from a dissipating pertinence, are such causal determinants; that which we may call history. Yuval Noah Harari, as both philosopher and global historian, calls the notion *historical chance* in which historically causal events transpire and a descendant may either be on the benefitting end, or suffers.[181]

In the British empire's expansive campaign, explorers and other European competitors shared the *telos* of outshining predecessors and competitors alike, but it was the Portuguese, who, akin to previous advances and successful coastal expeditions, were the first to land African shores south of the Sahel.

The Epistemological Quest

Northern European countries Norway, Denmark, Sweden, and Finland, often lionised for liberal policies germane to stances on immigration, education, egalitarianism, and so on, are often compared in political virtue (most especially immigration policies). But if one were to people Norway, Sweden, Denmark, Finland, while add Iceland, Estonia, Latvia, and Lithuania together, the sum total wouldn't even make up half of the UK's population. So the question is begged, what were the causes that lead Great Britain and Northern Ireland to people more than double the population of such nations? When the Nazis campaigned across Europe in the 1930s and 40s, where were those countries to ally and thwart the success of Hitler's terrorist campaign? From Roman expansion to Vikings, Jules, Angles, Saxons, Normans, temperate isolation, a second world war, colonial independence, and reparatory naturalisation, we may name some of those causes of the last two millennia. The Nordic north and Scandinavia are relatively small with distinguishing histories and have thus become a product of Harari's *historical chance*, most especially in terms of reparations for their historical dearth of imperial success.

It was adventures of figures such as Henrik Carloff across the 17th century, rising to become the Commander and Director of the Dutch West India Company and later joining both the Swedish then Danish Africa Company to laboriously induct them in slave trade-offs with other European explorers. Ultimately, Carloff founded his own Swedish Gold Coast on the Gulf of Guinea as one recognises the coast of Ghana today (Blake, 1950). The difference between minorities of adventurers of a Nordic north and Scandinavia, as opposed to powerful minorities of adventurers within the British isles (notwithstanding varying populations), is simply that the British outshined in maritime success.

Progressing in time from the 17th century, in with what some historians may define as an expensive enlightenment movement for slave traders and imperial beneficiaries, the British accrued debt in imperial reparations by the twenty-first century. Reni Eddo-Lodge adumbrated a political history of black Britain so well, in encompassing a political history of British racism across the twentieth century, such as significant events as the Murder of Charles Wooton, ingenuity of Harold Moody, and the Windrush. All that would fall consequent to ex-colonies returning (or arriving) to their colony's *motherland*, became the gratuitous struggle for co-existence in imperial relations, met with both vitriol and the newly articulated birth of racism (Eddo-Lodge, 2018). Going on to manifest itself in both intricate and ontological form, the British were coerced in redefining what it means to be British, inhered in contentious political philosophies that still trouble British parliamentary discourse today.

Numbers

The Census of both England and Wales by ethnicity (the most recent), held 13-14% of the over 56 million to be none-white. It is 7.5% of the 13-14% that were Asian, being dominated by Indians and Pakistanis as Indians made up 2.5% and Pakistanis, 2%. Blacks made up 3.3% of the 13/14%, as Africans dominated 1.8% of the population while Caribbeans sported 1.1%. Meanwhile, 2.2% would be mixed between white and Asian, white and black African, and other mixes, with the predominant mix being white, and black Caribbean coming in at 0.8%. Other ethnicities extraneous to Asians, blacks, and whites would make up 1% of the population of England and Wales (gov.uk 2011 Census, 2019).

If one focuses on black ontic inscription,[182] the issue corroborates itself the world over as something that sits in the *id* of British society.

The Epistemological Quest

As I said earlier, the *id*, for the most part, is that which one could call a body of biological truth. Often instantiated proverbially in the English language, is that one may show their true colours when inebriated (or simply that someone lightens up if slightly so), which I say is true of the *id*. One is more susceptible to carnal and bodily desires and the hyperbole of *aggressive* or *sexual* emotion. Implicitly, it's the loss of the *superego* and the *ego*, which cedes its power to the *id* on more and more inebriation. Once a minority of a particular environment approximate the *id*, they reveal what may lie in the majority of individuals (if not all). The differences will be that other individuals harness their *id* with such upholding *egos* and *superegos*.

It was the summer of 2021, on the 11th of July, when the footballer and son of Nigerian parents, Bukayo Saka, would take a penalty at the UEFA Euro 2020 finals that would reveal the *id* of such environments, ranging from England and its isles and beyond. Along with online racist abuse targeted at Manchester United players, Jadon Sancho and Marcus Rashford, the Arsenal prodigy, Bukayo Saka, also became a victim on social media platforms (Ahmad and Tucker, 2021).[183]

Even more peculiar, but yet indicative of reality in so far as it pertains, it was a man in his early twenties from Saudi Arabia that sent a number of monkey emojis on social media, leading to a 24-hour suspension on his social media account. So it may not have been intoxication that was necessary for the *id*, but for this Arab, *anger*. I remember my political science professor, David Tyrer, say in a lecture he wasn't an advocate of '*first* generations' and '*second* generations' terminology when referring to British citizens, as what would be instantiated is a hierarchy of Britishness onto ontological subjects, which as a matter of fact, would be akin to the Adlerian complex. This Saudi man claimed to have been supporting England's campaign, and during the final and was "angry" when the team lost penalties. Germane to his *id* and what I must admit have come to learn and induce as a British citizen, is that extraneous to the shores that form the British isles, the idea of Britain or England is phenomenologically white for a lot of the world, and one's cognitive dissonance ensures that.

With footballing stars having become more involved in British politics since the 'taking the knee' stance (rendered by the killing of George Floyd and adopted by the American footballer Colin Kaepernick), Marcus Rashford's activities off the pitch had been lauded by many across Britain. Rashford, on referring to his own experiences as a child, campaigned for children receiving free school meals as he had suffered from hunger when a child. A backlash saw conservative MPs state, 'it's a parent's job to feed their children', and another claim it's attributable to the ever-available trope of absent black fathers and broken homes, which would be akin to truths of personal responsibility (Hirsch, 2020). Be that as it may, why an innocent child should not suffer from the ramifications of their parent's actions is a feature of human reasoning within the species that has engendered welfare as a concept, rendering politics behind such conservative confutations as a frail moral compass.

It was the British Nationality Act in 1948 that imparted the status of citizenship of the United Kingdom and its colonies (CUKC status) to all British subjects connected with the UK or a British colony, bringing Caribbean communities to establish themselves in areas such as Brixton in south London, St Pauls in Bristol, Handsworth in Birmingham, and St Anns in Nottingham. The arrival of the Empire Windrush that brought those with such status at Tilbury docks on June 22, 1948, would be a watershed moment for the reconstruction era of Britain after the end of the second/world war in 1945. But the reception and subsequent discrimination in public spheres such as work, accommodation, and schooling, would unfurl the realities of racism in the motherland for those first venturesome Caribbeans (Pennant and Sigona, 2018).

But a *Windrush Scandal* that became salient in British politics 70 years later (2017-18), unpacked a complex system of racism in the UK, as, both the elderly who crossed the Atlantic to reconstruct Britain, or were of the Windrush generation, were suddenly compelled to provide evidence of their legal status. It led to many losing their jobs, homes, bank accounts, NHS services, while policed with detention and even deportation for septuagenarian blacks and older (Bloom, 2018). Irrespective of whatever political game one chooses to play in British politics, such mortification of one's grandmother and grandfather, parent, aunty or uncle, both regresses and adulterates the necessary epistemological trust that is being carefully constructed since colonialism and the abolition of slavery (since 1833) across Great Britain and British empires. Such scandals are only evident of why many blacks in Britain seem to carry recalcitrant, resentful, and disaffected attitudes towards authorities in the UK, even making it difficult for all British citizens to at times cooperate on the most menial of tasks.[184]

The political left claimed the political mishap of the *Windrush scandal* was ascribed to the 'hostile environment' set out by Theresa may in which one could say was both recklessly and negligently targeted at illegal migrants. It was the inception of the year 2015 that saw new operations purge the streets of homeless non-citizens through detention and deportation as a hasty response to asylum seekers filling the crevices of what would be perceived as the UK's problems. The left saw Theresa May's 'hostile environment' as catastrophising such aspects that did not play any part of the UK's political issues.

Chris Allen, another former professor of mine who altruistically gives up his time in the aid of refugees and those who seek asylum (through his organisation of a football team), shed light on the numbers of such realities subsequent to the 2016 Brexit vote and the spiralling of resulting racist hate crime:

'In the month immediately after the Brexit vote, in June 2016, it [racist hate crime] had increased by 49%. The Institute for Race Relations directly linked this rise to political and media influence, which had constructed immigration as a problem that was out of control.

Yet, the reality could not have been more different. In the year to March 2017, over 1 million people had sought safety in Europe. Britain received just 36, 846 of these asylum applications, which equated to 3% of all asylum claims made in the European Union (hereafter 'the EU') that year. It had taken a similarly low number in 2016, when, in contrast, Germany received 722, 265 applications. Less than half of applications for asylum in Britain would succeed anyway.

At the time, Syria was one of the major refugee producing countries, yet Britain had received only 10,626 asylum applications from Syrians since the beginning of the civil war. That equated to a miniscule 0.2% of the total number of Syrian refugees

worldwide, which, at the time, totalled 5.5 million. In fact, far from being an 'out of control' British 'problem', the UN's refugee agency estimated that developing countries were sheltering nearly 9 in 10 displaced people in the world. In two weeks of 2016 alone, Uganda had offered refuge to more people than Britain in that entire year.
If you read the British media, you would not think so. The media talk was about 'swarms' of people crossing the English Channel and the terrorist threat they posed. Even worse was to come. The Manchester Bomb occurred in May 2017, just before our AGM. Islamophobia hate crime increased by 500% in the immediate aftermath of the bomb. Life for my new friends was dire. However, I had an idea' (Allen, 2021, p. 32).

After reading a report about the shared sense of belonging football engenders, Allen's idea was football.
Conversely, appertaining to issues away from refugees and those seeking asylum, stories of migration in the UK variegate. This I would attribute to net migration and may approximate the untold issues of multiculturalism, and as this part of the book duly concerns racism, I will not hesitate in hanging fire and unfurl both the latent and manifest issues that may govern a western civilisation.
Demographic changes and net migration in the UK are mechanically calculated over three components necessary to make sensible reckonings of population change; that being fertility, mortality, and migration.
Instead of a languid approach predicting for a comprehensive, or holistic approach on migration from today to the future, it has been fragmented by the constituents that compose the United Kingdom, using a prediction for natural change (zero net migration), natural change (due to net migration), net international migration, and net cross-border migration between the years of 2018 to 2043. Such migration is predicated on the natural and not so natural processes of net migration.
So while Northern Ireland has a high natural change (zero net migration) coming at around a 4% increase,[185] and a net international migration rate of around 2.5%, natural change (due net to net migration) and net cross-border migration fall less than a 0% increase. While Wales' natural change (due to net migration) will increase by a rate of 1% and net international migration the driving factor for population growth at around 5%, the natural change (zero net migration) decreases at the rate of -5% while net cross-border migration tilts to -2%. Scotland's natural change (zero net migration) would decrease to -8% if it weren't for natural change (due to net migration) increasing at a rate of 2%, and both net international migration and net cross-border migration increasing at 4%.
Conversely, and instantiated in the issue I will elucidate, England would as expected differ from the other constituents as England entails of a larger population while holds the capital city, London. While England's natural change (zero net migration) increases at a rate of less than 1%, and net cross-border migration would decrease the population rate to slightly less than 0% by 2043, natural change (due to net migration) would increase at a rate of slightly less than 2% while net international migration increases at a colossal rate of 8% (Cangiano, 2021).
As political stories from the left and right can polarise themselves by the shed of light on distinguished numbers of migration patterns, one factor remains true, that England is expected to become even more multicultural attributable to natural change (due to net migration) and net international migration being the factors that predominate England's population increase. In fact, Cangiano points out that in the UK 'the cumulative net flow of new migrants after 2018 directly accounts for 73% of total population growth by 2028, and 84% by 2043'.

The Epistemological Quest

As of January 2020, the United Kingdom's withdrawal agreement was ratified by the Parliament of the United Kingdom and then European Parliament consequently. Notwithstanding one's political position in Brexit, two-thirds of the regions in the EU are projected to decline by 2050, and most of those regions comprise Southern and Eastern Europe, which one would admit is far from sociologically ironic, as a European layman would most probably infer correctly.[186] In observing population increase between urban and rural regions, Latvia, Lithuania, Romania, Bulgaria, and Croatia seem to have over 50% of their regions decreasing by equal to or more than -20%. The rest of the eighteen countries, in regional order, Slovenia, Portugal, Hungary, Finland, Poland, Slovakia, Estonia, Greece, Czechia, Italy, Germany, Spain, and France, show the population is set to decrease in over 50% of nationally regional areas which are falling below a 0% increase; hence decreasing.

Malta, Luxembourg, Ireland, and Iceland find that over 50% of their regions will increase by 20% or more, while Austria, Netherlands, Belgium, Denmark, Sweden, Switzerland, Cyprus, Norway, and Liechtenstein will find over 50% of their regions increasing by at least 0% or more (Eurostat, 2021). The evident trend with countries increasing by population is clearly of western nations in Europe.

When one, from a politically philosophical perspective, observes the issue of racism akin to those countries that decrease in population in the EU, pointing out countries such as Bulgaria, Hungary, Slovakia, and Greece, for instance, one cannot go on implausibly denying the nature of racism in such countries. As reverberated, when a Bulgarian family slowly immigrates the UK, or a Hungarian family slowly immigrates, along with what I would call the psychopathology of racism, the family brings along with them principles. Such principles manifest themselves by an adherence to *sins of omission*, rather than the *sins of commission* many still think racism manifests itself into. This would have also been principally relative to the contortions of the Schengen area across Western Europe.

As time, tumult, and generational vicissitudes have distorted and contorted the psychology of white British nationals since blacks first arrived in the UK, leading to tolerance and *interraciality* in the UK, England has in turn become one of the most comfortable places to live in the UK for blacks, in as far as such problems continue to be laboriously challenged in the UK. The psychopathological issue of racism and intolerance thereof, across Eastern Europe and many a distaste of other European nations, is set to then adulterate the progress made, most especially in England as opposed to the other constituents of the UK. The five recognised candidates for EU membership are North Macedonia, Montenegro, Albania, Serbia, and Turkey. When black football players spend less than even 72 hours in what one could infer to be three out of those five countries, they unsurprisingly seem to be subject to racist abuse.

As per *sins of omission*, racism may manifest itself in a wince, a languid interaction, avoidance, phlegmatic character, and more minutia individuals may be either cognisant or incognisant of, and it seems that racism in such countries across Eastern and Southern Europe come at a start-from-scratch approach when newly arrived migrants reside in the UK. Such interactions come at what is epistemologically induced by the black victim and may be implausibly denied by the white apologist in her idealistic endeavour, rendering such to publish books such as *Why I'm No Longer Talking to White People About Race*.

The issue of multiculturalism is thus complicated from a black phenomenological perspective in the UK or the west, in as far as black subjects have sufficiently ventured away from the British isles to truly understand how their ontological status fluctuates across the world; and not for the better by any stretch of the imagination.

It seems, and what should manifest itself to many as merely evident, are the proclivities and amenability to and of stories! The instantiation of stories is the very magic that adds meaning to the burden of suffering in our peculiar species, and our neurological faculties have only evolved to appreciate them. One finds such stories in one's own life, TV shows, novels, games, businesses, genealogies, mating, and so on. This is akin to the cognitive revolution and a use of fictive language we experienced 70,000 years ago, and this should be tacitly understood as the case.

One such story that has become ubiquitously distorted is that of nationalism. As a politically philosophical enterprise, one must deliberate what a nation truly is. As we find the recursive issues of intersubjective borders still problematised across countries such as China and Tibet, China and Taiwan, South and North Korea, Kashmiri Pakistan and India, Israel and Palestine, Kosovo and Serbia, and ongoing pervasive secessionist movements, stories seem to be the manifest corollary of divorce. A secular west may excoriate the stories told that abounds to *evil* across the East, coated in fundamental religionism, myth, and self-worship, but what stories abound to *evil* in the west?

One such story by a radical left that render individuals to feel guilty about their ontological status is that one must not feel *proud* of their own sense of identity appertaining to such borders. Attributable to imperialism, 'historical chance', tribalism and so on, one is expected to be always cognisant of such privilege. Moving away from a myopic perspective of privilege, and turning to a philosophical truth about claims, it seems perpetrators are almost incognisant of the chasm of privilege they have with billions of others in the world, which they ironically do not necessitate in rectifying.

The backlash has found those in the west in an endeavour to salvage the little of what is being dispelled of in their identity by a radical left, and is thus manifesting itself into tribalism, nationalism, racism, and the golden ages of history.

The colossal elision of stories being told in the west is that thus far, refugees have always existed in some form across the west, and most evidently across the world. In the UK, linguistic replacements of 'the refugee' with 'evacuee' across the first and second world wars seems to play out in the ontology of what someone who seeks refuge is or looks like. The catastrophising of refugees and asylum seekers are prejudicial to the bigger picture, conducive to leaving refugees in inhumane conditions with such parsimonious approaches to their conditions in the UK, and the cruellest of welcoming across Mediterranean countries sharing seas with the escape routes for African refugees:

'The reality is that the number of asylum applications in Italy are relatively miniscule. According to the European statistics agency, Eurostat, Italy received 128,850 applications in 2017, which is a rise on previous years. The UK, for its part, received a paltry 33,780, which constitutes a decline on previous years. Suffice it to say, not all these applications succeed.

Refugee Council statistics suggest that only half of all UK asylum applications succeed. In the case of unaccompanied children, only 55% of asylum decisions rested in refugee status being granted in the first 3 quarters of 2018. To translate some of these percentages into actual numbers, the UK government gave 'leave to remain' to a miserly 14,767 people in 2017. That amounts to one person every half an hour.

Let us put that figure into a global context. Each day the UK **accepts** a mere 0.1% of the world's new refugees, which are produced at a rate of one person every 2 seconds or, if you prefer, 900 people every half an hour. According to the Refugee Council, and the UN High Commissioner for Refugees, 86% of the world's 68.5m displaced persons are hosted by developing countries, such as Lebanon, rather than wealthy countries, such as Italy and the UK, which host less than 1%. Therefore, Italy and the UK are parsimonious, to say the least, when it comes to accommodating the world's displaced persons' (Allen, 2021, p. 192).

The narrative of refugees and those who seek asylum agglomerated with immigration 'floodgates' are mistakes made both from the left and right in British politics. One encompassing the franchise on having its fair share of discourse on net migration and the other to protect humans from persecution and wrongful repatriation seems to both warrant dissection in the purpose of human reasoning. In approximating realism without the intention of appealing to emotion this story underlies the costs of crossing the Mediterranean:

'I was on a small wood boat with 1000 people. I was a lucky one because, you know, when we crossed they stopped the engine. The people, the agents, they are just like a company, and they just want your money. It's a big sea and there are big waves and the boat can sink, so they need help from Italy. There is a border on the sea. When they reach there, they leave us. When they leave us, the boat was small. Most of the people were below deck. I was below deck. There were some people dead inside the boat because of lack of air. Everyone is fighting for their life, you fighting. There are 1000 people in a small boat in the middle of the sea with no engine. But, fortunately, the Italian came to rescue us because we are fighting for our lives. The boat was moving side to side so I went into the water. I stayed in the sea. Terrifying. I was thinking, "Is this my last day?" I just prayed oh my God. You don't want to hear the babies crying. And, when you are swimming in the sea, they rescue you. But, when I am in Italy they say there were seventy something people dead in the sea. And, there were some dead in the boat with the lack of air' (Allen, 2021, p. 232).

In 2019, I accompanied my mother to a British historian's inaugural lecture at the University of Manchester. A young British white woman, appearing to be in her late vicenarian years, confessed to David Olusoga she was a primary school teacher preoccupied with what seemed to be an insurmountable challenge. On divulging her challenge to the black historian and broadcaster, Olusoga, she asked, 'how must I participate in rooting out racism as a white woman in education?' Olusoga was suddenly tasked with giving a plausible and succinct response, while simultaneously tasked with evading the esoteric nature of his work. Let's say the answer given did not leave her any better off than before first bravely raising her hand.*[6]

As a historian and author of his phenomenal *Black and British: A forgotten history*, Olusoga believes that the United Kingdom is *losing slices of our past if we don't root out racism in our universities*. He invokes the report on 'Race, Ethnicity, and Equality in UK History' by the Royal Historical Society (RHS), based on surveys and interviews with more than 700 UK historians, fundamentally examining what is taught throughout tertiary education. He points out while the overall UK academic workforce seems to be 15% black and minority ethnic (BME), history departments change as the figure collapses to 6.3%. As 85% of all academics are white, for historians the figure is 93.7%. It means there is a mere representation of 0.5% of academic blacks in the contours of history departments. This then should be paralleled with blacks boasting a meagre 3% of the UK population (Olusoga, 2018). In response to these figures, the issue of implausible deniability presented itself when a respondent of the Royal Historical Society report wrote, "Whenever I tried

---

[6]* The answer to her great and courageous question, which he may have been too humble to retort, is that the start of the solution would be by reading his book!

to discuss it with my colleagues (all of whom were non-BME), I was told unequivocally that I was imagining it" (Olusoga, 2018).

As we slowly become cognisant of virtue-signalling motifs, expressed through social justice raids with battalions of disparity and injustice that protest in impositions of aggressive left-wing mobs, also elided is a form of virtue-signalling that can also be seen from the right. It's predicated on a world in which virtue stems from society in the west as contemporarily good, or more than reasonably fair, and thus failure is only attributable to excuses and a lack of personal responsibility achieved through conservative change.[187] Signals are akin to the virtue of appreciating the wars that allow people to revel in luxuries of the twenty-first century, along with science and political revolutions they allude to, leading to personal responsibility attributable to ignominious failure. It thus polarises itself with victim mentalities of the left and so it seems an individual can either, rest on a victim mentality, or take personal responsibility. In as far as personal responsibility is the only medicinal antidote we're left with as a **solution**, attributable to the *time arrow* and unfeasible solutions (such as economic reparations), it by no stretch of the imagination means it can be thus plausibly denied when dismantling the **problem** in its due perspicuity.

The oscillation between victimhood and personal responsibility manifests itself in the burden of what we gratuitously endure; life. Human reasoning is conclusive of having compassion for this burden we fall victim to, while sees that adopting 'personal responsibility' is the way we find agency. That which is akin to an inured victimhood may be injurious to the ears of he who in his daily endeavours succeeds in harbouring personal responsibility, while that which is akin to an evasion of the *injustices* that victimises another is also injurious to the ears of he who may necessitate such injustices. Maslow's hierarchy of needs are a fair imposition of when such a yardstick can tell for averaging 'injustice'.

In as far as one desires the necessitation of personal responsibility, relinquishing any sort of resentment from the past in order to move vigorously towards the future (agency), it quite evidently leads to the elision of an accurate history and its necessary problems. If one is driving her car on a four-hour journey and being twenty minutes away her car suffers a serious technical problem, she must stop the car. She cannot fall into a *sunk cost fallacy* attributable to being only ten minutes away, she must endure the problem at hand.

The oscillation of this pendulum that leans towards personal responsibility, in an effort to forego resentment, is only conducive to history being implausibly denied and thus the elision of a comprehensive narrative of what's epistemically *known*. What is epistemically known is the engine of our species cooperating in large and flexible numbers, and so the obtrusion of an asymmetrical ontology of black subjects only leads them further away from cooperation, approximating philosophies of conspiracy, snitching, and those that alleviates itself from a struggle being implausibly denied. If over 3% of the population of the UK are black,[188] over 2 million people, and 99.5% of the history departments are of historians who implausibly deny, history only continues to be elided as I write these words.[189]

It's also believed black history is viewed through the prism of American history when concerning both primary and secondary education. It thus leaves the children of the more than 3% of the population rather conflicted as primary school and high school teachers do not teach black British history. Frankly, because they do not know it.

With racism a philosophical issue as well as a geopolitical issue, the burden on schoolteachers abound. As suggested earlier, and in order to answer Olusoga's challenger, is that there isn't a shortage of literature that may pertain to colonialism, racism, geography, and philosophy. If a teacher is embarking on teaching in an environment, it might be necessary to know the demographics of the subjects you are teaching. With a fair share of literature that appertains to said histories and why those children may be there in the first place, a teacher will overcome her fears and add to her experience.

For instance, a whole 2.5% of the British population are Indian while 2% are Pakistani, it might be necessary to know about (a) the partition in 1947 (b) British colonial India across the nineteenth century and (c) the famous Mughal Empire. In so far as the British empire colonised many countries in Africa and the Americas, we know economic reparations are both infeasible and a suboptimal solution for imperialism. Academic reparations would reach as far as British explorers traversed the oceans, which would principally mean an edification of ex-British colonies that *predominate* the demographics of the UK, and anything superfluously relevant, as they become your neighbours, students, friends and family. This would be again akin to the facet of literature I express in the human prism of agency.

The elision of information in history doesn't only elide *truth* from the past. The ignorance of information at present thus leads to an elision of *truth* in both the present and the future, which is the work of history. This was and is attempted by communist regimes. The USSR had gone decades with the intent of keeping the rest of the world incognisant of what transpired in the gulags. It was thus George Orwell's fable *Animal Farm* which was the ingenious encapsulation of what was transpiring across the gulag archipelago, until Solzhenitsyn's work found his fable vindicated. Existence of ex-British colonies in sub-Saharan Africa may be conducive to academic reparations that will allow the UK to progress against the anathema of racism that instantiates itself upon British politics and the 'hostile environment' many believe has been pervasive. The White British that form over 80% of the population in England and Wales will have a curriculum predicated on academic reparations, giving a comprehensive curriculum in engaging on the critical solutions progressing the country further. Racism in schools are only corporeal corollaries of *sins of omission* from the educational board, as well as *sins of commission* that are for the most part necessitated, but history and a racist philosophy and psychology cannot continue to be implausibly denied, as hostility will abound for minorities who do not form a part of the 80%.

Any argument that racism may be irreparable due to in-group/out-group literature that targets skin colour may also play much less of a cause. Racism, I believe to be for the most part, germane to ontic inscriptions on subjects as we find signals that allow us to distribute ourselves into hierarchies, and so the influence on *blacks* to become agents in their lives, while inform ourselves, may help. But cognitive signals which most logically confuse the subject to exceptions, that which we call 'to stereotype and discriminate', engenders the risk for the issuing party. Profile and secure or strip her of what she's entitled to. As Michelle Obama (2022) wrote, 'I don't see you as being entitled to what you've got. I doubt that you belong with this thing that's made you proud' (p. 114).

Blacks and an Ontological Capitalism

A miracle in the ontological world for blacks in the west, which has distorted, contorted and almost alchemised both the black ontology and psychology for a significant part of history, is capitalism. As we see incommensurate representations of blacks in academia as the corollary of bad ideas such as *affirmative action*, the cooccurrence of a technological revolution with capitalism has been the miracle conducing the best place in the world to be black being many parts of the west outside of sub-Saharan Africa, and that may even been objectionable. In so far as we have necessitated blacks suffering from the lower ends of capitalism (being

Churchill's capitalism as the best of the worst choices we have for a flexibly cooperative system of sharing goods), black people have flourished.
The very idea that such a thing can be monetised on a 'multitude' demand, irrespective of how one feels of such demands, has led to the corollary of an unspeakable cornucopia of wonders for many blacks both freeing and morphing a manifest psychopathology in the west, through both sports and entertainment.
As ontological security is clearly predicated on what one may contribute to the species, their value thus rises. Fame is thus intraspecifically endemic to the human species in as far as fame and value do not necessarily render equality. And those inequalities can even be ramified into its fragmentations of why they exist, akin to such multifarious psychologies. Since the fifteenth century, English had been the *lingua franca* and in the twentieth century became the lingua franca of international business, education, science, technology, entertainment, radio, seafaring, aviation, and since the First World War, replaced French in international diplomacy. Then along with a tacit ability to read and write approximating the twenty-first century, blacks were able to monetise their proclivities for talent, becoming millionaires since the first black Jeremiah Hamilton, capitalising on sports, radio, music, and so on.
Be that as it may, and as written in the first part of this book, the ontologies of blacks must be isolated from the female. The ontic subsets of a working power of sexual selection renders value in its most specific forms. Feminist arguments, which are unfortunately prevailing in the west, incessantly endeavour to inscribe value upon women where reality does not coincide, and is achieved abusing a number of fallacies, whether it be *appeal to emotion, circular reasoning, slippery slope, false dichotomy*, and now even prevalent in universities, the *bandwagon appeal*.
Being that value is relative upon so many factors and as I drive this point home, is biologically pertinent, one cannot arrogate the authority to inscribe value where they see fit. Being that we are animals, anything, ethologically speaking, that can increase one's chances of survivorship and reproductive fitness is what renders one valuable. Gold, for instance, will never be valuable to a chimpanzee, and since males and females hold different sex cells within distinguished phenotypes in sexual dimorphism, such things lose value where the other sees fit.
Thus it had been the black male who first raised ontological value a lot quicker than the female, germane to the male ontic subsets I lay out in part I, being beauty, physicality, wealth, likeability, and intelligence, akin to the force of sexual selection in entertaining the female. Albeit a significant subset of beauty comes consequently to all others, while are comprehensively hostage to *assimilation* and *accommodation* in the entire working power of sexual selection. For instance, beauty is developmental, so comes after all or some of the others that have been achieved, being that they are psychological means for beauty to form during operational stages of a child assimilating and accommodating useful phenomena. This is attributable to a black male becoming first ontologically significant, while imputed to the black woman's lag of ontological security, as she become ontologically valuable not when rich, physically strong, or likable, but when beautified and represents the dominant male's position in a given hierarchy.
Programmes in the UK reverberated by liberals and incessant to conservatives are the campaigns that endeavour to purge racism particularly in sports, but also in school environments. With racism being psychologically operational, the onus is on the nation to tackle such a complicated matter as opposed to the child and it manifests itself in questions such as, 'can a child be truly racist?', or 'to what measure can a child take full responsibility for their own racism?' To look at an instance, Kehinde Andrews pointed out a fourteen-year-old white boy being sentenced to eleven years for stabbing his teacher in a racially motivated attack in 2015. Andrews

invokes charities such as *Show Racism the Red Card* and the policies that go without being made to deal with institutional racism across Britain, as he wrote:

'Britain is desperate for an education that provides a broad basis of knowledge and learning, which equips students for contemporary society' (Andrews, 2015).

The proclivities we have as a species of otherness may be simply defused with a few simple facts. Most especially in the second decade of the twenty-first century, and now onto the third, Indians, Nigerians, Pakistanis and Jamaicans have been a part of Britain long enough for school curriculums to necessitate their history and accentuate a miniscule political history of what Great Britain is today. The elision of such history and its pusillanimous nature, leads to its effects on tourism and beyond. A non-black tourist may not understand contemporary Great Britain and will often be surprised with what's paid for.

So, to what extent should the fourteen-year-old who stabbed his teacher be held accountable for his thoughts and actions, legally, 100%. Akin to his free will, 0%. Be that as it may, it must be understood that it was the same year as the end of the First World War that education became compulsory in the UK under the 'Fisher act' in 1918, meaning compulsory education hadn't even been a centenarian for the young offender committing his crime in an institution of education. Thus looking at this through the Platonian prism of his *Republic*, by no means is the nation protecting British adolescents from incendiary thoughts and practices of outgroups when sufficient literature on science, biology, psychology, philosophy, and even politics is at hand for teachers and parents to instantiate upon the *will* of the child. As I often remind my own adolescent students, 'we adults do not entirely trust you. We believe that if we were to give you complete *agency*, you would expend your time sleeping, playing games, eating, and sleeping some more, then by the age of twenty-one would have little to contribute to society. So we are happy to take away large parts of your *agency* for compulsory studying, at your timely expense. So while your responsibility as the adolescent may be to study as little as you can, our responsibility as teachers and parents is to ensure you study sufficiently and complete your homework. Thus I happily give you detention.'

To conclude on racism, in which I truly hope the reader does not accuse me of *appealing to emotion* in the defence mechanisms it inevitably incites, we may approach racism logically. Starting from the foundation of the epistemological hierarchy, or cake, we unanimously agree we're animals. Namely, the same species within the same genus, moving to variability within the same species. Founded on intraspecific relations (what we may call contractual philosophy), all cooperation is predicated on trust, and can only be. As we start from the smallest relations we experience or endure in our lives, such trust begins with the mother. It then subsequently moves to the institution of the family, friendship, community, and ultimately society. When one is overcome by virtue to cooperate and leave such disaffection behind, once maturing, he is rejected by his most reliable organ, his brain, or all that would encompass his intuition. It's as if a person were to be informed of their partner's infidelities many years ago. In so far as he intends to move on from such infidelities, his body rejects him. This I call inductive reasoning, and being simply philosophical, not all that are ontically inscribed as black are cognisant of the racism they thus experience (abductive reasoning). Just as in the same way all who believe not to be racist are this way until they try their hand in an institution such as real estate investment in the US. For instance, in which one would have to choose a *reliable* tenant who was more likely to ensure a positive cashflow against the loss of damages to their property, and given the realtor's right to discriminate, blacks may find themselves filtered out. I mean why take the chances.

Credit score, employment tenure, deposits, and so on do play as the essential variables, meanwhile, in many segregated areas in the United States, blacks are recent descendants of those who suffered Jim Crow housing laws before 1968. It would mean that one would have to do wonders to invest against the racial realities in many parts of the US, which only renders more distrust in the world of epistemology for *blacks*.

Predicaments extend to most institutions in which trust becomes the substrate of contractual philosophy. That which we may call latitude. In professional sport, for instance, the variables for *Homo sapiens* to cooperate are distilled. Technology has distilled variables even further in which intraspecific trust is not needed, only the ability to scientifically deduce. The more we abstract variables, the less distilled becomes human cooperation, in which *ontic inscriptions* do prove deleterious to the latitude that implausible deniability undermines.

Relative Privation and the Parable of the Workers in the Vineyard

"For the kingdom of heaven is like a landowner who went out early in the morning to hire men to work in his vineyard. He agreed to pay them a denarius for the day and sent them into his vineyard.

"About the third hour he went out and saw others standing in the market-place doing nothing. He told them, 'You also go and work in my vineyard, and I will pay you whatever is right.' So they went.

"He went out again about the sixth hour and the ninth hour and did the same thing. About the eleventh hour he went out and found still others standing around. He asked them, 'Why have you been standing here all day long doing nothing?'

"'Because no-one has hired us,' they answered.

"He said to them, 'You also go and work in my vineyard.'

"When evening came, the owner of the vineyard said to his foreman, 'Call the workers and pay them their wages, beginning with the last ones hired and going on to the first.'

"The workers who were hired about the eleventh hour came and each received a denarius. So when those came who were hired first, they expected to receive more. But each one of them also received a denarius. When they received it, they began to grumble against the landowner. 'These men who were hired last worked only one hour,' they said, 'and you have made them equal to us who have borne the burden of the work and the heat of the day.'

"But he answered them, Friend, I am not being unfair to you. Didn't you agree to work for a denarius? Take your pay and go. I want to give the man who was hired last the same as I gave you. Don't I have the right to do what I want with my own money? Or are you envious because I am generous?'

"So the last will be first, and the first will be last" (New International Version, 2000, Matthews. 20:1-16).

It's quite easy for one to become disheartened and fettered by the forces of pessimism when addressing the issue of racism, such that inclinations likewise do the same within the reality of evolution instantiating itself upon survival and reproduction. But realism only forgoes the problems of the future as they are first identified. That being said, there are many aspects of the twenty-first century in the west to be highly optimistic about.

One will understand relative privation to be the case, sometimes used as a defence mechanism to deny the philosophical and psychoanalytical intricacies of racism, be it plausible or implausible (as per the parable of the workers in the vineyard). The parable evokes avarice, ingratitude, unfairness, and the like. But it is *righteousness* concluding the parable, ensconced in what we call a political philosophy of today. Blacks may be compared to those that had toiled the field all day long. The fact that the Atlantic and Indian Ocean along with a vast Saharan desert isolated sub-Saharan Africa from unlost wars, millions lost in diseases, and a necessary cultural diffusion, are significant causes to the unpeeling of determining factors for racism and economic disparity of today. In as much as history shows we live in the safest society ever experienced, we must endeavour to compel rights, or policies, to be as fair as possible, which would be its advocate of an *equality of opportunity* (which respects variability of the species), in order to expect a nation state to cooperate as it maximally and possibly can. In terms of psychological racism, that would be in as so far as one may practice gratitude in their private abode. Like the workers of the vineyard, one learns the focus is on the agreement made relative to your own privation.

The Epistemological Quest

On a Conclusion to Knowledge

'An Evening School'

'This small night scene extols the virtues of teaching (light passed on) and practice (the sharpened pen), which the much admired Aristotle recommended as paths to knowledge. The subject would have appealed to many in Dou's city of Leiden, home of a famous university' (Dou, ca. 1655-1657).

The Epistemological Quest

Chapter 8

Science, Sexual Selection, and an Homage to Korea

A spectrum of knowledge that deal with both a priori and a posteriori knowledge render the ranges of petty skirmishes all the way through to wars and genocide (Pinker, 2011). The latter stages of the twentieth century recall the necessary recapitulation and revision of organising information, instantiated upon a human twenty-first century epistemology. Thus with such a recapitulation, the organisation of how we come to *know* a thing must now be optimally fathomed by the lay person be it woman or man, as franchise and compulsory education *seems* to inscribe the onus we have on our species to know a thing, which has served as a truly remarkable asset within our species, allowing us to cooperate largely within society. If one is sceptical about the agency of knowledge on the laity, picture the simplicities we inscribe on what one should know, such as speaking the native tongue of the land in which you reside, or how to send a letter at your local post office, how to vote, or how to send a message on your mobile phone.

The ubiquitous nature of our disorderly knowledge is most certainly an issue in the twenty-first century, and a competent compatibilist thinker of the future will certainly convey we simply can't be blamed, attributable to the torrent of information of our, our parents, grandparents and for many, great grandparent's generation. In as far as we cannot take responsibility for both causes in the environment as well as our own decisions of the past, we can be agents of the endeavour to *cause* others to act, and thus contribute to cooperation. If you were to travel back to the fifteenth century, stand in front of King Henry VIII and tell him you were sent from the future in order to make changes to fifteenth century England, you may quickly realise you cannot explain to King Henry how to make penicillin, vaccinations, a computer, mobile phone, or modern armaments. You will realise that, in so far as you may have experienced such twenty-first-century phenomena, you are rather useless standing alone as an individual subject.

We see both the twenty-first century philosopher and scientist often lock horns on the bigger questions that derive from the smaller, and somehow condone the conclusion, 'we have to agree to disagree'. It leaves those who feel vindicated frustrated, peevish and often deprecated on grounds of knowledge. In as much as one may claim one has the right to disagree, what many will know is **when** one has the right to disagree as opposed to **if** one can disagree with such a thing, akin to the fallacy of the *middle ground*.[190] This is the realm of a reasonable scepticism contra conspiracy. Can one truly conspire successfully without being sufficiently educated in such a thing? What does it mean to be sufficiently educated in such a thing? It's why I find it necessary to travel back to the drawing board, in a take-one-step-back-to-move-forward approach towards what we know today. Our biological sexes divide themselves into half the population of the species, and it seems today that a battle of the sexes in many parts of the west remains belligerent, which can be principally imputed to the sheer ignorance, denial, and elision of the biological sciences. One has the incessant propensity to reduce the biological sciences to the ethological practices of animals holding lower faculties of intellect in a survival-of-the-fittest-understanding, by claiming humans to be too far advanced with reason to be compared to. This is only indicative of a vacuously erroneous grasp of the nature of the life sciences. One must start from the very small of biology, and to achieve this one is lead back to millions upon billions of years or to a laboratory microscope of today, which then manifests itself into a philosophy of what one must

The Epistemological Quest

call teleology of the first adenine, cytosine, guanine, and thymine molecules. With this we become slaves of our endocrine system, endocrinology, and quite peculiarly one's genes, genetics. We ask ourselves, 'what was the *purpose* of my complicated action', and the discerning will find the complicated act to be attributable to a biologically rooted teleology, which qualifies itself as ethology, given that we are fundamentally animals.

I often ask my adolescent students to first think of biology by looking at their hands. I then get students to compare their hands with their classmates and ask them if they are the same or different. On being the same is clearly conducive to a vast and large gene pool that provides us with the necessary organs and features to survive. That being the dexterity nature provided us with our hands and thumb ratio, an ability to grasp, pull, pat, clinch, slap, twist, push, along with many more features of the human hand. I then ask students if they can do the same with their feet. The answer is, 'of course not!'. It seems the environment compensated a once quadrupedal aptitude towards hand dexterity and brain cells, which, if I may say so, was a good call from natural selection. Take this scenario I also often present to students:

Nine-year-old Lisa is standing face to face with a hungry lion around ten meters apart. Who will win?

The answer is Lisa. Attributable to how our species had evolved to become the most social species to exist on earth, along with the evolution of hand dexterity and the vocal tract thus attaining linguistic abilities to communicate incomparable to other species, with technology, innovation, and cultural diffusion, we came to necessitate our own intraspecific survivorship. Relative to the English language, this would appertain to linguistic terms such as, 'risk assessment', 'danger', 'safety hazard', and so on. The legal sciences have evolved well enough that humans can travel long distances around the planet to see animals such as lions with careful precautions on safety hazards. Lisa won as there will most probably be a fence between the lion and Lisa, or that Lisa is within a safe means of transport with a professional in her vicinity. The professional would have been taught risk assessment on how to competently fulfil his duties by other humans, the transport would have been built by other humans, and so the experience. Frankly, the chances of nine-year-old Lisa standing face to face with a lion with no recourse for safety isn't even worth thinking about.

When we start from the minute, especially with humans, we must understand before anything, that we are products of eukaryotic cells. As opposed to prokaryotes such as bacteria, we have a legion of organ-like structures that work for the cell sitting in a gel-like substance we call cytosol or cytoplasm (exterior), called *organelles*, such as mitochondria, responsible for energy production, vesicles and vacuoles, responsible for transportation and storage, and the classic ribosome principally responsible for protein synthesis. Scientists have been able to garner fossil records of organelles, taphonomy, harking back millions of years in order to understand the nature of eukaryotic evolution, creating more than a tree of life for the animal kingdom, but a tree of life for microscopic life as well (Carlisle et al., 2021). So while it's tacitly understood that chloroplasts are organelles conducting photosynthesis and chlorophyll converts sunlight into its green-like energy, being that plants cannot escape its principal nature, animals cannot escape the role of reproductive sex cells that in turn form our teleologically ethological behaviour. There is nothing merely reductionist about these facts nor should it dishearten the perspicacious ruminator. We're simply a product of the natural world. Such reproductive cells hold features that other cells do and so goes on the fundamental foundation of the epistemological hierarchy of knowledge.

Mould Juice

It was such microorganisms that changed the fate for humans, giving us the never-before-seen robust population of our species we see on earth today, residing on 29% of the earth's surface, land. On a Petri dish sitting next to an open window in 1928, bacteria started to die. Mould spores contaminating the Petri dish were conducive to the change of an agar gel. On isolating the mould, Alexander Fleming identified the colonies of mould as a member of the *Penicillium* genus, which he subsequently found to be deleteriously effective against diseases such as scarlet fever, pneumonia, gonorrhoea, meningitis, and diphtheria. They were the very diseases that killed our ancestors and could not simply run away from, prior to such knowledge of microbiology. On discerning what was killing the bacteria, he found it was not the mould itself, but the 'juice', as Fleming suggested:

'When I woke up just after dawn on September 28, 1928, I certainly didn't plan to revolutionize all medicine by discovering the world's first antibiotic, or bacteria killer. But I suppose that was exactly what I did' (Tan and Tatsumura, 2015, p. 366).

The recipient of many awards Alexander Fleming subsequently came to be, including his appointment as the Emeritus Professor of Bacteriology at the university of London in 1948. But it was 1945 in which Fleming was honoured with the Nobel Prize in Physiology/Medicine.
Speaking of Nobel Prizes in Physiology/Medicine, in 2016, a break-out cell biology under the concept of *autophagy* was put forward by Yoshinori Ohsumi, a cell biologist at the Tokyo Institute of Technology's Frontier Research Center. Within mammalian systems, it was known since the 1950s and 60s that some animal cells use autophagy to recycle proteins and important organelles. As a physiological process that responds to starvation and disease, Ohsumi used a single-celled organism, yeast (as yeast also carries out many of the same biochemical processes as animal cells). He made sure to choose strains that lacked the key enzymes that could play out the role of autophagy. On starving the yeast, they found vacuoles (the organelles responsible for recycling and storage) had become atypically large.
Vacuoles were originally too small to see under a microscope and then, attributable to such starvation, the mutant yeast showed enlargement of such vacuoles. Genes were also important for the strains of yeast to successfully mutate and comprise the type of vacuoles necessary for autophagy with successful research. Ohsumi and his colleagues were able to portray autophagy as significant to embryo development, cell differentiation, and the immunology. A functional autophagy system is conducive to diseases such as Huntington disease (a process of the breakdown of cellular nerves), cancer, and diabetes (Science News Staff, 2016).
As cellular differentiation transpires, which so happens numerously during the development of multicellular eukaryotic organisms, as cells develop into a complex system of tissue and then organs, autophagy plays a part in its likewise differentiation. It means the teleological functions of small organelles, such as vacuoles, are an adaptational feature to one's environment within time and space, as are the large cells such as nerve cells that have helped us to form deliberating machines such as the prefrontal cortex.
The issue of cellular structures and functions involve methodological naturalism yet again, touching on the realm of theological philosophy and subsequently, politics.

The Epistemological Quest

Modern philosophers, such as Sam Harris, have reacted to the anathema of stem cell research by religious authorities, believing that, fundamentally, they would not hold the *epistemological purchase* to involve themselves in such matters.

For a religious type, it is akin to when the 'soul' enters the human, or anything appertaining to the ethics of this debate. During the first week of pregnancy, embryonic development inside of a female begins with the fertilisation of the two haploid cells, the sperm, and the egg, fusing into that multicellular eukaryotic cell we know as the zygote which carries a complete set of chromosomes. Blastocyst cells come after approximately four to six days of pregnancy. Significantly, for what it may be relatively worth, it's the *inner cell mass* of such blastocyst cells subsequently forming embryonic development that would in turn lead to the development of a foetus. Stem cells are classified into three types; embryonic, umbilical and adult stem cells. The scientific study of stem cells, if being free to research, could cure neurodegenerative diseases, doing away with the likes of Parkinson's disease, Alzheimer's Disease, along with tumours such as leukaemia, brain tumours (such as anaemias), autoimmune diseases (such as Multiple Sclerosis and Systemic Lupus Erythomatosus), along with cardiovascular diseases (Agius and Blundell, 2007).

Thus the *epistemological purchase* which instantiates itself upon the hierarchy of a pyramid of knowledge and the cherry-picking of methods to address it is the problem that philosophical naturalists may invoke. Today, scientists publish research without the initial financial cumber, and publishers charge universities and other institutions such fees which in turn become the subscriptions that an institution may pay for. What was called "Plan S" was devised to help pay publishers and render scientific journal articles free, becoming the first mandate of an international coalition of funders to tackle paywalls in the scientific community and beyond, in an effort to create *Open Access* for all (Brainard, 2021). Ironically, the article I cite this point from was initially Open Access.

Would the act of theologians or of those of the religious communities be branded audacious for such impositions onto the hierarchy of knowledge, or the epistemological cake? This is exactly where Hobbesian contractual philosophy complicates itself within the world of the contradictorily methodological naturalism. It imposes upon contradictions of free will and human rights, as article 18 stipulates:

'Everyone has the right to freedom of thought, conscience and religion; this right includes freedom to change his religion or belief, and freedom, either alone or in a community with others and in public or private, to manifest his religion or belief in teaching, practice, worship and observance' (Roosevelt, 2001).

Should religious authorities have the right to thwart a field such as stem cell research? Well, according to the article, no. Whereby an individual may choose not to involve themselves, their offspring, or family members in stem cell research attributable to religious beliefs, religious authorities have simply not achieved the epistemological purchase to impose themselves on such significant matters. In fact, I would go as far as to say scientific institutions should also present their belief of the cellular verbiage of blastocyst cells and inner cell mass as purely biological to the public, with the complete elision of anything supernatural, steering clear from linguistic terms of a 'soul' for the pure purpose of professionalism (or philosophical naturalism). What one can and may believe about the implications of the supernatural is purely private, or at least secondary to scientific training in practice. And this is all written with the full consideration of *methodological* naturalism, as opposed to naturalism in its *philosophical* sense.

Flat earthers brandish their right to conspire in the light of freedom of thought, as long as conspiracy steers clear from scientific training in institutions such as, for instance, pilot school. If such conspiracy were to enter scientific training and all science's epistemic tumult, conclusively declaring the earth to be flat, using science and reason, then my hat goes off to such prescience!
As it stands, such languid acceptance, which would dismantle the function of trust (agreement) in the epistemic world of science, should be reviewed, and has thus far been successfully achieved through *peer review:*

'The benefits of peer review are real, whereas the alternative - giving up peer review in favour of scientific 'freedom of expression' – would create many problems of its own. Novel findings or ideas might not move into the mainstream of our understanding of biological processes if they are viewed as simple statements from the discovers, since peer review adds additional weight to claims that challenge our current understanding' (Gannon, 2001, p. 743).

Epistemological purchase may run further in science as we have political institutions, the British Broadcasting Corporation (BBC), who work concertedly with scientific communities by filtering sections of news outlets to the public via payrolls of science journalists. This way the laywoman has the material opportunity to edify herself to a plausible degree on the political outlet of such findings, in as far as such findings involve themselves to philosophical schools of thoughts.
We find ignorance of significant work, most especially in vast fields of physics,[191] leads to the corollary of erroneous predicates for religious and political thought while ideas on prescient ideas that may update the public find themselves subsided.[192]
I see in its absolute form that political commentators and the laity comprehensively understand science to obtrude in all thought. That's not just to say it is, but to understand it! Thus one would come to understand exactly the extent of the causal factors that created anything near the society in which we live in the west. When one ponders on the sciences of the individual, one must make the necessary inquiries as she quadrupedally moves as a living organism. She toils through space and time with this somewhat neurological expense of self-consciousness, and even worse, achieving the level of consciousness inquiring into the *purpose* of existence As the clinical psychologist Jordan Peterson articulated best:

'(People often get basic psychological questions backwards. Why do people take drugs? Not a mystery. It's why they don't take them all the time that's the mystery. Why do people suffer from anxiety? That's not a mystery. How is it that people can ever be calm? There's the mystery. We're breakable and mortal. A million things can go wrong, in a million ways. We should be terrified out of our skulls at every second. But we're not. The same can be said for depression, laziness and criminality)' (Peterson, 2018, p. 148).

That high serotonin and low octopamine characterises victors in skirmishes between lobsters is reason why he suggests readers to take on life by standing up tall with one's shoulders back,[193] as one has reason to understand their own diffidence. As classical relativity suggests, there is no such thing as absolute rest or absolute motion, all living beings are in a constant toil against both space and time. Neurochemical research challenges epistemological battles between what we may call agency in the world. A simple boost of serotonin helps Obsessive Compulsive Disorder (OCD) and Attention Deficit Disorder (ADD) in humans, meaning low dopamine in a system raises the inhibitory neurotransmitter, gaba, seeming to help

The Epistemological Quest

with the irrational feeling of anxiety too. When infusing oxytocin into the brain of a non-pregnant female rat, she rapidly induces maternal behaviour towards her young pups. Meanwhile, on endeavouring to understand the mammalian relationship between oxytocin in stimulating maternal behaviour and sexual activity in rats, voles, and humans, scientists agree to the term *bonding* (Herbert, 1994).

All the causal factors (or influences) render the illusion of the will. When scholars unfurl tribe, culture, nation, and group, they find it dissipates beyond the agricultural revolution around ten thousand years ago. That would continue to be endogenous to the Holocene epoch approximately 11.7 thousand years ago, the Quaternary period 2.6 million, the Cenozoic era 66 million, and the Phanerozoic aeon approximately 541 million years ago. Animals have both come and gone, endured a collision of an asteroid on their planet, and changed in evolutionary competition and Milankovitch cycles (thus adapting to their environment). Given the position of terrestrial zones of the earth and the forces creating continental composition of the earth's crust against the seas, we find causation to double on itself once again, withdrawing the act of the *will*.

When compelled to look at the evolution of both even-toed ungulates, artiodactyls, and their odd-toed cousins, perissodactyls, for instance, one apprehends the extent of evolutionary pressures that lead to such ultimately distinguished animals such as our popular domesticated horses and pigs, even when sharing ancestors in the Paleogene. Perissodactyls, who bear weight on three or one of their five toes, evolved from a type of chalicothere, the chalicotherium, which evolved creating a repertoire for high browsing in natural selection, eating from high trees while bearing weight on hind legs. What was strange to palaeontologists was how chalicotheriums had a miraculous feature of carnivores, claws, but had teeth of a herbivore, until realising such claws were used for high browsing. Some of those chalicotheres would eventually lose their claws to become the odd-toed ungulates we see today, which is its own fascinating story. But chalicotheres, eventually evolving into both ungulates from the common ancestor, condylarths, still belonged to the Cenozoic era, evolving into both even and odd-toed ungulates. Given the ongoing disputes between ancestry of a significant period across the Paleogene, postcranial research evinces the possibility artiodactyls evolved from a family of arctocyonids, attributable to the hindlimb and the morphological diversity with many of the genera of this family (Rose, 1996).

Sexual Selection, a Recap

I will return to a point I made in part I of this book, that of the most common misconception of biology. As a philosophical naturalist, all human behaviour must be biological, and so social constructionism can only be second to biology. Behaviour can only be a product of physiology or the environment, and the environment is the causal natural selection that rendered all animal's decision-making organs exist in the first place. This includes ethology, microbiology, evolution and so forth. As I mentioned in part I briefly, it tends to be ironically akin to the misused notion of *reductionism*, and when looking at phenomena through this biologically misconceived prism, there's no wonder one may see such theories as reductionistic. A most unjust proclivity is for the layperson to think of animals of the lower faculties of intellect. What follows as an ethological confutation of behaviour in many politically philosophical spheres is something along the lines of, *we are animals of higher faculties and therefore we can move beyond what we are biologically predisposed to enact*. What's reduced is the idea of intraspecific action which we cannot take responsibility for. If the self is more of a stream of consciousness between principal parts of our brain, what decision can be free from causation? In the sense of biological grandeur, does one truly believe we are as free as one could suspect? Literature on both the neuro and cognitive sciences clearly evince the right to be libertarian in incompatibilist free will belief, meanwhile, and epistemically, we must come to unanimously fathom from which point we can start to believe freedom to commence. Thus the ontic subsets of a working power of sexual selection I explain in the first part of this book I hope is not met with languid refutations of reductionism, and so I will further explain how the epistemic framework of teleology principally appears.

As humans are by far the most complicated animals to have existed on earth, which one may impute to a social predisposition, it may be optimal for one to become one's own critical anthropologist utilising a philosophy of teleology in order to fairly observe one's own actions, as there is not a species more intellectually advanced that will render such observation objectively. We may ask the simple question, what was the end *purpose* of my action? Or, in so far as I may be incognisant of the majority of my actions, is there a higher objective or goal far along in the future? Well, cognisant or incognisant, there certainly is.

The end goal as a member of the species is to acquire sufficient resources commensurate to the purchase of all of one's own *time*, and time is the most valuable asset in so far as a thing can sit as both an asset or a liability on one's own 'balance sheet'[194] as it may take away time or give you it. I quite confidently say no human has achieved this, meanwhile many have and do approximate the state of an acquisition of their own time compared to what others must expend contributing to the species. We're quite simply too social to achieve this.

As far as sexual selection, it must be perspicaciously understood females and males have different strategies to achieve this teleological end goal, in so far as many will hold similar strategies. It manifests itself in what we may call power, which also instantiates itself on the hierarchies for better or worse.

As Alfred Adler understood quite well, the inferiority of females to their male counterparts is well manifest and present within sexual dimorphism as, on average, males are faster and stronger than their female counterparts. Be that as it may, by no means is our velocity and strength a feature in the animal kingdom, as one could brainstorm animals faster and stronger than males rather quickly. Brain and hand dexterity are certainly unparalleled features in the kingdom, and there isn't any evidence of one sex being smarter or more dextrous than the other. Thus one cannot ascribe the most *powerful* of our species to being the fastest or the strongest.

Considering the head-start males have over females in some of those aspects, another burden strikes the female towards this teleological goal as viviparous species; pregnancy. It takes a whole nine months away from the female's objective of achieving the goal of an apprehension of resources, with the *gene's* goal quite happily and successfully achieved.

But as females have been disadvantaged, so too have males (what Maynard Smith calls *assymetries*). But what may this be? Females are not the weaker and slower versions of men, otherwise females would appear as small and weak versions of men. Instead, females have the advantage of neotenous features that are rather beneficial to they eye, even though they don't share as much genes as a mother or sister, and this would quite significantly go along with an endocrine system and neurological differences that would ultimately induce the male towards her. This holds as significant because in as much as the male achieves resourcefulness, his neurochemical, endocrine system and the like cannot reject this figure that ultimately his genes are attracted to, the female. That would be peptides secreted by

the pituitary gland such as endorphins and oxytocin that manifests itself into the highest of male pleasures.
Males lack such neotenous features females are advantaged with, thus on a female achieving resourcefulness, there is no real proclivity for the female to feel so altruistic to this male stranger. A male who feels necessitous of her resourcefulness may inversely be unattractive while she is bioengineered to value resourcefulness as a biological shortcut to allow her and her offspring to subsist. Thus she profits from the economics of being able to sexually select.
This significant difference between males and females for the teleological goal of resourcefulness most naturally renders between males and females a competing conflict of interest. Thus cooperation into such a monogamous substrate had been thus far ensconced in institutions such as marriage, in societies where more than 90% of the population endured extreme poverty while enduring tyrannically political and motivated systems. With the advent of reason being conducive to feminism, and also contraception, population increase, and a technological revolution, hypergamy prevails while marriage declines across the west, accentuating what seems to be a biological conflict of interest into a vicious war of the sexes.

You Can't Have Your Cake and Eat It! ... Oh Yes I can! And the Conflict of Interest

Sexual selection and the necessary ontic subsets of sexual selection I do believe to have manifested themselves within the twenty-first century, in so far as sexual selection sociologically continues to distort itself and has done so since the influence of feminism onto our earlier generations of men and women. It also intensifies itself in the elision of our scientific understanding, along with the elision of technological and thus sociological understanding of most of our human history. Given the access of birth control as opposed to abortion (a suboptimal experience for the female), we learned *empowered* the female and is also beneficial for the male, but how much so? As biology has it, given we are sexually reproducing species, one of the highest of pleasures, if not the highest, is to sexually reproduce. As we may come to unanimously agree, intense short-lived pleasures, or rewards, come with immediate costs, but what if they don't?[195]
The cost in our pastime was the burden of a new living being on the mother and one could say the long-term reward being the living being reaching adulthood, becoming an asset. Along with this self-explanatory idea is the essence of our primatological proclivities, quite specifically of the male. Dominant males in evolutionary psychology and as territorial primates, are programmed to assimilate behaviours germane to female chastity or abstinence, chartered towards the evasion of taking care of a child that is not his own.[196] One may say that there hadn't been so much of a conflict of interest between males and females in societies in which females were *coy*, most especially when coming to monogamously pair, as male interest would be in attaining a reasonably chaste female, and female interest in an evasion of impregnation by a *philandering* male without a desire to commit to her by taking care of offspring.[197]
Attributable to such an alleviation of the cost of pregnancy for the female, most especially across late denarian and vicenarian years, it seems females may now choose to enjoy all three privileges until menopause:

I.   The unhindered pursuit for resourcefulness.
II. Extramarital sex.
III. A male who should commit to her. The *faithful* male.

But as for the third privilege, which compels the male to act away from what I would contend is part of his freedom, one may say that you can't have your cake and eat it. The female who one may claim squanders her prime may come to an unfortunate cul-de-sac. In the event that her unhindered pursuit of resourcefulness fairs well, she may slowly have to come to terms with two issues. The first, hypergamy, which would be the simple proclivity within sexual selection for a female to select across and up within the tenacious pursuit for resourcefulness, rendering legions of men falling out of her criteria. The second issue, ageing. As she ages, she may find and intuit she cannot impose the same *conditions* on males that she could in her earlier years that would be willing to buy the entire cow, instead of feigning to eventually buy the cow in order to continue getting his milk for free whilst foreboding how long he can achieve this for. His success in achieving this supply of free milk is only attributable to his repertoire for sensing vulnerability, or a depreciated stock in the market, that continues to suffer from her timely unchasteness. She may utilise preservation techniques for those that pass by, but her unchaste philosophy is already well and truly ensconced, in which her philosophy may give herself up within a paltry amount of time within such encounters. Conclusively, and anathema to a concept of hypergamy, she may be compelled to lower her initial *asking price*.

Why Buy the Cow When You Can Get the Milk for Free? A Turn in the Emerging Mating Strategy for the Selectable Male

Like the stock market, one cannot arrogate the will of the gods in prognosticating the future. Science, technology, philosophy, and reason has tended to impede the ways of apocalyptic prognosticators of the past. However, as I will convey with a perusal of the stats on marriage, the decline in marriage and the tumult in science, technology, politics, and philosophy of the post 20th century female, may exacerbate the tumult to larger extents as I believe such conflicts of interest between the sexes to be incessantly and continuously endured. I again contend this from a natural standpoint.

Cooperation in marriage we can testify to experiencing personally, or somewhat vicariously in the 21st century and understand success to be short of a miracle. That's if the marriage does not end in divorce. As I often profess, in so far as we argue, skirmish and endure disarray with our parents, exceptions not being the rule, the chances of sustaining some type of relationship in each other's lives will continue until the day that either they die, or the day that you do. The same can be said of brothers and sisters, and its clearly attributable to the fact of one sharing fifty percent of the genetic makeup with one's parents and siblings. It halves itself as relatives become distant from that nuclear familial set up of having all that contains fifty percent of one's genetic makeup within the household. Thus on seeking sound advice or guidance from family members will be distinguished, which can be merely imputed to interests being little conflicted.

When effaced with a stranger in the hope of marriage, one does not share near as many as the same genes as a nuclear family member does. This manifests itself significantly because an investment becomes so much more elephantine for a prudent male when the risk of being estranged on one's death bed exponentially increased.[198]

As it will turn out, males, after sufficient experience in such encounters, once approximating later vicenarian and early tricenarian years, will assimilate such conflicts of interest and reassess. Males of reasonable ontological value, akin to what emboldens the male or what he contributes to the species, thus raises his *asking price*. How a male does this is in fact quite injurious to the epistemological cooperation of the species. Once sensing that the philosophy of his female counterpart intends to pursue resources unabatedly while alleviating herself of her unchaste history, he may endeavour to have his cake and eat it also. The corollary could be offset and bifurcate into two surreptitious and uncanny ways, whilst all being commensurate to his value on the market:

I. Acquire as much milk as he possibly can without fulfilling the objective of ever buying the cow.
II. Behaving mendaciously through unfaithfulness in his myriad encounters.

Why should he be compelled to purchase the unchaste cow if he can get the milk for free? Like real estate, fundamentally, it's not about the quality of the cow that the cow puts on itself. An owner of a house cannot simply choose her *asking* price of a property without comparison to others. He will act against what he can achieve against the other cows in the market (environment), and if those other cows also share the same taxing conflicts of interest, he may feel reduced to offset such interests in the two surreptitious ways to fulfil his pleasure and have his cake and eat it.

His teleological goal, whether he is cognisant of this or not, is to acquire a reasonably chaste female, valuable enough to appropriate him completely from the market by compelling him to buy the entire cow. It's the purchase of the cow that renders the female the victor, notwithstanding anything transpiring before the marriage or after.

On the Origin of Pleasure; Marriage

Plato believed those who indulge in pleasures do so like sheep with heads bent over pasture. And he believed such pleasures fundamentally insatiable. But a question is, most especially for a revamp of the twenty-first century, what is of the discourse of pleasure and its origin?

This question is most pertinent for the lay person who in the west has more access to the cornucopia of pleasures more than any human in history, both legally and illegally. Meanwhile, as many may have personally endured, or at least vicariously endured the ramifications of that which we may call a drug, a precise answer seems to linger with sex. We invoke scientists who give us healthy neurological retorts on matters of neurotransmitters, hormones, a dopaminergic system, and an involving mesolimbic dopamine pathway. But as discourse on philosophy suggests, instead of how such pleasures are experienced, there seems to be not enough import of *why* such pleasures have evolved, thus in part I of this book, I necessitate how necessary sex and the process of sexual selection may play in our own lives, both consciously and most especially, subconsciously.

I hope to encounter not a soul who would deny the proclivities of *Homo sapiens* to separate the ideal world, of what we may call romance, against the real world, manifesting two spirits, the idealistic and the real when on the topic of mating strategy. The youngster will dance in naivety of such meagrely experienced dalliances while the dyspeptic sceptic winces at the painfully endured realities of his pastime. I guess what I seem to be homing in on as a proposition, is that pair-bonding, that's notwithstanding what we may consciously or arrogantly believe against subconscious decision-making, is absolutely necessary for the human experience. Maybe as necessary as it is for a fish to feel the waters, birds to spread its wings, and caterpillars to metamorphosise. I do not for an instance reduce such

pleasures to a sexual act, but to the teleological pursuit, and the ongoing vicissitudes that such cooperation levies for the many times the earth comes to orbit one of the many the stars, as such a pair bonds while slowly yielding their resources onto offspring as those offspring likewise do to theirs.
As natural selection has had the gene for over 99% of human history, most humans have not had access to anything near approximating the power in stultifying human fertilisation up until recently, and so a resulting war between the sexes and thus unfortunate plummeting of marriage would only be unfortunate for the species if it interminably continues to do so. At least one could say our ability to cooperate is a colossal feature for our species in its manifestation of practical hierarchies while taxing for welfare for those who endure the bottom. And so, for instance, if one chose to limit, instead of entirely forego contraception, it can only be offset with the politicisation of female issues in policies that may protect the female, liberating females from the most essential experience in the *pursuit of resources*, instead of childbirth being only branded a burden in her teleological endeavour.[199]

Looking at a cohort of men at 30-years-old in England and Wales between 1940 and 1980, 83% of men born in 1940 were married, declining to 79% of those born in 1950, 64% in 1960, 41% in 1970, then down to 25% in 1980. This is a decline in which over ¾ of men were married by 30 in 1940, down to a meagre ¼ for those born in 1980. This is all while women's average age for marrying has increased to over 30-years old for those in countries such as Sweden, Norway, and the United Kingdom by 2017, which had initially been in women's later vicenarian years in 1990.
The causes and circumstances that lead to single-parent families across the world do variegate, meanwhile, Ortiz-Ospina and Roser found three recursive patterns in their data:

1. 'Women head the majority of single-parent households, and this gender gap tends to be stronger for parents of younger children. Across OECD countries, about 12% of children aged 0-5 years live with a single parent; 92% of these live with their mother.'

2. 'Single-parent households are among the most financially vulnerable groups. This is true even in rich countries. According to Eurostat data, across European countries 47% of single-parent households were "at risk of poverty or social exclusion" in 2017, compared with 21% of two-parent households.'

3. 'Single parenting was probably more common a couple of centuries ago. But single parenting back then was often caused by high maternal mortality rather than choice or relationship breakdown; and it was also typically short in duration, since remarriage rates were high' (Ortiz-Ospina and Roser, 2020).

Looking at the percentage of marriages ending in divorce in England and Wales since 1963, 1.5% of couples divorced before their fifth anniversary, 7.8% before their tenth, and 19% their twentieth. Conversely, the percentage of divorces has inversely halved since its peak in the 1990s, whilst those who married in the first 10 years of marriage has also fallen significantly.
With complications of marriage ramifying, Ortiz-Ospina and Roser show how difficult it is in the attempt to understand marriage and divorce rates globally. What can be positively assimilated from their work, is, that humans under intensified circumstances seem to find ways to work things out, and hopefully by the twenty-second century, an inevitable myopic perspective subsides while the more holistic

perspective supersedes, accounting for rapid changes in technology and science of both the twentieth and twenty-first century.[200]

A Conclusion on Marriage and a Conditional Love

As long as we brandish the right to discriminate in the west, we must understand love to be irrational, sexist, racist, homophobic, elitist, and all that renders disparity. On turning to marriage and on more contractually philosophical terms (most especially of ontology and essentialism), I believe it necessary to unfurl agreement into that which comes to allow a functional institution of marriage. There are six ontic agreements that take place consciously, but I contend far too often take place subconsciously. The first two agreements are the agreements one makes, or already has, with oneself (this could be germane to health, nutrition, finance, leisure, work, sex and so on). The second two agreements are the agreements one has with their future self, as Victor Frankl in his logotherapeutic practice understood most as that which may give life meaning. Attributable to the ongoing vicissitudes of life, many may neglect such agreements as they go unpunishable and manifest themselves in self-reproach. The last two ontic agreements are the agreements one has with the ontology of the other. Those agreements may be multifaceted and yet again manifest themselves throughout the distinguishing contexts of space and time in a venturing marriage. What becomes of the six ontic agreements is that conditions are ubiquitously disorderly, representing itself as a clear recipe for chaos and destruction in the marriage.

Many simple desires may not be materialised into agreements and will ultimately be imputed to feelings such as shame, fear, and so on, if unaddressed.[201] With the advent of technologies implicating itself on the twenty-first century, it holds the proclivity to complicate itself even further. Is love conditional or can it be unconditional? On the condition that the other is the same species can you love her, that you remember, that she remembers, that you think, that you phenomenologically exist. Might it be wise to think of what those conditions could be?

A Homage to Korea

I have written this manuscript sitting on one of the strangest peninsulas in the world. Even now, I do so while Christmas songs emanate the air, aware that it will be far from the type of indulgence many experience north of the Demilitarized Zone (slightly over 100 kilometres from where I sit). It has seemed, for countries outside of North Korea, the west, and even my Korean students, that through sheer incognisance of what transpires in the North, what occurs in this strange country is farcical, amusing, strange, and simply uncanny. Accentuated by instances of the strangest of associations with imperceivable relationships between Kim Jong Un and ex-Lakers' basketball star Dennis Rodman, who appeared on the popular American show, *The Late Show with Stephen Colbert*, as he talks about Kim Jong Un's strange attraction to the Sport's star, the situation exacerbates.[202] What is happening across the border is beyond the comfort of pleasantries Rodman experiences of the capital, Pyeongyang, and communist history seems to convey a repetition of its strategy of shrouding the evil transpiring in the nation, if one may even call it one.

Yeonmi Park's book, *In Order to Live*, recently stormed western literature through the successful translation to English for the western world. She unfurls the horrors of crossing the Yalu river while constantly suffers starvation, only to be received by

villainous Chinese human trafficking gangs, exploiting Koreans as they seek asylum, healthcare, food, and hope. While losing her father and becoming estranged from her mother and sister in the acrimonious evil of trafficking, Park unfurls the calamities of her life until reaching South Korea (Park, 2016).[203]
As Arnold H. Fang noted of North Korea through Amnesty International in 2016:

'In February 2014, the United Nations Commission of Inquiry on Human Rights in the Democratic People's Republic of Korea (the Commission) issued a 372-page long report documenting violations spanning the full range of human rights, many of which amounted to crimes against humanity.

Among other violations, the report detailed arbitrary detentions, torture, executions, enforced disappearances and political prison camps, violations of the freedom to leave one's own country and of the right to food and related aspects of the right to life. The report found an "almost complete denial of the right to freedom of thought, conscience and religion, as well as of the rights to freedom of opinion, expression, information and association"' (Fang, 2016).

The use of foreign movies, communication with the outside world, enaction of things of the "enemy state" (most specifically South Korea and the U.S.A), along with the upholding of neighbourhood units (*inminbans* that hold mandatory weekly meetings) ensure full submission to both the state and Kim family. The direct ramifications of being found to even 'languidly' submit to the Kim family is hard labour in camps, and even death. As a former resident of the capital told Amnesty:

'*The inminban leader knows the neighbourhood and how many people are in each household. He polices around the neighbourhood. If there are other people staying in the house. He will check if the people have reported to the police office. This was done to find out if there is someone spying. These days, the practice is used also to check if there are illegal electronic appliances being used in the household, including radios and foreign DVDs*' (Fang, 2016, p. 42).

As inconceivable as north of the DMZ may be, most especially as the west was critical of the Soviet ending in the early nineties, yet while acquainting oneself with Orwellian literature, conversely, the south of the DMZ greatly differs.[204]
Sitting almost literally and figuratively equidistant between both China and Japan, having belligerently endured occupation and wars with both countries, Korea champions the remarkable Joseon dynasty that lasted half a millennium, with its dying embers towards latter stages of the nineteenth century. With a two-thousand-year history encompassing the famous three kingdoms of Korea, namely Goguryeo, Baekje, and Silla, Korea (Confucian in nature), became entangled with politics of Soviet expansion and the West, conclusively dividing Korea after a three-year war ending in 1953, where up to 5 million people died.
Since then, the two divided countries bifurcated, as North Korea were subject to tyrannical submission of the Kim family, while South Korea endured a rather perilous road towards the *Hanryu* (the Korean Wave) of the nineties. Albeit critical of Park Chung-Hee's authoritarian and presidential campaign subsequent to the Yushin constitution in 1971,[205] Park was conducive to rapid economic growth, known as *the miracle on the Han River* transpiring after the Korean war, simply turning Korea into its developed country one observes today. Propelled by Summer Olympics in 1988, co-hosting the Fifa World Cup in 2002, and conglomerates such as Samsung, LG, and Hyundai technologically rising, Korea economically paved its

The Epistemological Quest

way into the *Hanryu*, in which K-pop, K-drama and Kimchi popularised the nation into a Korea of today (Asakura, 2016; Bae, 2009).
Notwithstanding the ever-prevalent wonders of South Korea, like any nation, its downfalls have a certain unique and philosophical aftertaste I do believe the world can learn from. As Confucius says best in his Analects:

'The Master said, 'If one learns from others but does not think, one will be bewildered. If, on the other hand, one thinks but does not learn from others, one will be in peril' (Confucius, 1979, p. 65).

The peril in this case is that as of 2020, the OECD reports 24.2 out of 100,000 commit suicide in South Korea, compared to 20.3 of Lithuania coming in second, Slovenia third with 15.7, and Estonia fourth with 15.2 out of 100,000 people ending their own lives (OECD, 2020). That's already an 8.7 difference between first and fourth place out of 100,000. As with the case of suicide, attributable to suicide may be endogenous to the practices of therapists, in as much as they display such unbiased competence. But as far as Frankl's logotherapeutic analyses go, a breakdown of meaning with the future can be hard hit in such a laborious and classist nation that sports South Korea.

The *Li* and the *Ren*

Confucius saw there were two principal virtues upon the onus of the citizen within a society, two virtues he saw as the *Li*, or the *Ren*. Without a clear translation into the English language, the *Li*, one would purport to be specifically deontological in its essence, instantiated upon the personal conduct of the citizen, the duty to conduct himself in such a way that one may call propriety in the Western world. The *Li* was more than propriety in its sense, as Dubs said quite succinctly, '*Li* included matters of politeness, court etiquette, religious ritual, governmental practices, and the state constitution, codes of conduct, and ethical principles' (Dubs, 1951, p. 33).
The *Ren*, however, found itself on the onus of individual virtue, and also used by the founder of the state Lu, the Duke of Jou who described himself as graciously kind to his subordinates, was so much more akin to what we would certainly reveal as humanity, or humanitarian. That would be a sense of graciousness, kindness, and love, within reciprocity you would hope from such counterparts, being the highest of human conduct for Confucius (Dubs, 1951).
It's believed the founding of Korea to be as long ago as 2333 BC by heavenly design, as it was the Lord who sent his son, Hwanung, down to earth, descending to the Paektu mountain in which he would encounter a woman, previously a bear, to wed. She would beget a child, Dangun, who would engage in the early example of nation-building, which would go on to become the long-lasting Korea that exists today, notwithstanding its division.
It seems, albeit being a historically rooted Confucian society,[206] that Korea was pulled in many directions by supranational influence from China, Japan, and ultimately the U.S.A, given its interwoven political relationship and ongoing tension with North Korea. But what of the *action* can be achieved without more tensions? The British journalist, Tim Marshall, said best:

'All the actors in this East Asian drama know that if they try to force an answer to the question at the wrong time, they risk making things worse. A lot worse. It is not unreasonable to fear that you would end up with two capital cities in smoking ruins, a civil war, a humanitarian catastrophe, missiles landing in and around Tokyo and

another Chinese/American military face-off on a divided peninsula in which one side has nuclear weapons. If North Korea implodes, it might as well also explode, projecting instability across the borders in the form of war, terrorism and and/or a flood of refugees, and so the actors are stuck. And so the solution is left to the next generation of leaders, and then the next one.
If world leaders even speak openly about preparing for the day when North Korea collapses, they risk hastening that day; and as no one has planned for it – best keep quiet. Catch-22' (Marshall, 2015, p. 217).

It's estimated 10,000 artillery weapons lie north of the DMZ. The weapons are located on the hills, while dug in and fortified by bunkers and caves as the South Korean capital, Seoul, lies a short 35 miles south of the DMZ, as opposed to Pyeongyang, North Korea's distant capital. It's evident that the harm North Korea could inflict upon the dense population of Seoul and its areas would lead to gratuitous involvement from neighbours and the U.S.A. As Marshall puts it, given what transpires with the Kim family and its plethora of possible human rights violations (which would be subsequently acted upon on capturing North Korea), if North Korea is willing to implode, then it might as well do as much damage as it possibly can, which was also akin to Hitler's obstinacy towards the latter stages of the second world war.[207]
While Korea both quickly and exponentially pulled itself up from the lesser developed country it was, all while germane to its philosophy of *danil minjok*,[208] it took with it classist ramifications that daubs the sovereignty of the individual we find in many of the other less developed countries of the world. That being academic education, materialism, and familial ties as ontically valuable in Korean society.[209]
Through classism, as both age and position call for the traditional Korean bow, the sovereign of each individual seems to dissipate on enumeration, as one of the approximately fifty-one million championing a rather dense population of the peninsula subsides in the collectivist world.
And so with such staggering suicide rates of a developed country such as South Korea, one is only led to inquisitiveness. Causes are classically biosocial. Many will ascribe dark reasons of suicide to another, albeit it seems throughout Korea the issue of *relative privation* most certainly ensues. Like many places we encounter in the world, one can have gratitude of wielding the bare necessities of life, and be comfortably safe of Maslow's hierarchy of needs, but once such necessities are achieved, one will, through the *environment*, encounter social others with a proclivity to compare one's assets with those within the vicinity of such an environment. One would certainly be compelled to have an effectual understanding of resources others hold in an environment if one wishes to successfully pair-bond, bequeath sufficient resources onto offspring, and so on, as the forces of sexual selection are at play.
Glands produce endocrine and neurotransmission of biosocial responses to one's own ontological position through the comparison of one's own status to the rest in an environment. This could be via the secretion of dopamine, GABA, serotonin, or the other plethora of peptides and neurotransmitters to choose from.[210] As *meaning* in a classist South Korea inscribes itself upon its materials, founded on financial muscle and status, those extraneous to such success become ever more susceptible to meaninglessness,[211] coerced into long working hours to offset misfortune.[212]
As part III of this book concerns, South Korea, like many countries, distils an insidious form of racism to its own unpalatable essence, at least, unpalatable for me. With high surveillance and stringent immigration laws rendering citizens intransigent, violent acts end up with, even in as little as a fortnight, one receiving a

surprise phone call from the National Police Agency adumbrating compensation to his victim.
So, of *sins of commission*, one would not expect to see acts such as hate crime, but, in as far as hate crime does not tax the victim, only the fatuous would purport racism to not exist. One must return to *sins of omission* to understand how racism wields itself in a nation that changed so rapidly over the last eighty years.
With abrupt changes, one finds stark differences within generations (in as much as generations are similar), and so what women did and still often endure is now disclosed by women reaching tricenarian years. The writer, Cho Nam Joo, a former scriptwriter for TV programs who wrote the international bestselling fiction novel *Kim Jiyoung, Born 1982*, accentuated such issues. Cho,, along with other issues, picks out the compulsory mandate of new-borns taking the patriarch's last name under the *hoju* system:

'In the late 1990s, the dispute over the *hoju* system (the traditional family registration scheme, in which all members of a family must be registered under the patriarch) began in earnest with the emergence of organisations arguing for its abolition. Some people publicly used both of their parents' surnames, and a few celebrities revealed their painful childhood memories of being picked on for having a different family name to their fathers. At the time, a very popular TV show about a single mother at risk of losing custody of her child, whom she'd been raising all on her own, to a deadbeat dad taught Jiyoung about the absurdity of the *hoju* system. But there were still those who thought its abolition would turn blood relations into strangers and make Korean society savage' (Cho, 2020, p. 118).

The *hoju* system, like other Korean practices, is outdated,. Thus, as Cho points out, the *hoju* system was abolished in 2008 as it did not conform with the South Korean constitution's gender equality clause in February 2005. Be that as it may, the matter of the *hoju* system being merely obsolete should not run any further than it needs to, as through evident technological and scientific changes over the last century, which can be quite easily extended to two, has witnessed significant changes in which men would slowly be freed from conscription to war, and opportunities away from the first and second economic sectors. The frustration for women in recent times has been a sentiment of both the lack of need for men, whether true or untrue, and some men exploiting such a position.
I often claim the death penalty, still officially legal in South Korea, albeit *de facto* abolished, is also a great yardstick seeing reason play its course within a nation.[213] As well as literature, I have discussed this issue with thousands of students and have come to find that, irrespective of political inclination, it seemed to be the most ethically conflicted topic. This is only when given the worst types of crimes (the Socratic method), including a number of killed victims, child sex offence, infanticide, parricide, dismemberment, and so on, given Asia as a continent that continues to gravitate towards the death penalty:

'The practice of the death penalty remains most entrenched in Asia, where more than 90 percent of the world's judicial executions take place' (Bae, 2009, p. 2).

An End

As we have sailed the epistemological flagship and hopefully progressed towards a better approach to knowledge, it may suffice to begin with the tenet that one quite easily can learn or know, and so safeguarding logic and an eye to cognitive biases

The Epistemological Quest

may be a solution for a populace. Thus one finds that in day-to-day conversations today logic has a more than likely tendency to slip, rendering listeners having to consult with others and themselves by often turning to the question, *'How does he/she know?'*, attributable to what may be akin to the commission of the legions of logical fallacies one would claim is committed far too often:

- "My cousin is a scientist and he told me so that's how I know". *Appeal to authority*
- "It's all made from natural ingredients so that's why it will be better for you". *Appeal to nature*
- "My grandmother had never received so much as an injection in her life and she's 97 years old!". *Anecdotal*
- "The next thing you know, I'll be able to identify myself as a unicorn". *Slippery slope*
- "We know this because the word of God is infallible, he *is* what's necessary in all possible worlds". *Begging the question*

One can only train oneself in becoming the competent sceptic against the deluge of knowledge and information we continuously endure from the toboggan of sciolism today.[214] An approach to knowledge that may alleviate oneself in order to cooperate is thus intricate, as one is seldom willing to forego one's principles in the event that a better one takes place, which I believe to be akin to the metaphysics of being, ontology. As per the fate of the fallible cousin of the chimp, hardwired with the ACTG molecules driving us teleologically to our determination, a brief quest of sexual selection, morality, and racism was pending. We have thus been only left with the burden of finding our own way, which some purport to be achieved by asking ourselves the conundrum that has troubled philosophers for time immemorial, *how do you know that you know that you know you are knowing?*

Bibliography

Adler, A. (2010). *Understanding human nature*. Martino Publishing.
Agius, C. M., Blundell, R. (2007). What are stem cells? – Review paper. *International Journal of Molecular Medicine and Advance Sciences*, *3*(4), 145-150.
Ahmad, A., Tucker, M. (2021). *Euro 2020 racism: File on 4 confronts Saka troll*. BBC News. https://www.bbc.com/news/uk-58466849
Alberts, B., Johnson, A., Lewis, J., et al. (2002). *Molecular biology of the cell*. 4th edition. New York: Garland Science.
https://www.ncbi.nlm.nih.gov/books/NBK26842/
Alexopoulos, G. (2005). Amnesty 1945: The revolving door of Stalin's Gulag. *Slavic Review*, *64*(2), 274-306. https://doi.org/10.2307/3649985
Allen, C. (2021). *Football without borders*. Twenty9 Publishing.
Andrews, K. (2015). Why Britain's schools are failing to tackle racism. The Guardian. https://www.theguardian.com/commentisfree/2015/aug/12/racism-schools-government-reforms-targets
Angelou, M. (2009). *I know why the caged birds sings*. Ballantine Books.
Anti-Defamation League. (2016). *Tattered robes: The state of the Ku Klux Klan in the United States*. Anti-Defamation League.
https://www.adl.org/sites/default/files/documents/assets/pdf/combating-hate/tattered-robes-state-of-kkk-2016.pdf
Aristotle. (2009). *The Nicomachean ethics*. Oxford World's Classics.
Aronson, J. K. (2015). Five types of skepticism. *BMJ Clinical Research*, 350. http://dx.doi.org/10.1136/bmj.h1986
Asakura, T. (2016). Cultural heritage in Korea – from a Japanese perspective. *Reconsidering Cultural Heritage in East Asia*, In A. Matsuda & L. E. Mengoni (Eds.), *East Asia* (pp. 103-120). Ubiquity Press.
Attenborough, D. (2020). *A life on our planet: My witness statement and a vision for the future*. Witness Books.
Aurelius, M. (2020). *Meditations*. Macmillan Collector's Library.
Austen, J. (2014). *Pride and prejudice*. Penguin Classics.
Bae, S. (2009). South Korea's de facto abolition of the death penalty. *Pacific affairs*, *82*(3), 407-425. https://doi.org/10.5509/2009823407
Bahcall, N. A. (2015). Hubble's Law and the expanding universe. *Department of Astrophysical Sciences*, *112*(11), 3173-3175.
https://doi.org/10.1073/pnas.1424299112
Baluch, P. (2011). *Incomplete metamorphosis has three stages: Egg, nymph, and adult*. Arizona State University School of Life Sciences Ask A Biologist.
https://askabiologist.asu.edu/incomplete-metamorphosis
Barrowclough G.F., Cracraft J, Klicka J, Zink R.M. (2016). How Many Kinds of Birds Are There and Why Does It Matter?. *PLoS ONE, 11*(11).
https://doi.org/10.1371/journal.pone.0166307
Beard, M. (2016). *SPQR: A history of ancient Rome*. Profile Books.
Beevor, A. (2016). *Ardennes 1944*. Penguin Books.
Bennett, C. (1993). *We the Americans: Blacks*. U.S. Department of Commerce: Economics and Statistics Administration.
Bialey, S., Loeb, A. (2018). Could solar radiation pressure explain 'oumuamua's peculiar acceleration?. *The Astrophysical Journal Letters, 868*(1).
http://dx.doi.org/10.3847/2041-8213/aaeda8
Ali, M. B., Sudiman, M. S. A. S. B. (2016). *Salafis and Whahhabis: Two sides of the same coin?*. Nanyang Technological University. https://www.rsis.edu.sg/wp-content/uploads/2016/10/CO16254.pdf

Blackburn, D. (1999). Viviparity and oviparity: Evolution and reproductive strategies. *Encyclopedia of Reproduction, 4*, 994-1003.
Blake, J. W. (1950). The study of African history. *Transactions of the Royal Historical Society, 32*, 49-69. https://doi.org/10.2307/3678477
Blaustein, A. R. (1981). Sexual selection and mammalian olfaction. *The American Naturalist, 117*(6), 1006-1010.
Bloom, T. (2018). *Windrush generation latest to be stripped of their rights in the name of 'migration control'*. The Conversation. https://theconversation.com/windrush-generation-latest-to-be-stripped-of-their-rights-in-the-name-of-migration-control-95158
BMG Research. (2017). *BMG research poll: Two-thirds of people don't read political manifestos*. BMG Research. https://www.bmgresearch.co.uk/bmg-research-poll-10-people-dont-know-manifesto/
Borgmann, A. (1999). Gender, nature, and fidelity. *Ethics and the Environment, 4*(2), 131-142. https://doi.org/10.1016/s1085-6633(00)88416-8
Brain, C. K., Prave, A. R., Hoffman, K., Fallick, A. E., Botha, A., Herd, D. A., Sturrock, C., Young, I., Condon, D. J., Allison, S. G. (2012). The first animals: ca. 760-million-year-old sponge-like fossils from Namibia. *South African Journal of Science, 108*(1-2), 83-90. https://doi.org/10.4102/sajs.v108i1/2.658
Brainard, J. (2021). Open access takes flight. *Science, 371*(6524), 16-20. https://doi.org/10.1126/science.371.6524.16
Brinkbäumer, K., Höges, C. (2006). *The voyage of the Vizcaína: The mystery of Christopher Columbus's last ship*. Harcourt.
Brown, B. (2015). *Daring greatly*. Penguin Life
Brown, B. (2018). *Dare to lead*. Penguin Random House: Vermillion
Brown, L. (2020). *BLM mob beats white man unconscious after making him crash truck: video*. New York Post. https://nypost.com/2020/08/17/blm-mob-beat-white-man-unconscious-after-making-him-crash-truck/
Buis, A. (2020). *Milankovitch (orbital) cycles and their role in earth's climate*. NASA: Global Climate Change. https://climate.nasa.gov/news/2948/milankovitch-orbital-cycles-and-their-role-in-earths-climate/
Burke, D., Nolan, C., Hayward, W. G., Russel, R. (2013). Is there an own-race preference in attractiveness?. *Evolutionary Psychology, 11*(4), 855-872. https://doi.org/10.1177/147470491301100410
Cangiano, A. (2021). *The impact of migration on UK population growth*. The University of Oxford: The Migration Observatory. https://migrationobservatory.ox.ac.uk/resources/briefings/the-impact-of-migration-on-uk-population-growth/
Carlisle, E. M., Jobbins, M., Pankhania, V., Cunningham, J. A., Donoghue, P. C. J. (2021). Experimental taphonomy of organelles and the fossil record of early eukaryote evolution. *Science Advances, 7*(5). https://doi.org/10.1126/sciadv.abe9487
Chang, R. L. (1999). *A gesture life*. Riverhead books.
Chaudhry, R. Bordoni, B. (2022). Anatomy, thorax, lungs. In *StatPearls*. StatPearls Publishing. https://www.ncbi.nlm.nih.gov/books/NBK470197/
Cho, N. J. (2020). *Kim Jiyoung, born 1982*. Simon & Schuster: Scribner.
Chu, J. (2021). *Physicists observationally confirm Hawking's black hole theorem for the first time*. Phys.org. https://phys.org/news/2021-07-physicists-observationally-hawking-black-hole.html
Clark, R.D., Hatfield, E. (1989). Gender differences in receptivity to sexual offers. *Journal of psychology & Human Sexuality, 2*(1), 39-54. https://doi.org/10.1300/j056v02n01_04

Clark, S. (2008). Deconstructing the laws of logic. *Philosophy, 83*(1), 25-53. https://doi.org/10.1017/s0031819108000296
Colescott, R. (1987). *Knowledge of the past is the key to the future: Upside down Jesus and the politics of survival.* Portland Art Museum, Portland, OR, United States. http://portlandartmuseum.us/mwebcgi/mweb.exe?request=record;id=25718;type=101
Confucius. (1979). *The Analects.* Penguin Classics.
Corbey, R. (2012). "Homo habilis's" humanness: Phillip Tobias as a philosopher. *History and Philosophy of the Life Sciences, 34*(1-2), 103-116.
Darwin, C. (2017). *On the origin of species.* Macmillan Collector's Library.
Darwin, C. (2017). *The descent of man and selection in relation to sex.* CreateSpace Independent Publishing Platform.
Dawkins, R. (2016). *The extended phenotype.* Oxford Landmark Science.
Dawkins, R. (2016) *The God delusion.* Black Swan.
Dawkins, R. (2016). *The selfish gene.* Oxford Landmark Science.
Descartes, R. (2008). *Meditations on first philosophy.* Oxford World's Classics.
Diamond, J. (1993). *The third chimpanzee: The evolution and future of the human animal.* Harper Perennial.
Diamond, J. (2017). *Guns, germs, and steel: The fates of human societies.* W.W. Norton & Company.
Doležal, S. (2019). The political and military aspects of accession of Constantine the great. *Graeco-Latina Brunensia,* (2), 19-32. https://doi.org/10.5817/glb2019-2-2
Dominey, W. J. (1984). Alternative mating tactics and evolutionarily stable strategies. *American Zoologist, 24*(2), 385-396. https://doi.org/10.1093/icb/24.2.385
dos Reis, M., Donoghue, P., Yang, Z. (2016). Bayesian molecular clock dating of species divergences in the genomics era. *Nature Review Genetics, 17*(2), 71-80. https://doi.org/10.1038/nrg.2015.8
Dostoyevsky, F. (1991). *Crime and punishment.* Penguin Classics.
Dou, G. [ca. 1655-1657]. *An evening school.* The Metropolitan Museum of Art, New York, NY, United States. https://www.metmuseum.org/art/collection/search/436209
Douglass, F. (1995). *Narrative of the life of Frederick Douglass.* Dover Thrift Editions.
Dubs, H. H. (1951). Confucius: His life and teaching. *Philosophy, 26*(96), 30-36. https://doi.org/10.1017/S0031819100019185
Dumas, A. (1997). *The count of Monte Cristo.* Wordsworth Editions Limited.
Dwyer, P. G. (2010). Napoleon and the foundation of the empire. *The Historical Journal, 53*(2), 339-358. https://doi.org/10.1017/s0018246x1000004x
Earman, J. (1993). Bayes, Hume, and Miracles. *Journal of the Society, 10*(3), 293-310. https://doi.org/10.5840/faithphil19931039
Eddo-Lodge, R. (2018). *Why I'm no longer talking to white people about race.* Bloomsbury.
Effingham, N. (2018). *To argue with flat earthers, use philosophy not science.* Quartz. https://qz.com/1264453/to-argue-with-flat-earthers-use-philosophy-not-science/amp
Elder, L. (2020). Preface. In C. Owens, *Blackout* (p. xv). Threshold Editions Hardback.
Eurostat. (2021) *Population projected to decline in two-thirds of EU regions.* Eurostat. https://ec.europa.eu/eurostat/web/products-eurostat-news/-/ddn-20210430-2

Fadelli, I. (2021). *Researchers observe stationary Hawking radiation in an analog black hole*. Phys.org. https://phys.org/news/2021-02-stationary-hawking-analog-black-hole.html
Fang, A. H. (2016). *Connection denied: Restrictions on mobile phones and outside information in North Korea*. Amnesty International. http://dx.doi.org/10.13140/RG.2.1.1327.5126
Farlow, J. O., Dodson, P., Chinsamy, A. (1995). Dinosaur Biology. *Annual Review of Ecology and Systematics, 26*(1), 445-471. https://doi.org/10.1146/annurev.es.26.110195.002305
Fisher, R. A. (2018). *The genetical theory of natural selection*. Franklin Classics.
Forgan, D. Rice, W. K. (2010). Numerical testing of the Rare Earth Hypothesis using Monte Carlo realization techniques. *International Journal of Astrobiology, 9*(2), 73-80. http://dx.doi.org/10.1017/S1473550410000030
Frank, A. (1993). *The diary of a young girl*. Bantam Edition.
Frankl, V. E. (2008). *Man's search for meaning*. Rider.
Freud, S. (1997). *The interpretation of dreams*. Wordsworth Editions Limited.
Gannon, F. (2001). The essential role of peer review. *EMBO reports, 2*(9), 743-743. https://doi.org/10.1093/embo-reports/kve188
Gates, B. (2021). *How to avoid a climate disaster: The solutions we have and the breakthroughs we need*. Alfred A. Knopf.
Goodall, J. (2010). *In the shadow of man*. Mariner books.
Gov.uk. (2019). *Population of England and Wales*. Gov.uk. https://www.ethnicity-facts-figures.service.gov.uk/uk-population-by-ethnicity/national-and-regional-populations/population-of-england-and-wales/latest#full-page-history
Gramlich, J. (2019). *The gap between the number of blacks and whites in prison is shrinking*. Pew Research Center. https://www.pewresearch.org/fact-tank/2019/04/30/shrinking-gap-between-number-of-blacks-and-whites-in-prison/
Grant, M. (2020). *Gravitational wave echoes may confirm Stephen Hawking's hypothesis of quantum black holes*. Phys.org. https://phys.org/news/2020-01-gravitational-echoes-stephen-hawking-hypothesis.html
Gregg, T. (2007). Intelligent design: Jonathan Wells and the tree of life. *Journal of College Science Teaching, 36*(7), 10-11.
Gualtieri, L., Brigham-Grette, J. (2001). The age and origin of the little Diomede island upland surface. *Arctic, 54*(1), 12-21. https://doi.org/10.14430/arctic759
Guindon, M. H., Green, A. G., Hanna, F. (2003). Intolerance and psychopathology: Toward a general diagnosis for racism, sexism, and homophobia. *American Journal of Orthopsychiatry, 73*(2), 167-76. https://doi.org/10.1037/0002-9432.73.2.167
Haggard, P. (2008). Human volition: towards a neuroscience of free will. *Nature Reviews Neuroscience, 9*(12), 934-946. https://doi.org/10.1038/nrn2497
Haggard, P., Clark, S., Kalogeras, J. (2002). Voluntary action and conscious awareness. *Nature Neuroscience, 5*(4), 382-385. https://doi.org/10.1038/nn827
Hale, B. (2012). What is Absolute Necessity?. *Philosophia Scientiæ*, (16-2), 117-148. https://doi.org/10.4000/philosophiascientiae.743
Hamilton, W. D., Zuk, M. (1982). Heritable true fitness and bright birds: A role for parasites?. *Science, 218*(4570), 384-387. https://doi.org/10.1126/science.7123238
Haney, C., Banks, C., Zimbardo, P. (1973). Interpersonal dynamics in a simulated prison. *International Journal of Criminology and Penology*, 1, 69-97. https://doi.org/10.21236/ad0751041
Harari, Y. N. (2015). *Sapiens: A brief history of humankind*. Vintage.
Harari, Y. N. (2017). *Homo deus: A brief history of tomorrow*. Vintage.
Harari, Y. N. (2018). *21 lessons for the 21st century*. Random House.

Harari, Y. N. (2020). *Yuval Noah Harari: 'Will coronavirus change our attitudes to death? Quite the opposite'.* The Guardian. https://www.theguardian.com/books/2020/apr/20/yuval-noah-harari-will-coronavirus-change-our-attitudes-to-death-quite-the-opposite
Harris, S. (2012). *The moral landscape.* Black Swan.
Hatch, R. A. (1998). *Correspondence Networks.* The Scientific Revolution. http://users.clas.ufl.edu/ufhatch/pages/03-Sci-Rev/SCI-REV-Home/resource-ref-read/correspond-net/08sr-crrsp.htm
Haug, C., Haug, J.T. (2017). The presumed oldest flying insect: More likely a myriapod?. *PeerJ, 5*, e3402 https://doi.org/10.7717%2Fpeerj.3402
Hauserman, S. (2013). *Thomas Henry Huxley (1825-1895).* The Embryo Project Encyclopedia. https://embryo.asu.edu/pages/thomas-henry-huxley-1825-1895
Hawking, S. (2016). *A brief history of time: From the big bang to black holes.* Penguin Random House.
Hawking, S. (2018). *Brief answers to the big questions.* John Murray.
Herbert, J. (1994). Oxytocin and sexual behaviour: A new chapter in psychopharmacology may be beginning. *BMJ: British Medical Journal, 309*(6959), 891-892. https://doi.org/10.1136/bmj.309.6959.891
Herndon, A. W., Glueck, K. (2021). *Biden apologizes for saying black voters 'ain't black' if they're considering Trump.* The New York Times. https://www.nytimes.com/2020/05/22/us/politics/joe-biden-black-breakfast-club.html
Hobbes, T. (2008). *Leviathan.* Oxford University Press.
Hinton, A. R. (2018). *The sun does shine: How I found life and freedom on death row.* St. Martin's Press.
Hirsch, A. (2020). *Racist responses to Marcus Rashford's campaign for children are no surprise.* The Guardian. https://www.theguardian.com/commentisfree/2020/sep/09/racist-responses-to-marcus-rashfords-campaign-for-children-are-no-surprise
Holze, H., Schrader, L., Buellesbach, J. (2021). Advances in deciphering the genetic basis of insect cuticular hydrocarbon biosynthesis and variation. *Heredity, 126*(2), 219-234. https://doi.org/10.1038/s41437-020-00380-y
Horwitz, S., Wade, M. (2021). *Regulation and the perpetuation of poverty in the US and Senegal.* The Center for Growth and Opportunity at Utah State University. https://www.thecgo.org/books/regulation-and-economic-opportunity-blueprints-for-reform/regulation-and-the-perpetuation-of-poverty-in-the-us-and-senegal/
Huber, R., Smith, K., Delago, A., Isaksson, K., Kravitz, E. A. (1997). Serotonin and aggressive motivation in crustaceans: Altering the decision to retreat. *Proceedings of the National Academy of Sciences of the United States of America, 94*(11), 5939-5942. https://doi.org/10.1073/pnas.94.11.5939
Jablonski, N. G., Chaplin, G. (2010). Human skin pigmentation as an adaptation to UV radiation. *Proceedings of the National Academy of Sciences of the United States of America, 107*(2), 8962-8968. https://doi.org/10.1073/pnas.0914628107
Jefferson, T. (1997). *Declaration of independence.* Applewood Books.
Joel, D., Berman, Z., Tavor, I., Wexler, N., Gaber, O., Stein, Y., Shefi, N., Pool J., Urchs, S., Margulies, D. S., Liem, F., Hänggi, J., Jancke, L., Assaf, Y. (2015). Sex beyond the genitalia: The human brain mosaic. *The Rockefeller University, 112*(50), 15468-15473. https://doi.org/10.1073/pnas.1509654112
Jung, C. G., Von Franz, M. L., Henderson, J. L., Jacobi, J., Jaffé, A. (1968). *Man and his symbols.* Dell Publishing.
Kant, I. (2007). *Critique of pure reason.* Penguin Classics.
Kant, I. (2021). *Metaphysics of Morals.* Anastic Press Edition.

Kids Count Data Center. (2019). *Children in single-parent families by race and ethnicity in the United States*. Kids Count Data Center. https://datacenter.kidscount.org/data/tables/107-children-in-single-parent-families-by-race-and-ethnicity?loc=1&loct=2#detailed/1/any/false/1729/10,11,9,12,1,185,13/431
Kiyosaki, R. (2017). *Rich dad poor dad*. Plata Publishing.
Knoll, A. H., Javaux, E.J., Hewitt, D., Cohen, P. (2006). Eukaryotic organisms in Proterozoic oceans. *Philosophical Transactions: Biological Sciences, 361*(1470), 1023-1038. https://doi.org/10.1098/rstb.2006.1843
Kravitz, G. (2014). The geohistorical time arrow: From Steno's stratigraphic principles to Boltzmann's past hypothesis. *Journal of Geoscience Education, 62*(4), 691-700. https://doi.org/10.5408/13-107.1
Lacourse Jr, R. D. (1997). *Three strikes, you're out: A review*. https://www.washingtonpolicy.org/publications/detail/three-strikes-youre-out-a-review
Laine, P. E. (2016). The role of cosmic rays in the origin of life. Conference: Evolution of Chemical Complexity: From simple interstellar molecules to terrestrial: http://dx.doi.org/10.13140/RG.2.2.20794.59848
Lehtonen, J., Okasha, S., Helanterä, H. (2020). Fifty years of the Price equation. *Philosophical Transactions the Royal Society Publishing, 375*(1797), 20190350. http://dx.doi.org/10.1098/rstb.2019.0350
Lewis, C. L. E. (2001). Arthur Holmes' vision of a geological timescale. *Geological Society, London, Special Publications, 190*(1), 121-138. https://doi.org/10.1144/gsl.sp.2001.190.01.10
Library of Congress. (2021). *General Information*. https://www.loc.gov/about/general-information/#year-at-a-glance
Iida, A., Arai, H. N., Someya, Y., Inokuchi, M., Onuma, T. A. Yokoi, H., Suzuki, T., Hondo., E., Sano., K. (2019). Mother-to-embryo vitellogenin transport in a viviparous teleost Xenotoca eiseni. *The Jackson Laboratory, 116*(44), 22359-22365. https://doi.org/10.1073/pnas.1913012116
Lieberman, P. (1986). On Bickerton's review of the biology and evolution of language. *American Anthropologist, 88*(3), 701-703.
Limbong, A. (2020). *Fired over anti-Semitic comments, Nick Cannon wants 'wild 'n out' ownership*. NPR. https://www.npr.org/sections/live-updates-protests-for-racial-justice/2020/07/15/891422809/fired-over-anti-semitic-comments-nick-cannon-wants-wild-n-out-ownership
Lipka, M. (2014). *The Sunni-Shia divide: Where they live, what they believe and how they view each other*. Pew Research Center. https://www.pewresearch.org/fact-tank/2014/06/18/the-sunni-shia-divide-where-they-live-what-they-believe-and-how-they-view-each-other/
Livingstone, M. S., Harris-Warwick, R. M., Kravitz, E. A. (1980). Serotonin and Octopamine Produce Opposite Postures in Lobsters. *Science, 208*(4439), 76-79. https://doi.org/10.1126/science.208.4439.76
Locke, J. (2014). *An essay concerning human understanding*. Wordsworth Editions Limited.
Lomborg, B. (2006). *How to spend $50 billion to make the world a better place*. Cambridge University Press.
Long, B., Lynch, S., Loehr, G. T., Jacquin, K. (2019). INCELs, misogyny, and mass violence. *Conference: American College of Forensic Psychology*.
Loving v. Virginia, 338 U.S. 1 (1967). https://www.law.cornell.edu/supremecourt/text/388/1
Mahl, D., Zeng, J., Schäfer, M., S. (2021). From "nasa lies" to "reptilian eyes": Mapping communication about 10 conspiracy theories, their communities, and

main propagators on twitter. *Social Media + Society, 7*(2), 205630512110174. https://doi.org/10.1177/20563051211017482
Malan, G. (2016). Mythology, *Weltanschauung*, symbolic universe and states of consciousness. *HTS Theological Studies, 72*(1). https://doi.org/10.4102/hts.v72i1.3243
Mansur, U. [ca. 1610]. *Peafowl*. The Metropolitan Museum of Art, New York, NY, United States. https://www.metmuseum.org/art/collection/search/76023
Mardlife. (2018). *Not my business – Niyi Osundare*. Mardlife. https://www.mardlife.com/not-my-business-niyi-osundare-a-poem/
Marshall, M. (2020). *Charles Darwin's hunch about early life was probably right*. BBC. https://www.bbc.com/future/article/20201110-charles-darwin-early-life-theory
Marshall, T. (2015). *Prisoners of geography: Ten maps that tell you everything you need to know about global politics*. Elliot and Thompson Limited.
Marx, K., Engels, F. (2015). *The communist manifesto*. Penguin Books Ltd.
Maslow, A. H. (1943). A theory of human motivation. *Psychological Review*, 50, 370-396.
Mathews, G. A., Fane, B. A., Conway, G. S., Brook, C. G. D., Hines, M. (2009). Personality and congenital adrenal hyperplasia: Possible effects of prenatal androgen exposure. *Hormones and behavior, 55*(2), 285-291. https://doi.org/10.1016/j.yhbeh.2008.11.007
Matzel, L. D., Sauce, B. (2017). IQ (The intelligence quotient). In J. Vonk & T.K. Shackelford (Eds.), *Encyclopedia of Animal Cognition and Behavior* (pp. 1-9). Springer International.
Mauer, M. (2018). *Long-term sentences: Time to reconsider the scale of punishment*. The Sentencing Project. https://www.sentencingproject.org/reports/long-term-sentences-time-to-reconsider-the-scale-of-punishment/
Maynard Smith, J. (1982). *Evolution and the theory of games*. Cambridge University Press.
Mayr, E. (1992). The idea of teleology. *Journal of the History of Ideas, 53*(1), 117-135. https://doi.org/10.2307/2709913
McPhee, J. (2000). *Annals of the former world*. Farrar, Straus and Giroux.
Menkhorst, E., Nation, A., Cui, S. Selwood, L. (2009). Evolution of the shell coat and yolk in amniotes: A marsupial perspective. *Journal of Experimental Zoology, 312B*(6), 625-638. https://doi.org/10.1002/jez.b.21235
Michod, R. E. (2005). John Maynard Smith. *Annual Review of Genetics, 39*(1), 1-8. https://doi.org/10.1146/annurev.genet.39.040505.114723
Mileham, R. (2020). *First Animals*. Oxford University Museum of Natural History Research Team. https://www.oum.ox.ac.uk/firstanimals/
Milgram, S. (1963). Behavioral study of obedience. *The Journal of Abnormal and Social Psychology, 67*(4), 371-378. https://doi.org/10.1037/h0040525
Mill, J. S. (2002). *On liberty*. Dover Thrift Editions.
Mill, J. S., Bentham, J. (2004). *Utilitarianism and other essays*. Penguin Books.
Miller, G. F. (1998). How mate choice shaped human nature: A review of sexual selection and human evolution. In C. B. Crawford & D. L. Krebs (Eds.), *Handbook of evolutionary psychology: Ideas, issues, and applications* (pp. 87–129). Lawrence Erlbaum Associates Publishers.
Morgan, F. (1999). The hexagonal honeycomb conjecture. *Transactions of the American Mathematical Society, 351*(5), 1753-1763. https://doi.org/10.1090/s0002-9947-99-02356-9
Narayan, Y. (2021). *On histories of policing, academic reconstruction and reparation*. Identities: Global Studies in Culture and Power.

Natoire, C. J. (1740). *The rebuke of Adam and Eve.* The Metropolitan Museum of Art, New York, NY, United States.
https://www.metmuseum.org/art/collection/search/437180
Newman, J. H. (2012). *Darwin's star orchid.* NYBG.
https://www.nybg.org/blogs/plant-talk/2012/03/exhibit-news/darwins-garden/darwins-star-orchid/
Nicholson, O. (2000). Constantine's vision of the cross. *Vigiliae Christianae,* 54(3), 309-323. https://doi.org/10.2307/1584644
Nicoll, R. A., Roche, K. W. (2013). Long-term potentiation: Peeling the onion. *Neuropharmacology,* 74, 18-22.
https://doi.org/10.1016/j.neuropharm.2013.02.010
Nietzsche, F. (2018). *Beyond good and evil.* Bookk.
Nowak, M. A., Krakauer, D. C. (1999). The evolution of language. *Proceedings of the National Academy of Sciences,* 96(14), 8028-8033.
https://doi.org/10.1073/pnas.96.14.8028
Obama, B. (2016). *Barack Obama: Dreams from my father.* Canongate.
Obama, M. (2022). *The light we carry.* Penguin Random House.
OECD. (2020). *Suicide rates.* OECD. https://data.oecd.org/healthstat/suicide-rates.htm
Olusoga, D. (2017). *Black and British: A forgotten history.* Pan Books.
Olusoga, D. (2018). *We risk losing slices of our past if we don't root out racism in our universities.* The Guardian.
https://www.theguardian.com/commentisfree/2018/oct/21/we-risk-losing-slices-of-our-history-if-we-dont-root-out-racism-in-our-unversities
Oman, C. W. C. (2008). *The Byzantine empire.* Westholme Publishing.
Ortiz-Ospina, E., Roser, M. (2020). *Marriage and divorces.* Ourworldindata. https://ourworldindata.org/marriages-and-divorces
Orwell, G. (2003). *1984.* Longman: Penguin Readers.
Owens, C. (2020). *Blackout.* Threshold Editions Hardback.Pace v. Alabama, 106 U.S. 583 (1883). https://www.law.cornell.edu/supremecourt/text/106/583
Park, Y. (2016). *In order to live.* Penguin Books.
Parker, A. A. (1937). History of Carlism in Spain. *Studies: An Irish Quarterly Review,* 26(101), 16-25.
Parker, G.A., Lehtonen, J. (2014). Gamete evolution and sperm numbers: Sperm competition versus sperm limitation. *Proceedings of the Royal Society,* 281(1791), 20140836. http://dx.doi.org/10.1098/rspb.2014.0836
Pennant, A., Sigona, N. (2018). *Black history is still largely ignored, 70 years after Empire Windrush reached Britain.* The Conversation.
https://theconversation.com/black-history-is-still-largely-ignored-70-years-after-empire-windrush-reached-britain-98431
Peterson, J. (2019). *12 rules for life.* Random House Canada.
Peterson, J. (2021). *Beyond Order: 12 more rules for life.* Portfolio/Penguin.
Peterson, J., Flanders, J. L. (2005). Play and the regulation of aggression. In R.E. Tremblay, W. W. Hartup & J. Archer (Eds.), *Developmental Origins of Aggression* (pp. 133-157). The Guilford Press.
Piaget, J. (1962). *Play, dreams and imitation in childhood.* The Norton Library.
Pinker, S. (2000). *The language instinct.* Penguin Books Ltd.
Pinker, S. (2012). *The better angels of our nature.* Penguin Group.
Pinker, S. (2019). *Enlightenment now: The case for reason, science, humanism, and progress.* Penguin Books.
Pinker, S. (2021). *Rationality: What it is, why it seems scarce, why it matters.* Viking.
Plato. (2007). *The republic.* Penguin Classics.
Plato. (2012). *The trial and death of Socrates.* BN Publishing.

Pomiankowski, A., Iwasa, Y. (1998). Runaway ornament diversity caused by Fisherian sexual selection. *Proceedings of the National Academy of Sciences, 95*(9), 5106-5111. https://doi.org/10.1073/pnas.95.9.5106

Puts, D. A. (2010). Beauty and the beast: Mechanisms of sexual selection in humans. *Evolution and Human Behavior, 31*(3), 157-175. https://doi.org/10.1016/j.evolhumbehav.2010.02.005

Quigley, C. A., De Bellis, A., Marschke, K., El Awady, M. K. (1995). Androgen receptor defects: Historical, clinical, and molecular perspectives*. *Endocrine Reviews, 16*(3), 271-321. https://doi.org/10.1210/er.16.3.271

Ratnieks, F. L. W., Foster, K. R., Wenseleers, T. (2011). Darwin's special difficulty: The evolution of "neuter insects" and current theory. *Behavioral Ecology and Sociobiology, 65*(3), 481-492. https://doi.org/10.1007/s00265-010-1124-8

Rogers, J. A. (2014). *100 amazing facts about the negro: with complete proof.* University Press of New England.

Roosevelt, E. (2001). *Universal declaration of human rights.* Applewood Books.

Rose, K. D. (1996). On the origin of the order artiodactyla. *Proceedings of the National Academy of Sciences of the United States of America, 93*(4), 1705-1709. https://doi.org/10.1073/pnas.93.4.1705

Roser, M. (2020). *The short history of global living conditions and why it matters that we know it.* Our World in Data. https://ourworldindata.org/a-history-of-global-living-conditions-in-5-charts

Rousseau, J.-J. (1968). *The social contract.* Penguin Books.

Sadam, R. (2021). *Riemann hypothesis: 161-yr-old math mystery Hyderabad physicist is waiting to prove he solved.* The Print. https://theprint.in/science/riemann-hypothesis-161-yr-old-math-mystery-hyderabad-physicist-is-waiting-to-prove-he-solved/692466/

Sands, P. (2016). *East west street.* Weidenfield & Nicolson.

Sapolsky, R. (2018). *Behave: The biology of humans at our best and worst.* Vintage.

Schadewald, R. J. (1980). *"The flat out truth".* Science Digest. In Tfes. http://library.tfes.org/library/the_flat_out_truth.pdf

Schultz, W. (2010). Dopamine signals for reward value and risk: basic and recent data. *Behavioral and Brain Functions, 6*(1), 24. https://doi.org/10.1186/1744-9081-6-24

Science News Staff. (2016). *Nobel honors discoveries on how cells eat themselves.* Science. https://www.science.org/content/article/nobel-honors-discoveries-how-cells-eat-themselves

Shackelford, T. K. (1999). Facial attractiveness and physical health. *Evolution and Human Behavior, 20*(1), 71-76.

Shenton, D. (2009). *"In defense of the flat earth".* In Tfes.org. http://library.tfes.org/library/daniel_shenton_flat_earth_essay.pdf

Shultz, S., Maslin, M. (2013). Early human speciation, brain expansion and dispersal influenced by African climate pulses. *PLoS ONE, 8*(10), e76750. https://doi.org/10.1371/journal.pone.0076750

Solzhenitsyn, A. (2003). *The gulag archipelago 1918-56.* The Harvill Press.

Soon C. S., Brass, M., Heinze, H. Haynes, J. (2008). Unconscious determinants of free decisions in the human brain. *Nature Neuroscience, 11*(5), 543-545. https://doi.org/10.1038/nn.2112

Smith, T. (2017). Methodological naturalism and its misconceptions. *International Journal for Philosophy of Religion, 82*(3), 321-336. http://dx.doi.org/10.1007/s11153-017-9616-3

Smyth, A. (1995). *The Hamilton-Zuk hypothesis: Finding its rightful place among models of the evolution of female mate preferences.* Berkeley. http://ib.berkeley.edu/courses/ib160/past_papers/smyth.html

Snowden, R. (2019). *Jung: The key ideas.* John Murray Learning.
Stowe, H. B. (1995). *Uncle Tom's cabin.* Wordsworth Editions Limited.
Striker, G. (2001). Scepticism as a kind of philosophy. *Archiv für Geschichte der Philosophie, 83*(2), 113-129. https://doi.org/10.1515/agph.83.2.113
Sutterer, M. J., Koscik, T. R., Tranel, D. (2015). Sex-related functional asymmetry of the ventromedial prefrontal cortex in regard to decision-making under risk and ambiguity. *Neuropsychologia, 75,* 265-273. https://doi.org/10.1016/j.neuropsychologia.2015.06.015
Tan, S. Y., Tatsumura, Y. (2015). Alexander Fleming (1881-1955): Discovery of penicillin. *Singapore Medical Journal, 56*(7), 366-367. https://doi.org/10.11622/smedj.2015105
Tassaert, O. [ca. 1850]. *Heaven and Hell.* The Cleveland Museum of Art, Cleveland, OH, United States. https://www.clevelandart.org/art/1980.287
The Holy Bible, New International Version. (2000). The Gideons. (Original work published 1979)
The Holy Quran. (M. M. Ali, Trans.). (2002). Ahmadiyyah Anjum Isha'at Islam Lahore Inc., U.S.
Tory, S., Blankenbuehler, P. (2017). *How private prisons became a booming business.* High Country News. https://www.hcn.org/issues/49.8/how-the-booming-business-of-private-prisons-came-to-be
Trisel, B. A. (2015). Does death give meaning to life?. *Journal of Philosophy of Life, 5*(2), 62-81.
Tuskegee University Archives Repository. (2021). *Lynchings stats year dates causes.* Collection: Lynching Information.
http://archive.tuskegee.edu/repository/wp-content/uploads/2020/11/Lynchings-Stats-Year-Dates-Causes.pdf
Tymowski, M. (2014). Death and attitudes to death at the time of early European expeditions to Africa. *Cahiers d'études africaines, 215,* 787-811. https://doi.org/10.4000/etudesafricaines.17843
Tyrer, D. (2013). *The politics of Islamophobia.* Pluto Press.
Tyson, E. (2021). *Investing for dummies.* John Wiley & Sons Inc.
United Nations. (2015). *Transforming our world: The 2030 Agenda for sustainable development.* https://documents-dds-ny.un.org/doc/UNDOC/GEN/N15/291/89/PDF/N1529189.pdf?OpenElement
van Anders, S. M., Steiger, J., Goldey, K. L. (2015). Effects of gendered behavior on testosterone in women and men. *Proceedings of the National Academy of Sciences of the United States of America, 112*(45), 13805-13810.
https://doi.org/10.1073/pnas.1509591112
Voltaire. (1991). *Candide.* Dover Thrift Editions.
von Goeth, J. W. (1989). *The sorrows of young Werther.* Penguin Classics.
Wallace, A. R. (2022). *Darwinism: An exposition of the theory of natural selection, with some of its applications.* Independent Press.
Wallace, H. (2020). *Men without God: The rise of atheism in Saudi Arabia.* Free Inquiry. https://secularhumanism.org/2020/01/men-without-god-the-rise-of-atheism-in-saudi-arabia/
Warner, B. (2010). *The Hadith: The traditions of Mohammed.* CSPI, LLC.
Watson, J. (2015). *Report finds two-thirds of private prison contracts include "lockup quotas".* Prison Legal News.
https://www.prisonlegalnews.org/news/2015/jul/31/report-finds-two-thirds-private-prison-contracts-include-lockup-quotas/
Weale, S. (2018). *London's black cultural archives get £200,000 stopgap funding for survival.* The Guardian. https://www.theguardian.com/uk-

news/2018/dec/13/london-black-cultural-archives-gets-thousands-stop-gap-funding-for-survival-heritage
Wells, S. (2003). *The great leap.* The Guardian. https://www.theguardian.com/education/2003/jul/03/research.highereducation 1
Wessels, Q., Taylor, A. M. (2017). Anecdotes to the life and times of Richard Owen (1804-1892) in Lancaster. *Journal of Medical Biography, 25*(4), 226-233. https://doi.org/10.1177/0967772015608053
Williams, M. (2014). *What percent of Earth is water?.* Phys.org. https://phys.org/news/2014-12-percent-earth.html
Willink, J., Babin, L. (2017). *Extreme ownership: How U.S. navy seals lead and win.* St. Martin's Press.
Wimmer, G. E., Liu, Y., McNamee, D. C., Dolan, R. J. (2023). Distinct replay signatures for prospective decision-making and memory preservation. *Proceedings of the National Academy of Sciences, 120*(6). https://doi.org/10.1073/pnas.2205211120
Witte, J. (2015). Sex and marriage in the Protestant tradition, 1500-1900. In A. Thatcher (Eds.), *Oxford Handbook on Theology* (pp. 204-223). Oxford University Press.
Woodson, C. G. (2010). *The miseducation of the negro.* CreateSpace Independent Publishing Platform.
Young, L., Camprodon, J. A., Hauser, M., Pascual-Leone, A., Saxe, R. (2010). Disruption of the right temporoparietal junction with transcranial magnetic stimulation reduces the role of beliefs in moral judgements. *Proceedings of the National Academy of Sciences of the United States of America, 107*(15), 6753-6758. https://doi.org/10.1073/pnas.0914826107
Young, M. M. I., Winters, S., Young, C., Weiß, B. M., Troscianko, J., Ganswindt, A., Barrett, L., Henzi, S. P., Hingham, J. P., Widdig, A. (2020). Male characteristics as predictors of genital color and display variation in vervet monkeys. *Behavioral Ecology and Sociobiology, 74*(2). https://doi.org/10.1007/s00265-019-2787-4
Zahavi, A., Zahavi, A. (1999). *The handicap principle: A missing piece of Darwin's puzzle.* Oxford University Press.
Zahavi, A. (1981). Natural selection, sexual selection and the selection of signals. *Evolution Today,* 133-138.
Zimmerman, S., Ryan, L., Duriesmith, D. (2018). Who are Incels? Recognizing the violent extremist ideology of 'Incels'. *Women, peace and security + Gender, peace and security.*
Zou, Yawen. (2014). *Charles Darwin's theory of pangenesis.* Embryo Project Encyclopedia. https://embryo.asu.edu/pages/charles-darwins-theory-pangenesis
Zúquete, J. P. (2008). The missionary politics of Hugo Chávez. *Latin American Politics and Society, 50*(1), 91-121. https://doi.org/10.1111/j.1548-2456.2008.00005.x

Notes

On an Introduction to Knowledge
Chapter 1: ἐπιστήμη

[1] Necessary: As in true in all possible worlds.

[2] Which can be found in the 'Note on the Text and Translation' of Michael Moriarty's edition.

[3] Also a much more distinguished contemporary of Descartes at the time of his work being published.

[4] I wouldn't wish of being guilty of the type of languor that would render me accused of false cause reasoning, as the process of secularisation could be attributable to amalgams upon amalgams of causes, but as the historical consensus shows, and until a more prevalent cause can be attributed, the enlightenment prevails.

[5] The issue is most pertinent to fallaciously *appealing to authority* that many will naturally hold the proclivity to do. If one were to successfully appeal to authority, for what it's worth, I'm sure one would rather see information backed by a European university such as Cambridge or Oxford, than any university in Africa for that matter. Given the history between Europe and Africa in the latter half of the millennium, one would be foolish not to see the amenability of the African populace to step back from such authority, rendering them more amenable to the variegating conspiracy theories to choose from.

[6] I will elucidate further on exemplary limits of free will when looking at *intraspecific influence* in Part I of Sexual Selection.

[7] Compensatory in that we are truly inferior animals without the ability to cooperate socially. I touch on Alfred Adler's notion of the *inferiority complex* in part II and part III.

[8] Thus meddling with voluntariness within free will.

[9] Some of those thinkers across the enlightenment were Thomas Hobbes, John Locke, Montesquieu, Voltaire, Jean-Jacques Rousseau, Marquis de Condorcet, Mary Wollstonecraft, Jeremy Bentham, Immanuel Kant, David Hume, Adam Smith and many more (Pinker, 2019).

[10] As opposed to the myriad tools that have been used across history. This has been both an abstract and loaded claim concerning the enlightenment. Albeit reason may have been used, one could certainly ask, 'what has ever been incendiary in the politicisation or radicalisation of any issue throughout history that is truly free of tools such as identity politics, a raise in hierarchal structures, the Barnum effect, the framing effect, or the other number of biases the mind bestows upon the subjective nature of reasoning?'

[11] i.e. deductive reasoning.

[12] Which is also ironically akin to the spotlight effect within the human experience.

[13] https://www.youtube.com/watch?v=Iy2BcY0Drws&t=1s

[14] Well, when logically distilling the facetious nature of her argument, Phoebe would be guilty of *personal incredulity* in as much as she holds a plausible objection of there being but a small possibility, as there is superfluous evidence corroborating the theory of evolution.

[15] The *great leap forward*, coined by the American physiologist Jared Diamond in his book *Guns, Germs, and Steel*, suspects an environmentally strenuous period within our modern species to have happened between 50,000 to 100,000 years ago, thus leading them to disperse from Africa remunerated with a more than profitable set of neurological advances (Diamond, 2017). It's of significant importance to understand the unpeeling of free will through both an evolutionary and philosophical lens when concerning our socially geared evolution. Many ongoing theories about the causes of the *great leap forward* still proliferate since Diamond published his book in 1997. What was it that pushed those earlier species during the last glacial period? Was it an optimal mutation in a language gene, the FOXP2 gene found in language impediments today, or other genetic advances compelled to endure such cold climates? Whatever it was, epistemic contextualism ensued as a live or die phenomenon in our modern species as our brain size increased in cubic centimetres by the hundreds, reaching the over unprecedented thousand we flaunt today (Wells, 2003).

[16] Note 40 briefly explores these ideas of Yuval Noah Harari's subjective, objective, and intersubjective realities, best expressed in his bestseller *Sapiens: A Brief History of Humankind*.

[17] That which we call cognitive dissonance. Epistemic contexts can be easily perceived as two-dimensional circles, once circles contiguously overlap, the dangers of cognitive dissonance may thus prevail.

[18] Such information within varying epistemic contexts I refer to could be religion, political leaning, accents, racism, or general stereotypes that lead to the cognitively influenced biases consequently embedded in the decision-maker, thus the decision-making.

[19] *Apodictic truth* is the Aristotelean term that concerns itself with what is necessarily true (in all possible worlds), self- evidently true, or clearly established beyond dispute. Thus again it leads to the varying confirmation and belief biases we often both experience and observe, that being specifically *reactance* and the *backfire* effect!

[20] See note 94.

[21] As it's often argued hunter-gathering was more profitable for opportunities to be sedentary.

[22] The French astronomer and scientific correspondent Nicolas-Claude Fabri de Peiresc left as many as ten to fourteen hundred letters at his death bed!

[23] Such a besiege, as I reiterate, would be founded on enlightenment reason as I would contend that to focus on one reason alone approximates such false causes, i.e. *cum hoc* or *post hoc*...

[24] I will also come back to Roser's article for its quantitative significance throughout this work.

[25] Thanks to the exclusive correspondence network of intelligencers initially safeguarding the scientific community by those such as Samuel Hartlib and Henry Oldenburg in the seventeenth century, well, in so far as they could possibly protect science (Hatch, 1998).

[26] *Institutions* may use a comprehensive system for reference, citation, and peer review that may protect academia from collusion alleviated from conflicts of interest in allowing for the best possible science to ensue.

[27] Article 19 of Eleanor Roosevelt's draft on human rights stipulates that 'Everyone has the right to freedom of opinion and expression; this right includes freedom to hold opinions without interference and to seek, receive and impart information and ideas through any media and regardless of frontiers' (Roosevelt, 2001). Alexander Solzhenitsyn (2003) epitomised the dangers of disenfranchisement and censorship in his famous work *'The Gulag Archipelago'*. The political reins on the grey area of freedom of opinion and expression plagued the twentieth century with post Marxist ideologies that have led to millions of deaths. The political chronology on whether Marxism was solely the catalyst for Soviet atrocities will be explored in part II.

[28] Albeit states have their nuances on what constitutes First Degree murder, it generally alleges to carry both the intention to kill and premeditation on the defendant's part.

[29] As ought to be learned throughout, three stringent laws governing the filtering of any information I contend to be logic, scepticism, and probability. Thus we find ourselves often frustrated in health sciences, and reasonably so. I find it quite rare for a specialist to be *conclusive* about anything, not specifically in not *knowing* what they may be talking about. In such circumstances one ought to push for referrals, not managers. But as it may be the case for many, money talks.

[30] The base rate neglect, or the *prior probability*. Fundamentally the prior probability times the likeability of the data divided by the commonness of the data across the board. This would essentially equate its 'posterior probability'. As Bayes' theorem had all but pithily ratified David Hume's objection on miracles:

> 'So if the usualness of the event is reflected in the assignment of a low prior probability, then the usualness of the event does, other things equal, diminish the strength of the testimony' (Earman, 1993, p. 297).

And as Pinker also writes of the importance of statistics:

'Educational institutions, from elementary schools to universities, could make statistical and critical thinking a greater part of their curricula. Just as literacy and numeracy are given pride of place in schooling because they are a prerequisite to everything else, the tools of logic, probability, and causal inference run through every kind of human knowledge. Rationality should be the fourth R, together with reading, writing, and arithmetic' (Pinker, 2021, p. 314).

[31] Even eschatological stories of time, i.e. heaven and hell, offset the conundrum with the concept of an *eternal* state, never *temporary* or *limited.*

[32] Until Gottfried Wilhelm Leibniz's mathematical speculations eventually went on to influence Immanuel Kant.

[33] For most of Holmes' life, Holmes attempted to complete a reliable geological time scale. Contemporaneous to the publishing of *The Age of the Earth* by Holmes in 1913, was Frederick Soddy's discovery of isotopes along with Ernst Rutherford finding the transmutation of elements ascribed to radioactivity (Lewis, 2001). Whereas chemical reactions were the work of changing electrons, nuclear reactions were the changes that occurred to the nucleus leading to nuclear decay, thus leading to radiation. So with work on isotopes being rather too immature at such a revelation, Holmes knew it would take a lifetime to approximate anything close to a comprehensively reliable geological timescale.

[34] As for the consensus of time and conducive to antitheses of both science and scripture, the issue wasn't discrepant by a mere few years as it was the archbishop of Armagh, James Ussher, dating the beginning of genesis, precisely, around 6pm on October 23rd, 4004 B.C. (Lewis, 2001; Hawking, 2018).
Today, the geological time scale, comprehensively accepted, primarily contends with ages, epochs, periods, eras, and aeons, accounting for a distinguishing geology, ecology, geography, biology, and cosmology of both the earth's and universe's history.

[35] I refer to this issue in later chapters within the two types of naturalism that involve such frustrations, methodological and philosophical naturalism.

[36] Epistemic.

[37] Key to the notion of relativity is that it could almost be seen as a conduit to all other sciences, meaning in principle, it must be understood and epistemically accepted. For example, biologists deal with evolution, and evolution deals with time, and time is apparently space, so can a biologist who doesn't accept relativity really be a competent one? Can both evolution and relativity succeed as mutually exclusive doctrines?

[38] i.e. muscles.

[39] Which explains why so many fall into the trap of infinite regress.

[40] I use the 'world' for the purpose of simplifying.

[41] A successive synthesis is what Kant believed to be *manifolds*, i.e. a collection of categories that constituted representations (within a possible noumenal world), a key claim to his *Transcendental Deduction.*

[42] As it does not compel such consultation or experiences from the senses.

[43] Another example of Kant's synthetic *a priori* examples within mathematics would look like this: As we know two and two are equivalent to four, as *a priori*, what about the equivalence of three hundred and twenty-six and one hundred? The equivalence, four hundred and twenty-six, does not require such consultation or experience of the senses, nor is as easily countable (or obvious) as two and two, thus becoming synthetic *a priori* in judgement.

Chapter 2: Tierra! Tierra! Tierra!

[44] Shouted from one of Christopher Columbus' men on board the Santa Maria when spotting land in 1492, promised maravedis if successful. Columbus heartlessly contended he spotted the same land before him (Brinkbäumer and Höges, 2006). Tierra also means earth in Spanish.

[45] Duncan H Forgan and Ken Rice wrote a helpful juxtaposition of the *Rare Earth Hypothesis* and *Baseline Hypothesis* in 2010, conveying possible criteria thus needed for life to emerge. Such criteria consider: plate tectonic activity to regulate the composition of the atmosphere, the presence of a planet like Jupiter to control for asteroids and comets, a steady production of oxygen, sufficient raw materials generating amino acids, a critical mass range, a critical range of orbital radii, and low orbital eccentricity (which would preclude extreme temperature changes of the planet) (Forgan and Rice, 2010).

[46] All proteins on earth are made up of the same 20 amino acids!

[47] The four nitrogenous bases, thus building blocks in DNA, are adenine and guanine (purines), then cytosine and thymine (pyrimidines). RNA shares the two same purines and the same pyrimidine, cytosine, but instead of thymine has uracil. They are commonly known as the A, C, G, and T nucleotides that form the gene.

[48] A cistron is the closest means of proportionally measuring a sequence of nucleotides with a start and end symbol in a single protein chain.

[49] As the expression goes, 'you can lead a horse to water, but you can't make him drink it!'

[50] Stephen Hawking conveyed from both his works A *Brief History of Time* (1988), and his almost posthumous work, *Brief Answers to the Big Questions* (2018), how the face of determinism could change within a 'Grand Unified Theory', such that a Darwinian means of natural selection would be fervently reconsidered. In particle physics, scientists study the elements of time and space through careful observations of universal forces, i.e., the electromagnetic force, the strong nuclear force, the weak nuclear force, and gravity. The behaviour of these forces is believed to have derived from one singular force in earlier stages of the universe. Physicists closely observe positive and negative charges of electromagnetics, radioactivity of weak nuclear forces, how strong quarks become bound to form protons and neutrons, as well as astronomical gravity. In the world of probability, namely, quantum mechanics, scientists may be nullifying the compatibilist notion of libertarian free will, professing determinism as a highly plausible case in that one

may not be too far away from unfurling a 'Grand Unified Theory' than one may believe, leading to a would-be comprehensive understanding of the wiring of biological agency throughout the universe. Nonetheless, it is the notion of consciousness thinkers believe provides alternate philosophies on libertarian free will and agency.

[51] Max Roser showed how more than a third of the population lived in a colonial regime and almost everyone else lived in some type of autocracy in the nineteenth century. It was not until after the second world war political freedom slowly permeated the world, cohering with the rise in population, science, and technology (Roser, 2020). With such oppression at stake, anthropological fieldwork eager to compare intellectual faculties of such 'races' have been rendered rather unfeasible and thus unreliable, which also significantly appertain to issues effaced in geographic history that hark back to both proximal and ultimate causes from the first domestication of plants and animals. With an eagerness in apprehending the mechanisms that test for such intellectual faculties, many have been far too conclusive in such an effort. But the fact of the matter is, in as much as one can test for intellectual faculties in one given environment founded on the same worthwhile variables, variables change from environment to environment, so who gets to decide what those variables are?

[52] That human races are of different origins, as opposed to monogenism, of a single origin.

[53] Categorical syllogisms distribute middle terms in both premises (belief), two major terms which are predicates of the conclusion and in the first premise (uncertain), and minor terms which are subjects of both the conclusion and one premise (knowledge).

[54] An issue that appertains to *methodological naturalism* which I address in later parts of this book. If such belief is justified through faith other than reason and logic, irreconcilability ensues.

[55] As John McPhee said best, 'The human mind may not have evolved enough to be able to comprehend deep time. It may only be able to measure it' (McPhee, 2000, p. 90). It's quite well understood that for knowledge to be *synthetically a priori*, it has to both be *necessary* and *universal* while sport an aptitude to be *ampliative*. I would argue 'deep time' may be yet another instance of a Kantian *synthetic a priori* knowledge.

[56] It's not so obvious making presuppositions of uniform states that contend with such long lapses of time. Just imagine what one could infer about lepidopteran species from its larval and mature state without observing metamorphosis.

[57] A fascinating question springs to mind when concerning Abrahamic religions that seem to dominate the world's religious institutions today. For a moment, consider religious scripture to be the philosophical domain that controls for its propagation of both metaphysics and ethics. Barring ethics, if religious scripture had been written after the scientific revolution, would such scripture control for a more successful propagation of its metaphysics? Or, having been written long

before the scientific revolution, have such inconsistencies thwarted what would be ultimately compensated and therefore given itself away?

[58] Which pertains to a *Shell Model* of proportionate protons and neutrons in the strong nuclear force to help the nucleus from breaking apart.

[59] For instance, the uranium-238 isotope found in rocks endures both alpha and beta decay that alters to lead-206, in which uranium-238 is then known to display a half-life of 4.5 billion years!

[60] As the eccentric German philosopher Friedrich Nietzsche complained.

[61] A laity that includes various academics from fields outside of the life sciences.

[62] I refer to the philosophy of *telos* very carefully in this sense and will continue to use such a uniform sense of teleology throughout, as the discussion of teleology is subjected to various versions of its meaning pertaining to either *ends, means, goals, point* and *endpoint*, as the evolutionary biologist Ernst Mayr precisely pointed out:

> '*Telos means either endpoint or goal; they are the same.* By contrast, for the evolutionary biologist there is a great difference between *telos* as goal and *telos* as endpoint. If one asks whether natural selection and, more broadly, all processes in evolution have a telos, one must be clear which *telos* one has in mind' (Mayr, 1992, p. 123).

With Mayr's distinctions understood, the evolutionary sense of teleology is what I then present throughout axioms concomitant to the various ways of associating *purpose* and sought-after conclusive *goals*, rather than an ultimate *endpoint*. It is somewhat germane to a reverberated philosophy of mine in that there is no such thing as *just*, as for the sake of expedition, or convenience, the English language avails itself of the adverb to merely hold packets of reason that would otherwise be subjected to being epistemologically unpacked or unfurled. For instance, the interrogative would be, 'well, how do you know?', and such a reply would be, 'I just know'. Pendent is the justification of how one knows such a thing, or comes to know such a thing, and so I thus proposition we hold such teleology within these very packets of reason to expedite practicable communication as a concerted species. The biological sense of *telos* holds zigzag trajectories germane to a complexity of adaptation in its evolutionary sense, which I again proposition holds a pure teleological position in that such a gene thus mobilises the organism to act for its purpose to survive in the vast arena of its gene pool.

[63] It should be noted that it would take me less than 6 minutes and 36 seconds to encircle each block as I would only have to travel the alleyways once.

[64] Or raises the probability of being between its whole and its double.

[65] Due to the crossing of the alleyways that separate the blocks around the perimeter.

[66] Well fathomed by the sceptic is the ossification of one's bias being its ultimate authority. When all else fails it's the *appealing to authority* one is inclined to. Such foundational authority includes titles, history, ceremonies, and uncountable hours invested in beliefs that render the improbability of relinquishing such ideas. It goes

along with an *ad populum* that renders such an authority truly infallible. As satirical as it might seem for many North Koreans to worship the Kim family and in turn glorify an infallible entity, it wouldn't be as surprising if you were to dissect its very foundation. The same ingredients would be a successful *ad populum*, titles, history, ceremonies, uncountable hours invested, and a handsome veneer. Other ingredients and adornments notwithstanding, I would argue these principal ingredients could be applied to any human population today, along with the extant populations of hunter gatherers.

[67] For instance, the success in dating fossils back millions of years in time.

[68] Cnidaria, a notable phylum of remarkable marine animals such as jelly fish, are of organisms truly perverse to a human phenomenology as the organic make-up does not consist of a brain, a heart, bones, or the retinal organ, the eye. Annelids encompass earthworms and parasitic worms such as leeches, while chordates possess the notochord germane to all five vertebrate classes. Arthropods, encompassing the exoskeleton, are home to the popular classes of insects, arachnids, and the subphyla that evolved into crustaceans and myriapods. After the collision of an asteroid with planet earth around 66 million years ago in what was a truly significant event; the K-T boundary, it led to the remnants of Palaeozoic life that started afresh in the Cenozoic Era, in what we witness to be the interspersed life that surrounds us today, including us.

[69] Charles Darwin had a keen interest in both lancelets, fish-like chordates, and ascidians, which are essentially marine invertebrate chordates. Chordates, evolving around 535 million years ago as some of the first post-Cambrian forms of life, are rather paramount when compelled to think of, as Darwin put it, 'the lower stages in the genealogy of man'. Chordates branch off into three distinct subphyla; the tunicates, the cephalochordates, and the vertebrates. Vertebrates then evolved into the five popular classes of fish, birds, reptiles, amphibians, and mammals across the Palaeozoic and Mesozoic eras. Lancelets were therefore taken with keen interest as they belonged to the *cephalochordata*, while ascidians belonged to the *tunicata*, which gift taxonomists alike the great puzzle of the Palaeozoic tree of life:

> 'We should then be justified in believing that at an extremely remote period a group of animals existed, resembling in many respects the larvae of our present Ascidians, which diverged into two great branches – the one retrograding in development and producing the present class of Ascidians, the other rising to the crown and summit of the animal kingdom by giving birth to the Vertebrata' (Darwin, 1871 p. 102).

[70] A classic appeal to ignorance fallacy: '*No aliens have been proven to exist, therefore they exist*'. One can also shop from the causal fallacies committed by Jonathan Wells here as well!

Part I: Of Sexual Selection

Chapter 3: What If I Wanted to Get a Yoghurt on the Way Home, Would You Pay for That?

[71] A famous study was conducted by Russel D. Clark and Elaine Hatfield on *Gender Differences in Receptivity to Sexual Offers* with interesting results from both 1978 and 1982. Women and men were offered three requests by the opposite sex around university campuses, in which a man or woman would be offered (a) a date, (b) an invitation to their apartment, or (c), an invitation to bed by a relatively attractive man or woman. When men offered women the request, 56% of women agreed to the date, 6% to the apartment, and 0% to bed. When women offered men the request, 50% agreed to the date, 69% agreed to the apartment, and by no surprise, a whopping 75% to bed. Of the repeated study in 1982 when men offered women the request, 50% of the women agreed to the date, 0% to the apartment, and 0% to bed. While 50% of men agreed to the date from the woman, 69% agreed to the apartment, and 69% to bed (Clark and Hatfield, 1989).

[72] Geoff A. Parker and Lehtonen portray how the economics of both sperm competition and limitation within gamete evolution and anisogamy play out. It's more of an interesting rebuttal to the classical view that males produce large numbers of sperm to enhance the probability of fertilisation.

[73] As John McPhee reminds us of the greatest extinction event that ends the Paleozoic era, the Permian Extinction (299-252 million years ago):

> 'Whatever the cause, no one argues that at least half the fish and invertebrates and three-quarters of all amphibians – perhaps as much as ninety-six per cent of all marine faunal species disappeared from the world in what has come to be known as the Permian Extinction' (McPhee, 2000, p. 82).

And as Attenborough reminds us of the danger of atmospheric carbon:

> 'A radical change in the level of atmospheric carbon was a feature of all five mass extinctions in the Earth's history and a major factor in the most comprehensive annihilation of species – the Permian extinction, 252 million years ago. The exact cause of that change is disputed, but we do know that one of the longest and most extensive volcanic events in Earth's history had been growing in strength over a period of a million years, covering what today is Siberia with 2 million square kilometres of lava. This lava may have spread through the existing rocks and reached vast beds of coal, igniting them and discharging sufficient carbon dioxide into the atmosphere to raise the temperature of Earth 6°C above today's average, and increasing the acidity of the entire ocean. The warming of the ocean put all marine systems under stress and, as the waters became more acidic, marine species with calcium carbonate shells – such as corals and much of the phytoplankton – simply dissolved. The collapse

of the entire ecosystem was then inevitable. Ninety-six per cent of the marine species on Earth disappeared' (Attenborough, 2020, p. 88).

[74] Bats are great examples of polyphyly. If you were to ask a child the vertebrate class bats belonged to, she may assume birds due to its evident airborne prowess, however, a bat's most common ancestry lies with primates and rodents, for instance, as they also evolved from those same Mesozoic mammals.

[75] A female thus calibrates her strategy to ensure the new embryo acquires the best possible genes it can acquire, against the odds of her reducing her own and the child's chances of survival and eventual reproduction, that being if the child does not achieve the rearing and stability the other child may advantageously achieve within intraspecific economics! Notwithstanding the evils of non-consensual sex, this gives the female both cognizant and incognizant strategies parallel to her ontological position of her environment she plays in, which result into what could be portrayed as a Venn diagram of vast similar strategies against her contingent or anomalous differences. I say anomalous differences, or mutations, in so far as evolution morphed our species into self-consciousness, which effectively opens up the debate on free will or an extended phenotype. Science is exemplary of how we have been able to accentuate anomalous differences for the female within sexual selection, inasmuch as the biological substrate for tendentious behaviour leaves the organism in her will.

[76] When one ruminates long enough on the notion of *deep time*, and on how genes had been carefully recycled molecule after molecule, on to sexually reproducing eukaryotes, being generation after generation, the entailing profitable characteristics within the environment should both reflect and retract a microscopic teleology. When we hark back to Triassic and Jurassic periods of the Mesozoic era over 200 million years ago, around a time of when vertebral mammals started to evolve, we should think about what part of any phenomena could disdainfully arrogate the authority to call itself 'the self'. For instance, let's call the currently construed notion of the self, Leo, who in this context is an organism, a human. If Leo unfortunately fell victim to a fatal accident and were to lose both of his arms, where would Leo *be*? Would Leo *be* his arms, *inside of his arms*, or the rest of his body? Of course, Leo would be the rest of his body, or *inside of the rest of his body*. Sure, he may despondently have to alternate a few locks to his ontological security, but we know where Leo *is*. If Leo were to lose his heart, where would Leo be? We know heart transplants are ongoing occurrences throughout the world today, and so we know Leo would continue being Leo without being preoccupied with 'loving' strangers after a procedural heart transplant. Maybe it could be agreed that the biggest part of Leo *is* his brain, inasmuch as the functions of his brain variegate into what would constitute the illusion of what Leo *is*. Before we look at a few exemplary parts and functions of those parts of the brain, we have to deliberate on what part of the brain we could now say Leo *is*, as there is no real ontological central headquarters to Leo, and we know that interaction among all of these signals may create a deceptive stream of consciousness. To take an example, could we say that Leo is the hippocampi, mostly responsible for short-term and long-term memory, or only **one** hippocampus? Could we say he's the amygdala, principally responsible for anxiety

and stress response? Or the entire cerebral cortex, which is responsible for multiple functions such as decision-making? There has been a plethora of instances in which damage to particular parts of the brain render the individual *behaving* in peculiar ways, which can be best sought in Robert Sapolsky's work when reasoning with this neurological conundrum, *Behave, the Biology of Humans at Our Best and Worst*. So let's anchor here and look at how the hippocampus may be quintessentially adduced to an evolutionary game theory model.

[77] The notion that the human mind can be reduced to computational algorithms leading to an almost sci-fi sense of superhumans usurping the world we live in has been a rather intrepid claim within the academic climate, as objections raised point out the *de facto* presupposition that abiotic computers are inhered with the evolutionary proclivity to compete as biotic life, which does appear to be somewhat tenuous.

[78] Of which adrenal glands are responsible for glucocorticoid, epinephrine, and norepinephrine secretion.

[79] Conversely, neural circuits of long-term depression (LDP) may lose synaptic excitability for what is inadvertent in sharpening gratuitous and ulterior signalling in the depressed. Likewise, research by G. Elliot Wimmer et al. (2023) show how 'replay' may be significant for sequencing planning and memory preservation in the hippocampus.

[80] Of the three essential cycles the Serbian Milutin Milankovitch demonstrated.

[81]
> 'In any case, the essential requirement is a cool summer. A little snow from one winter must last into the next. Every forty thousand years, the earth's axis swings back and forth through three degrees. Summers are cooler when the earth is less tilted towards the sun. The sun, for that matter, is not consistent in the energy it produces. Moreover, the relative positions of the sun and the earth, in their lariat voyage through time, vary, too – enough for subtle influence on climate' (McPhee, 2000, pp. 271-272).

[82] This leads us onto what I believe is a paramount aspect of sexual selection, which has had implications for sociological practices within our species, sexual dimorphism. Sexual dimorphism is currently proving to be controversial in aspects such as the cultural evolution of sports, as there is a poor epistemic endeavour to understand the intricacies of biological differences within our sexes as a species, however, I do not intend to venture or digress towards this inclining sociological controversy. Rather, sexual dimorphism should at least be adduced to glare through the prism of a collective ontology within human sexual selection, and yet again, a great way to edify oneself on this matter is to venture away from the extraneous perimeters of our taxonomical species. Sapolsky, also a primatologist spending over thirty-three years observing primates in east Africa, gives quintessential instances of where one could place their social philosophy within our species, playing out in both monogamy and polygamy, under what he polarises

as 'pair-bonded' and 'tournament' species, both internal and external to our primatological taxon. This should also be perceived through the ontological prism for what *is*, as opposed to what *should be* or what one *would like to be*, as it only becomes caustically deleterious in the long-term for a more stable and practical epistemological apparatus.

[83] However distinctions between men and women in the VMPFC is not so clear, albeit men with right-sided VMPFC damage and women with left-sided VMPFC damage seem to show lower aversion to risk and ambiguity in decision-making (Sutterer and Koscik et al., 2015). Dimorphism in the brain between the sexes has often been negligible, referring to the brain as a mosaic of differences between both males and females, which puts more emphasis on the brain being a blueprint for our distinguishing *telos* (Joel and Berman et al. 2015).

[84] Simians such as rhesus monkeys and mandrills with redder faces have been observed to gain more proceptive behaviour (pre-copulation) under the RWB display. Male geladas leading larger groups have been observed to display redder skin. Likewise, male snub-nosed monkeys have been observed to have redder lips when increasing in age and leading large groups of conspecifics (Young, et al., 2020).

[85] In 1982, a ground-breaking book in the field of evolutionary biology, *Evolution and the Theory of Games*, was published by the British biologist and mathematician John Maynard Smith. His ideas were laudably appropriated from the American mathematician John von Neumann, stealing the idea of game theory used in economics, and applying it to the field of biology in which Maynard Smith came to win the Crafoord prize for in later years (Michod, 2005). For Maynard Smith, it started out as a payoff matrix he laid out during the USA/Vietnam war while he was on holiday in Chicago, and so, as a result, named the two types of animals playing the payoff game Hawk, or Dove. To understand the payoff matrix in evolutionary games, let's suppose there are two types of animals playing the game. One animal bites, scratches, and kicks while the other animal shows more innocuous behaviour, maybe by displays of aggression, as opposed to actually enforcing it. The winner of the contest would be the Hawk, the former animal, as Hawks are animals who escalate and fight until either they win or the other is forced to retreat, and so would conclusively pass his genes on to the next generation, and so forth. The latter animal, the Dove, are animals that display, and if the adversary escalates then the dove will inevitably flee. So, according to the payoff matrix Maynard Smith devised, V, as a unit, would be the value of what the two animals are fighting over, a resource, and -C would be the cost of injury for the animal playing the game. Games can be positive, negative, and zero-sum. Positive being *beneficial* for both parties (e.g. a trade-off of resources), negative being costly for both parties (e.g. both parties getting injured), and zero (e.g. one party taking all) (Maynard Smith, 1982; Pinker, 2012).

[86] A cost of injury (-C) is greater than the value of resource (V) when playing the game. When pairing off animals at random in a population that may play one or the other game of Hawk and Dove as a strategy or play a bit of one strategy and a bit of the other, mixed strategies, let's say that after playing a certain number of games animals stop, weigh out the value of their payoff, and reproduce. In the case of

evolution, Hawks will produce Hawks, Doves will produce Doves, and those with mixed strategies will produce offspring with mixed strategies relative to the probability of the parents playing either the Hawk or Dove strategy. Then, the number of offspring reproduced is going to be equal to the payoff acquired within the matrix, and so the payoff becomes indicative of the changes that eventually manifest themselves into biological fitness, and over long periods of time the population-evolved strategies then manifest themselves into what will appear to be an *evolutionarily stable strategy* (ESS) in a particular environment. The evolutionarily stable strategy is a strategy in which, if all the members of a population play the given strategies, there would be no possible mutant strategy that could invade the population and do better, attributable to long periods of evolutionary changes, benefits, and costs stabilising themselves out within an environment. Also, augmented to the phenotypic behaviour of the ESS would be a DSS, a *developmentally stable strategy*, best understood as heuristics learned through individual trial and error, and then also a CSS, a *culturally stable strategy*, learned vicariously through other members of the species, extraneously but yet intraspecifically assimilated by the organism, or the survivorship machine.

[87] Asymmetries may be between male and female, old and young, small and large, owner and non-owner etc. An example of asymmetries within the extreme Socratic sense for those apprehensive by nature would be to look at heritability scores. Two monozygotic twins enter a physical bout in a contest. They are both right-handed, weigh exactly the same, had the same training in whatever martial art, and are both as indignant with each other relative to the reason they are about to fight. What would be of asymmetries in this context?
Phenotypic strategies amidst animals, whether under the telos to sexually reproduce or survive, may manifest themselves into an interoperable amalgam of an ESS, a DSS, or a CSS, creating probabilistic stochastic strategies in order to avail themselves of as much benefit as the organism can acquire against its cost, while all akin to its asymmetries and completion of information the survivorship machine yet manages to harbour.

[88] A key theory within the theory of sexual selection, after a long hiatus since Darwin originally proposed his theory in 1871, was of 'Fisherian runaway selection', implemented in his 1930 book, *The Genetical theory of Natural Selection* (Fisher, 2018). Fisherian runaway selection, established by the British geneticist and statistician, Ronald A. Fisher, also an acquaintance of Charles Darwin's son, Leonard Darwin, pushed population genetics to the forefront of post Darwinian evolution. A chasm between 1871 and the early 20th century could be readily ascribed to a scientific quarrel between Alfred Wallace and Charles Darwin about the supposedly spurious nature of Darwin's empirical claim amidst the theory of sexual selection. As Darwin wrote extensively throughout his book, giving examples even extraneous to the perimeters of the vertebral classes of how it may be that females select for aesthetics in a male, Wallace had remained inexorably incredulous:

> 'It will be seen, that female birds have unaccountable likes and dislikes in the matter of their partners, just as we have ourselves, and this may afford us an illustration. A young man, when courting, brushes or curls his hair, and has his moustache, beard, or whiskers in perfect order, and

no doubt his sweetheart admires them; but this does not prove that she marries him on account of these ornaments, still less that hair, beard, whiskers, and moustache were developed by the continued preferences of the female sex. So, a girl likes to see her lover well and fashionably dressed, and he always dresses as well as he can when he visits her; but we cannot conclude from this that the whole series of male costumes, from the brilliantly coloured, puffed, and slashed doublet and hose of the Elizabethan period, through the gorgeous coats, long waistcoats, and pigtails of the early Georgian era, down to the funeral dress-suit of the present day, are the direct result of female preference. In like manner, female birds may be charmed or excited by the fine display of plumage by the males; but there is no proof whatever that slight differences in that display have any effect in determining their choice of a partner' (Wallace, 2022, pp. 286-287).

What became significant about Ronald A. Fisher's theory is that his notion made Darwin's idea of female preference merely respectable. For the theory of sexual selection, the question always remerged, what became so substantive about Fisher's claim? Fisherian runaway selection is what is best understood as the manner in which sexual dimorphism evolves under an evolutionary substrate, accounting for the benefit of mate preference against the cost of mate choice within natural selection, picking itself out cyclically over long spans of evolution.

[89] Pomiankowski and Iwasa also claimed cyclic evolution under sexual selection may also explain for allopatric speciation in populations. Allopatric speciation concerns itself with species evolving separately due to being geographically isolated, as opposed to sympatric speciation, in which populations evolve separately irrespective of geographic proximity (Pomiankowski and Iwasa, 1998).

[90] You can witness Sir David Attenborough narrate the bizarre courtship of the twelve-wired bird-of-paradise and the black sicklebill (*epimachus fastosus*) on Our Planet, Youtube: https://www.youtube.com/watch?v=rX40mBb8bkU

[91] For the male, this would always play against his chances of ostentatiously, or for the part, meretriciously, displaying his merits until those merits summon a predator that costly devours the male in his ornamentation, not allowing the beautification to runaway genetically to the male and female offspring for both choice and beauty. A Fisherian model reconciles natural selection, in which a covariate between predator and prey instantiates upon survivorship for the male having to adapt to his environment, and sexual selection, in which female preference instantiates upon sexual selection for the female choosing the most ostentatious ornaments, against a covariate of the taxing process of choosing, which render stable states of equilibria within timely evolutionary models.

[92] A significant theory within the Darwin vs Wallace duel came when the Israeli biologist, Amotz Zahavi, came to the forefront with what appeared to be a first unfounded theory about the nature of sexual selection, in what he termed 'signalling'. His idea of signalling endeavoured to swing the pendulum to Wallace's objection that sexual selection couldn't only be ascribed to female choice, and so for Zahavi, natural selection for survival had to play a larger part within evolution.

The Fisher revelation suggests that as well as male anatomy, female choice was also under genetic control. We could say that for the Fisherian model this could principally transpire in two rather explicit ways. The first is that natural selection may work on genes in males for them appearing more beautiful, i.e. in ornamentation, and for females in selecting beauty. The second is that both the female and male young inherit genes from their father for beauty and also genes from their mother for selecting for beauty. This would all be conducive to secondary sexual characteristics that render something as spectacular as a peacock's tale, while appearing as what Zahavi calls a 'handicap' in terms of predation risk, or the twelve wired-like filaments in the twelve-wired bird-of-paradise (Zahavi, 1999). So, as for the instance of the twelve-wired bird-of-paradise and his wire-like filaments as he strokes against the hindquarters of the female, he may not only be displaying his beauty, but may also be displaying profitable aspects of his fitness. His fitness simply implies, 'I have this spectacularly beautiful ornament and I am still alive, which predators would have devoured by now if I didn't have such amazing genes!' As natural selection selects for characteristics which are potentially profitable for populations adapting to their environment, Zahavi perceived signals to be simply by-products of natural selection:

'Characters which have evolved as signals may contribute to the diversity of the morphological, anatomical, physiological and behavioural repertoire of the species. It is important to remember that this contribution of signal selection is a by-product of selection for reliable signals' (Zahavi, 2007).

[93] Yet again contributing to the theory of sexual selection and thus coming back to the secondary sexual characteristics of the twelve-wired bird-of-paradise, is the popular theory of parasites, which Adrianna Smyth spells out of the Hamilton-Zuk hypothesis (Smyth, 1995). Relative to Zahavi's handicaps being insufficient to the extent of animals signalling and displaying fitness, Bill Hamilton and Marlene Zuk's hypothesis shows how a display of health under Zahavian signals may resist parasites and thus show for 'good genes'. So secondary sexual characteristics of the twelve-wired bird-of-paradise may be indicative of the fitness to resist parasites. But why? As we have endured a Covid-19 pandemic*[93], we see how the behaviour of a virus has tested global medicinal science. Gene frequencies are the slowest to respond to antigens as molecules of microbes change from generation to generation. For sexually reproducing species without the advent of medical science, (namely, all of the others) plagues, viruses, and the quickly changing evolution of parasites are vehemently positioned throughout the prism of the Hamilton-Zuk hypothesis. Health is the primary virtue according to their claim, as an advertisement of the twelve-wired ornamental filaments for the bird-of-paradise is indicative of both robust health and useful resistance to coevolving parasites that reinvent themselves over short spans of evolutionary time. Be that as it may, health then becomes predicated on females choosing for secondary sexual characteristics limited to parasite infection. Why? They may opt for twelve-wired traits in order to obtain those especially resistant characteristics for future offspring, rendering heritable variation that coincides with the coevolution of the said parasites, allowing for their offspring to resist them over *cyclic* generations thus taking its turn (Hamliton and Zuk, 1982). This swings the pendulum towards Wallace while at the same time shows how the teleology for survivorship and

reproduction may be interoperable when performed, ensconced within the organism's genetic makeup.

[94] Theories on inheritance of acquired characteristics (IAC) are also brewing in biology. Darwin's theory of 'Pangenesis' in 1968 hypothesised that living cells could be sending information to germ cells by what he termed 'gemmules' (Zou, 2014). This controversially held Lamarckian approach may be correct and achieved through extracellular vesicles being transmitted into the germline from the soma.

[95] A more than plausible question is asked about the relevance of Darwin's work in modern science today, as it would be logical to enquire into the possible *epistemological purchase* his ideas still hold, as his work on both natural and sexual selection are as much as one hundred and fifty years old. Be that as it may, for Darwin, this is not the case, as Lewens stated precisely "Darwin is still a part of modern Darwinian biology in a way that Einstein is not a part of modern physics" (cited in Ratnieks et al., 2011, p. 481), and as Alcock claimed, "When biologists differ over issues in modern science, they often try to claim Darwin for their team. Darwin is still regarded as a quotable biological authority, and struggles go on between biologists over how his views should be interpreted" (cited in Ratnieks et al., 2011, p. 482). His relevance in modern science is rightly ascribed to the comprehensive scope of naturalist work that ranged from botanical fields across to sexual selection in entomology. Darwin first presented his theories by observing what should be significant to the epistemological consensus of where his truly incipient stages of sexual selection derived, by scratching the surface of entomology. Hence, it's important in the epistemic domain to see how far both natural and sexual selection expand across the animal kingdom, and insects are produce of millions of years of cyclic evolutionary mutations that have survived extinction events, endured aggressively fluctuating environments, and thus evolved truly remarkable phenotypes.

[96] Insects that molt and undergo the incomplete three-stage-metamorphosis of egg to nymph, then to maturity are true bugs, grasshoppers, cockroaches, termites, praying mantises, crickets, and lice (Baluch, 2011).

[97] Bees have even been found to show emotions by their distinguished tones of humming while varying in many colours across the taxonomical order.

[98] There is also the mathematical elision of George Price's equation on covariance of traits and fitness through alleles in the 1970s (Lehtonen and Okasha et al., 2020).

[99] Professor Thomas Hales of the University of Michigan showed how hexagonal tiling provided the 'least-perimeter' way to enclose infinitely many unit areas. Polygons with less than six sides do worse for perimeters, and those of more sides that approximate a circle do better for decreasing perimeters (Morgan, 1999).

[100] The research team at the Oxford University Museum of Natural History, written by Rebecca Mileham, makes for an excellent read of the first evolving animals. The museum now also uses an X-ray scanning technique which builds three-

dimensional models of fossils that you can now rotate and better envisage of what those first animals would have looked like (Mileham, 2020).

[101] Of sauropsids we find extant species of crocodiles, turtles, birds, lizards, and snakes. Of synapsids we find extant species of mammalian placentals, marsupials, and monotremes. And as Zahavi noted, some adaptations are likely irreversible given that some reptiles and mammals went back to the waters (i.e. our extant dolphins):

'After all, some adaptations are likely to be irreversible, since they involve a complex set of interconnected changes. Reptiles and mammals who left the land and returned to the water continue breathing through their lungs, even though in their new habitat it would benefit them to breathe through gills' (Zahavi, 1999, p. 148).

[102] Menkhorst and Nation et al. make for a good reading on the evolution of our synapsid shell coat and yolk in monotremes, placentals, and marsupials (Menkhorst and Nation et al, 2009).

[103] Thus we find the spongy pulmonary pleura and functional tissue (parenchyma) that help to reduce air pressure found in the lungs. The asymmetrical lungs consist of a right lung that has three parts: the right upper lobe (RUL), the right middle lobe (RML), and the right lower lobe (RLL). The left lung consists of both the left upper lobe (LUL) and left lower lobe (LLL) (Chaudhry and Bordoni, 2021).

[104] That entail of those organs such as the glottis and vocal folds that control for airflow.

[105] As Amotz Zahavi's work on warblers had shown, the misconception that warbler's warning-calls were only useful to communicate to other warblers that a predator was coming was confounded. After successful observation, those warning-calls were communicating to predators that they could already see them coming, thus communicating not to expend energy you could expend on food you are more likely to catch!

[106] * It may not be as far-fetched as it originally appears to be in imagining humans quacking for effectual communication. We see characteristics that are evolutionarily profitable across even distinct phyla. For instance, legs seem quite useful for locomotion, arms for prehensility and grappling, tongues for acoustics and eating, and so on, so a measure of profitability could be certainly cogent with the characteristics in themselves. Interestingly, this may also be why some scientists do not think it would be too far-fetched or anthropic to be too surprised to see extra-terrestrial life resembling humans, I guess it would all be contingent upon its measure of evolutionary profitability, in so far as their 'would-be' characteristics not entirely running afoul of the natural laws of the universe.

[107] The causes of such drastic speciation, and the marked encephalisation of our species, I find such a fascinating topic. Its causes share a certain concomitance with the notion of Creationism first contested by evolutionists such as Thomas Huxley across the latter stages of the nineteenth century, when it was inconceivable that monkeys could be related to our species via the then nebulous theory of natural selection. Since then, as oblique as the causes of speciation are, a number of

hypotheses stand on how speciation may have occurred. The first thing we must remember is that irrespective of the interspersion of hominins across Africa and Eurasia subsequent to two million years ago, the vast majority of hominin species originated in East Africa, and then subsequently ventured and colonised Africa and Eurasia, respectively. The *Savannah hypothesis* necessitates the long-term trends of expansion and a harshly arid climate that those hominins would have had to adapt to across east Africa. Conversely, the *variability selection hypothesis* necessitates the everchanging and intermittent vagaries across the environment that could play out in both behavioural and ecological effects of life within adaptation. A *turnover pulse hypothesis*, used to look principally at ungulate speciation, has also been summoned to understand hominin speciation which observe acute climate shifts, driving adaptation and speciation as a result. Then a final hypothesis is an alternative to the *turnover pulse hypothesis*, the *pulsed climate variability hypothesis*, which looks at how wet-dry cycles endogenous to East Africa could have propelled adaptation and speciation due to such a comprehensive threshold for survival. A correlation between the varying species within the sub-tribe *australopithecina* and the genus *homo* were observed to explain for possible encephalisation, migration, species, and turnovers found amid the formations and environmental conditions. To explain for such speciation would have to be rendered by ascribing speciation to the powers of environmental fluctuations proposed in hypotheses such as the controversial *turnover pulse hypothesis*. As mentioned earlier, Milankovitch cycles have such powerful spatial-temporal forces that can change the fate for life in distinguished environments on earth through such subtle changes of 'spacetime' between a motioning sun, as noted in lake cycles within these very hypotheses, 'Up to and including 2.6 Ma, these lakes track 400 ka eccentricity cycles' (Shultz and Maslin, 2013, p. 2).

[108] In mammals, it has been speculated that certain odours play out as secondary sexual characteristics that can determine for mate choice and has been speculated to be as seemingly equivalent to 'bird plumage, deer antlers, or the bowers of bower birds. Sexual selection would act upon these odours just as it would act upon visually conspicuous characters' (Blaustein, 1981, p. 1007).

[109] As we'll see in later parts, inductive reasoning is well found in the mind and often lacks its articulate spokesperson to its defence. As psychology is questionably tenuous in its scientific trajectory towards absolute certainty (of which we have heretofore been unable to attain), psychology opened itself to two issues in the political sciences. (1) It is hijacked by those with agendas to propagate false claims about the nature of the psyche. (2) One sticks to the legal sciences alone or on the other extreme acts as if we're only computational entities.

[110] Infanticide outside of mother's killing their own offspring has been observed in other primates. New male leaders killing the offspring of babies in new-founded territories in order to ensure that both or all children belong to them is common (e.g. rhesus monkeys).

[111] I say ontological security in a very particular sense which should be explicated and by no means be misconstrued. Ontology, coming from the philosophy of being, existence, and rooted by the infinitive 'to be', conjugated as 'is', and 'are', is the extremely sensitive foundation of the fluctuating psyche we forever endure as conscious organisms. I usually say that 'it's the reason why one chooses to get up in

the morning', notwithstanding the unfathomably numerous reasons we have not to. As the *time arrow* moves from past to future, fundamentally an *a priori* intuition on space and time, as organisms, we are only victims of spatial-temporal senescence, and so in turn, every waking moment is a struggle for existence, and we know this through processes such as homeostasis. In turn, ontological security is anything and everything that adds to the security of *being* and *existing*, which tends to fluctuate and has to at times be resynchronised, revisited and reconciled.

[112] I compare it to the paradox of the 'white woman' who claims the 'straight white male' is he who suffers nothing in this world. Do you honestly believe you inherited no privileges at all from ancestral white men (or those who selected for those white men in the first place)?

[113] In so far as religion has been tricky to dissect, akin to its evident role throughout history, gender and sex have also risen as contemporarily sensitive issues attributable to the disparate positions women have occupied throughout history, but nowhere near to the degree that so many narratives throughout the humanities claim today, being decontextualised, misguided, and unfortunately, manipulated, all but using facile univariate narratives and analyses to explain for some of the most complicated periods in human history. One only has to look at the epistemological contours of the comprehensive data surrounding women's suffrage throughout the occidental west, without having to enter 'feminist history', which is tacit in having transpired throughout the nineteenth century across to the former stages of the twentieth century. Coincidentally enough, the rise of women's suffrage coincided with the rise of democracy, health, population, prosperity and education (see Max Roser for a better insight) (Max Roser, 2020). What shouldn't be eschewed too impetuously, is that according to the United Nations, over 90% of the world lived in extreme poverty by the turn of the 19th century, living on less than $1.90 a day, which only less than 10% of the population in the world today endure. That being said, the population of our species was 7-8 times smaller, and two thirds of the world only became democratic towards the latter stages of the nineteenth century. So, in terms of women's franchise, most of the time throughout our history data shows that, as per democracy, most people were disenfranchised before an evolutionarily enlightened democracy. Do we really know what this means? Have we seriously meditated on what life could have been like for those humans? The 90% of the less than one billion people who existed before the nineteenth century were only compelled to cooperate by the extreme vicissitudes of everyday life, with that extreme privation. For a minority of women to have been without benefitting from positions held by such miniscule minorities of privileged men seems to be highly unlikely, notwithstanding the vast majority of the population's incessant plight. As for today, the data points out for most people in the west, notwithstanding naturalisation in western nations, our lives were so much better than that of our grandparents alone. For most people, crossing the first two thirds of the twentieth century, until 1970, more than half of the population fell victim to the harsh turmoil of extreme poverty. Albeit the other misconception amidst some gratuitous feminist claims, which I hope will not be unduly interposed in my approach to the theory of sexual selection, is that testosterone is in any part conducive to male power and competition, as the literature clearly suggests women are as equally susceptible to testosterone

secretion when in positions of wielding power (M. van Anders et al., 2015). But the question would be when, particularly predicating itself on the question of time.

[114] It must be remembered that psychology didn't even exist when Charles Darwin outlined his theory of sexual selection, as psychology initially came roughly a decade after him publishing in 1871.

[115] In order to evade being sadly misconstrued, I do believe that the structure of IQ testing is quite telling when we think about intelligence as a ontic subset of the power of sexual selection. The somewhat left-leaning analogy of multiple intelligences, in which 'you can't teach a monkey how to swim, you can't teach a fish how to fly, and you can't teach a bird how to climb', standing in for intellectual relativity, is not only confounded, but is also a concerning understanding of evolution. The approach to IQ testing currently being utilised, founded on psychometric centred testing, use three types of IQ tests which approximate different means for testing. The three types of IQ tests used are often disputed, but not as much as one may believe given the domain for intelligence, which naturally gravitate towards distinct predicates for the psychometrics in focus. The more culturally aimed IQ test, standing in for a test that endeavours to contextualise the cultural relevance of a school's system, is the Stanford-Binet IQ test, which, if you're an evolutionary biologist, might prove to be quite useful, as it could be effectively argued that adaptation to the organism's environment is primarily necessitated. Another test used, which is less culturally gravitated, is the Wechsler or WAIS IQ test, which includes categories of questions that are merely predicated on verbal comprehension, working memory, perceptual reasoning, and processing speed. Conversely, the Raven's Progressive Matrices, or the RPM, is surprisingly founded on one ability, approximating difficult tests that aim to understand perceptual or analogical reasoning, postulated on the idea that reasoning is the primarily necessitating attribute testing for one's intelligence. IQ testing is relative to the individual's performance compared to individuals of the same age, so the score should be the same throughout one's lifetime. Centred on both the tasks and the domain of ability together work for a psychometric substrate in testing IQ, which thus understands function of hierarchical models for intelligence. Matzel and Sauce continue to display such frameworks for IQ testing in play. At the pinnacle, branching itself down into more compartmentalised facets, is what can be seen as general ability for the individual tested. This they call level 3. Level 2 would entail of five possible domains of ability that travel through to level 1 of specific testing, fairly analysing the individual's domain of ability throughout. Mediating between level 1 (the testing) and level 2 (the domains of ability being tested) is effectively how intelligence then becomes detectable. Reasoning tasks test for a reasoning domain, speed tasks test for a processing speed domain, memory tasks then test for a memory domain, and spatial tasks test for a comprehension domain. The fifth domain, for Matzel and Sauce, they leave as a question mark, as they acknowledge that other domains certainly may exist (Matzel and Sauce, 2017). The dicta among IQ testing is that 'people who perform well on one domain tend to perform well on other domains'. If one closely looks at the structure of the Wechsler or WAIS intelligence test for a full-scale IQ (FSIQ), it may give an observer a better idea for what is being tested for, especially when audaciously applying intelligence to something as complicated as an empirical claim of a working power within sexual selection. The Wechsler intelligence test

divides the FSIQ into two scales of intelligence for its purpose in testing. One scale found on performance, and the other verbal. A verbal comprehension index is tested using vocabulary tests, similarities and differences tests, comprehension and informative passage tests, while testing for working memory (or short-term memory) using arithmetic, the sequencing of numbers and letters, and digit-spans for how many digits can be recalled. For performance in the FSIQ, a perceptual organisation index uses matrices, block design and picture completion while a processing speed index is tested for using digit symbol coding. It must be remembered that IQ testing first came about to detect disabilities, rather than testing for the most intelligent kind. And so as the test approximates the highest faculties of intelligence, the tests predicate themselves more and more on the dicta that if one performs well on one domain, then they invariably perform well on other domains, which ironically could be argued makes the test more and more unreliable for the highest faculties, in as far as the test proves useful.

[116] Interestingly enough, and in regard to those ontic subsets of a working power of sexual selection, the rise of what the online world now call 'Incels', or 'Inceldom', briefly emanated sociological literature and circles in previous decades. Incels (or involuntary celibates), claiming to be dispelled from competition for the female by superior other males, female promiscuous behaviour, and post-1960s feminist movements, are males seemingly disaffected due to the relative misfortune of a failure to consummate during any endured courtship. According to some, they contort scientific literature, namely sexual selection, and fortify misfortunes by engendering villainously resentful behaviour towards women that have even led to attacks on innocent females (Long and Lynch, 2019; Zimmerman and Duriesmith, 2018). What I find fascinating about the ideas of 'inceldom', and also giving the devil his worthwhile due, is that, as a male, many of the ideas Incels profess I find rather substantive. Meanwhile, on understanding a philosophy and the calamities within their own plight, the quintessential question touching the surface for those professing the ideas is... well what about the plight of women? In the extreme sense, the plight of women seems to have engendered what I see as analogous to the incel movement, radical feminism. Again, like that of inceldom, many ideas professed by radical feminists I find rather substantive, well, to the degree that I could possibly commiserate as a male myself. Males, as products of physiology (organs, tissue, cells, organelles, molecules, and atoms), fall into what have become extremely incessant organisms borne with proclivities heterosexual females have had to endure for time immemorial, having to encounter ways to cooperate and live with a sex females will never ontologically fathom. Likewise, the female organism, being a mere product of nature, consists of the very similar biological faculties that men have had to endure, having to encounter different ways to cooperate, live with, and be selected by a sex males will never ontologically understand in myriad environments.

[117] Such as the inclination a lower animal such as the peahen may select for beauty in the peacock.

[118] It can't be reiterated enough that every ontic subset is and will only be germane to one ontologically working power, one male organism, for sexual selection. The ontic subsets are only divided to confer in such a complicated and relative function for our complicated species in order to briefly articulate what could be ethereally

propositioned as true, and again taking responsibility for the heterosexism the theory is subjected to.

[119] An interesting television series broadcast on the popular production company Netflix, 'The One', plays on the fascinating idea of sexual selection played out in humanity. As of the first season, the television series uses the biological properties of 'CHCs' found in insects as a precursor for humans selecting for 'the one', leading to an unencumbered biochemical matching, destroying marriages, relationships and families as the entertaining element of the show. CHCs, or cuticular hydrocarbons, are chemicals found to be produced biosynthetically by insects, which may work through eusocial insects finding nestmates, as they may individually secrete their own CHCs through specialised secretory cells, associated with epidermal layers in the skin. The fruit fly genus *drosophila* have been closely studied for CHCS collating all types of CHC signalling in assertive mating, sexual communication, and prezygotic reproductive isolation to map out CHC genetics (Holze, Schrader, and Buellesbach, 2021). Like the television series, the attempt to understand the biological matching of phenotypic traits throughout sexually reproducing social organisms remains a fascinating phenomenon amidst the theory of sexual selection.

[120] This may be best displayed as his 'selectibility'.

Part II: Of Good and Evil

Chapter 4: *Of Contractualism and Objectionable Conditions*

[121] According to some scientists, the idea of extra-terrestrial life doesn't seem as fictively sci-fi as one may initially like to believe. Not too long ago, around the latter stages of 2018, Shmuel Baily and Abraham Avi Loeb wrote an article proliferating in the scientific community concerning an interstellar object travelling rather too quickly into the solar system for their liking. The object, according to Loeb, was the first object that came from interstellar space that didn't consist of a cometary tail for what it was originally inferred to be. Comets are rocks that have a high percentage of its mass as ice, meaning that as it becomes bound to a star, like the sun, ice begins to vaporise and hence formulates a cometary tail that can be tracked by observing scientists. Asteroids, on the other hand, principally entail most of its mass in rock, and has a very small percentage of its mass as ice. The cometary tail has an effect in which it pushes itself away from the sun for a short period as a form of propulsion as it evaporates. Meanwhile, this particular object has such a propulsive effect without having a cometary tail, which would have thus showed that it approximated the definition of an asteroid. The conundrum has been a corollary of ideas such that extra-terrestrial life could be testing or watching us. Loeb claims that like Sherlock Holmes, his job as a scientist is to follow the evidence irrespective of its anomalous explanations that may be anathema to one's own expectations. Other curiosities, such as the oblateness, or flatness of the object's tumbling trajectory, and the brightness of the object being changed by a factor of more than ten, led Loeb to write a more than interesting

article that disseminated amongst the astrophysical and cosmological community (Bialy and Loeb, 2018).

[122] As opposed to harbouring an illogical philosophy of necessitating an expedition of every death, for instance.

[123] If you have ever been to a hospital, and in being there been timely referred to distinct wards after being treated by more than a nurse or doctor, you have unwittingly inscribed the epistemological purchase onto uniforms and possible insignias of the individuals treating you.
In attending one of Yuval Noah Harari's lectures at the Central European University in Budapest, Hungary, he fruitfully elaborated on what he termed 'the Bright Side of Nationalism'. Harari articulated the miracle of him being a functional tool of his nation, Israel. He claimed to feel a sense of national pride for a country with a population of eight and a half million people, then noted that he probably knows less than a million of them, less than one hundred thousand, less than ten thousand, one thousand, and so on! It seems that as long as one agrees(or acts as if one agrees), epistemology prevails.

[124] The best examples epitomising nihilistic scepticism are the leading conspiracy theories engrossed in covid-19 vaccination in the current climate. Conspiracy theories suggest vaccines are poorly tested and that longitudinal effects can't be possibly *known*, and so people who accept quasi-government mandates to be vaccinated are only guinea pigs as the government targets the poor and ethnic minorities first. The question for the type of nihilistic scepticism would be, 'how can one ever know the longitudinal effects?', at some point we must cede epistemological purchase to scientists, unless one is poised to endure government restrictions, masks, and social distancing until the end of one's days.

[125] As a cohort of possibly five citizens you will be a member of, not all citizens in that cohort will be epistemologically complete in knowledge, knowledge enough to prosper from the epistemological knowledge you would thus acquire from all citizens in such distinguishing specialities.

[126] The benefit-cost boomerang quintessential for what we know as the Hobbesian *articles of peace* as opposed to extremity, in which we approximate a *state of war*, play as a foundation for political philosophy.

[127] For instance, maybe I am ill-fated in being the victim of a terrible accident and I prefer to die than endure such vegetative states. Unfortunately for me, such privileges are *extraneous* to civilization. As long as I am *contractually* a part of my civilization and am socially secured, 'living' takes precedence.

[128] Which would be one of the infinite objections that could be made.

[129] The history of Eastern Rome I found on the work of Charles Oman. Of Constantine, the Czech historian Stanislav Doležal. The history I refer to of Western Rome is founded on the work of Mary Beard exclusively, who, like many, in her erudite form, dedicated her life and work to understanding the inception of Roman history, up until the death of Commodus around 192 CE.

[130] It must be remembered that it wasn't until the nineteenth century that literacy had started to take such a toll in modernity.

[131] Also fascinating is the radical stance and rise of atheism in the Islamic capital, Saudi Arabia (Wallace,2020).

Chapter 5: A Balance of Good and Beyond...

[132] Being the most extreme circumstances to venture the problem at hand.

[133] After careful introspection, it then becomes one's responsibility to seek help from a specialist.

[134] As biology has it, competition for the most judicious*[134] of humans to reside in the safest of areas ensues, approximating Roosevelt's first Article. Many humans thus seek the congenial position respective of space and time to fulfil their laborious duties and achieve the *spirit of brotherhood*; civilisation. The congenial location the human reckonings may find is in what we call the suburbs, not too far from one's laborious duties, and yet not too close to the tumult of the city. So, how could I achieve such a life? Some may argue education. In as far as we find many who necessitate the education of the child, we can thus discern the issue of, 'how much responsibility does the state hold in educating the child against that of the parent?' This takes us straight to rule II, as we understand 'parents' don't simply appear from nowhere!

[135] Milgram, S. (1963). Behavioral study of obedience. *Journal of Abnormal and Social Psychology*, 67, 371–378.

[136] As we observe even throughout the parsing of scripture and classical literature, thematic ties of marriage and heartbreak reveal profound complexity of what scientists may call intemperate vagaries of sexual selection. It's the irreconcilable nature of the protagonist in the story of 'The Count of Monte Cristo' of the French novelist, Alexander Dumas. Edmond Dantes, who consecrates his revenge against those who betrayed and plotted against him, found himself embroiled in villainous and sordid machinations. Set in the early nineteenth century, an undertow of political hostility between the Jacobins and the Girondins come to their fruition as Napoleon's expected return on a brief spell known as 'The Hundred Days' eventually occurs. Dumas conveys the horrors of such stories pertaining to the Château d'If, a fortress and prison located and hidden on the smallest island of the Frioul archipelago. Edmond Dantes, the most magnanimous and yet conscientious of young men, is promoted captain by Pierre Morel, an employer who finds the countenance of Dantes' magnanimity most endearing. The resentful vigour of Dantes' conspecifics coveted the imminent success of Edmond Dantes' providence, as he was betrothed the most gracile and beautiful fiancé, Mercédès.
Upon the fulfilment of his captain's duties, Edmond Dantes meets Napoleon on the island of Elba, and is thus compelled to hand an 'innocuous' letter to an anonymous friend of Napoleon's. Such sordid machinations then duly manifested, as the

treacherous letter handed from Napoleon was ironically to the father of Gérard de Villefort, Monsieur Noirtier de Villefort, an anti-royalist and close ally to Napoleon, informed of the re-usurpation of Napoleon's throne. On Gérard de Villefort retrieving the letter and his revelation of who the correspondence was addressed to, his father, he resorts to the calculated plot with Fernando Mondego, a Catalan fisherman in love with Edmond's betrothed, by inculpating Edmond Dantes leading to his imprisonment of what became fourteen years in the hellish Château d'If. On miraculously escaping the prison, in which he was originally subjected to life imprisonment due to the undocumented nature of such political affairs, the complication of the story actually lies in the person who Edmond Dantes loves the most, his dear fiancé, Mercédès.

Dantes' providence abounds as fate enriches his estate to becoming the *Count of Monte Cristo*, and eventually plotting his own revenge on all who were complicit in sending him to the Château d'If. As his enemies acquired and accrued their fortunes at Dantes' expense, Mercedes is deceived into marrying Fernando Mondego who finds his own wealth and success. Even after avenging, Alexander Dumas portrays the lugubrious succumbing of the Count of Monte Cristo, Dantes' heart, as even after discovering that his betrothed continued to love him, notwithstanding a treacherous marriage, Dantes was unable to overcome the betrayal, and could not find the space in his heart to alleviate his trauma. As the Count of Monte Cristo concluded, 'She was dead?' – 'Worse than that she was faithless and had married one of the persecutors of her betrothed' (Dumas, 1997, p. 851).

[137] This would be merely representative of women's franchise, the First and Second World War, USSR, and the segregationist laws of interracial marriage laws that plagued the United States of America up to the Apartheid in South Africa in the early 90s.

[138] One cannot find contraception to neurobiochemical responses.

[139] In 1943, Abraham Maslow, the now prominent American psychologist, published a paper he titled 'A Theory of Human Motivation' outlining what many now know as 'Maslow's Hierarchy of Needs'. Maslow argues five sets of goals, which he calls basic needs, are briefly physiological, safety, love, esteem, and self-actualisation. Notwithstanding the physiological needs of hunger and shelter, akin to the physiological and safety basic needs, man is a perpetually wanting animal. That would mean in so far as she feels partially satisfied, she will also feel partially unsatisfied with such basic needs as desires are not altogether mutually exclusive. Maslow proposes the question that accentuates the intricacies of truly fathoming such basic desires anathema to the mere satisfaction of the perpetually wanting animal, 'Who is to say that a lack of love is less important than a lack of vitamins' (Maslow, 1943)?

[140] And as Steven Pinker suggested:

> 'Though Napoleon did implement a few rational reforms such as the metric system and codes of civil law (which survive in many French-influenced regions today), in most ways he wrenched the clock back

from the humanistic advances of the Enlightenment. He seized power in a coup, stamped out constitutional government, reinstituted slavery, glorified war, had the Pope crown him emperor, restored Catholicism as the state religion, installed three brothers and a brother-in-law on foreign thrones, and waged ruthless campaigns of territorial aggrandizement with a criminal disregard for human life' (Pinker, 2012, pp. 287-288).

[141] If a man succeeds the height of 190cm once reaching his vicenarian years, and the other man of the said age doesn't, then the former man is simply shorter than the latter. One cannot change the height of these men in order to espouse an egalitarian standing, but politically one may endeavour to render an approximation of uniformity. In so far as the literature for IQ testing becomes more and more nebulous the higher the faculties of intellect may run in a person, if one person has significantly higher faculties of intellect than another, this runs congruously with the laws of biology and is thus instantiated in both a genetic and archetypal history of evolution.

[142] All behaviour is biological, as long as behaviour comes from (a) physiology, and (b) the environment.

[143] I use the scientifically classified terms in order to harbour the notion of biological responsibility. We must constantly be reminded that thirty-three thousand years separates us from other species within our same genus. Hence rights concomitant with our genus would mean that if by a miracle another species of human were to bioengineered, which isn't as far-fetched as you may think as it has been conferred quite seriously within scientific circles, then that other species would technically hold the same *human rights* you also 'inalienably' hold.

[144] It's been said that if a rich man gave his over-one-hundred-billion dollars to the 7.8 billion people in the world, the rich man would still be remarkably rich. What people who profess such nonsense fail to comprehend is that, as much as it is economically nonsensical, most of the retrievers of such money would lose the money fairly rapidly. Why? That's because it's attributable to financial illiteracy on the behalf of the retriever. Many would increase their expenses, and thus their liabilities, eschew the generation of income and thus neglect the building of assets, and be in the same position they originally found themselves in (Kiyosaki, 2017). It would be just like giving an athlete's body to an obese person, the obese person wouldn't know what to do with it! So a person who wants more capital, or a lot of it, may not be necessarily greedy. It would depend on what their motive is to brand someone such a claim. Yes, it could be Brené's scarcity problem motivating them. However, if a person becomes rich, their motivation could be to buy enough time to be more altruistic along with their financial independence, offering their time, knowledge and money to the necessitous, more than anyone who isn't financially independent could ever dream of giving.
Taxing the rich highly may help with certain trivialities, meanwhile, teaching a man how to fish will also help him in so far as he's willing to learn. But him learning how to manufacture and distribute rods against his inclinations of morbid pleasures would be imputed to his creative mind. Taxing the rich highly ends up being so much more counterproductive than one professes, as the rich pay lower

¹⁴⁵ Akin to the existing hierarchies of epistemic-cooperative global warming issues was a recently produced 2021 Netflix sci-fi film, written and directed by Adam McKay, *Don't Look Up*. It stars illustrious actors Leonardo Di Caprio, Jennifer Lawrence, Meryl Streep, Jonah Hill, Tyler Perry, Cate Blanchett and many more in which the film satirises the epistemic tumult of handled information between scientists, politics, and the rest of the world. Two astronomers detect a comet heading directly for the earth, and upon warning the president, and attributable to the sceptical, or over-sceptical world lived in today, the matter seems to dissipate into the tumult of information services. The matter circulates TV journalism, capitalism, politics, social media, and more until it becomes too late to act on. Adam McKay talks of the origin of his idea to write his screenplay, being of global warming until having an epiphany when presented with the analogy. You can watch Adam McKay and some of the actors talk about the film here on Youtube: https://www.youtube.com/watch?v=77pyaEoT3dQ&ab_channel=EntertainmentWeekly

¹⁴⁶ Green Premiums are simply the greenest alternatives to the *de facto* options that lead to our emissions.

¹⁴⁷ I use the term *agency* as opposed to responsibility, as, being a causal determinist, I do not believe the latter to exist (which we will get to later). Given great ideas on transcendent forms of what I believe is agency, such as what ex U.S. Navy Seals Jocko Willink and Leif Babin term *extreme ownership*, I believe the term responsibility philosophically inaccurate (Willink and Babin, 2017).

¹⁴⁸ A worthy look at the perpetuation of poverty across the world albeit *The Great Enrichment* of the last two hundred years was noted by the Senegalese entrepreneur Magatte Wade and Professor Steven Horwitz (Horwitz and Wade, 2021). The answer to why extreme poverty continues to pervade sub-Saharan Africa is endeavoured as they ascribe such failure to what they call (a) 'permissionless innovation' and (b) 'trade-tested betterment'. The former predicated on the refrainment of regulation and the latter on the latitude of its own benefit/cost trajectory in the market.
They conclude three tenets for a prospering economy in entrepreneurship:

1. The respect of private property.
2. The adoption of the rule of law.
3. The maintaining of sound money (against the precariousness of both inflation and deflation) (Horwitz and Wade, 2021).

¹⁴⁹ I will revisit this idea later in the book.

¹⁵⁰ As I am a compatibilist free will believer, I believe everything is a mitigating circumstance.

¹⁵¹ i.e. anything that falls within subjective phenomenology of space and time, which one would certainly claim to be everything.

[152] Exceptions being the case of suicide.

[153] I say this without conjecturally concluding the cession of all that fortifies such an ontological apparatus, as the issue is complicated, being embedded in the evolution of RNA cycles for over millions of years.

[154] For those unfamiliar with Socrates' allegory of the cave you can read the abridged translation by Thomas Sheehan on the Stanford website at: http://web.stanford.edu/class/ihum40/cave.pdf or read the allegory directly in Plato's *The Republic* (see references).

[155] For instance when someone's life is at stake or in cases of emergencies.

Part III: A Brief Discourse on Racism

Chapter 6: The Negro

[156] I use the term 'black' as the English and interchangeable replacement of the word 'negro', considering its distinguished meaning but merely for the purpose of simplification and for the evasion of confusion.

[157] Or the 'Great Leap Forward' after around 100,000 years ago.

[158] A terrace on the small island, the *Little Diomede Island and Upland Surface*, was extensively studied and found to corroborate periglacial weathering moving from what would have been initially a marine terrace, to what would have become a gradually slow process of cryoplanation through the disintegration of rock from frost (Gualtieri and Brigham-Grette, 2001).

[159] One will reverberate the fact of humans before Columbus *discovering America*, such as that of the Vikings finding land or African exhumations found with American products. In as far as there may be evidence of this, Columbus' voyage was the first 'grand' conflation of the civilisations of the New and Old World dating back to around 16,000 years ago.

[160] I also stress this for better or worse in the case of identity politics today.

[161] I have often observed compensation of the dearth of general knowledge (countries in Africa) on at least two or three of the continent's nations outside of white South Africa to be in the concomitant phrase, 'my geography is really bad', or specifically, 'my geography of Africa is really bad'. Teaching over one thousand students in Spain over a course of five years, I could most probably recall one or two instances of university students knowing of the one country that officially speaks Spanish in sub-Saharan Africa... Equatorial Guinea.

[162] You can watch Dave Ruben and Larry Elder's discussion on Youtube here: https://www.youtube.com/watch?v=IFqVNPwsLNo&t=1521s

163 For instance, the strategy of deniability is used sordidly by my adolescent students in forgetting to do their homework. Students will complain of not seeing, not receiving, or not *knowing* about the homework while their classmates had completed it, in an endeavour to render their ignorance or deniability plausible!

164 I stress again the significance of literacy. Our general history is not frankly a hobby, interest, or anything akin to esoteric leisure, but a necessary condition for our franchise to actualise itself. While the adolescent is exonerated from branding the topic of the past boring, attributable to their developing faculties of intellect and their dearth of free will from education, the adult remains responsible, in so far as he can vote.

165 The argument that racism is biological is another misconstrued argument. I would contend the more unlearned the subject is, the more biological racism becomes, and hence the more learned an individual is, the less work the *ego* would need to do in alleviating the *superego* from its skirmishes with the *id*. There is evidence of racism being biological, and you can again return to Sapolsky's work on this. Be that as it may, in such studies of those giving racist evidence, the invocations of more questions touch the surface, such as, 'how acquainted with racially different people is the participant, how learned is the participant, and so on.

166 The same argument must be applied to those who must assiduously deal with feminism, most especially within gender inequality. We know, exceptions not making the rule, that females are inferior in strength and speed, which is most essential for space and time as an animal. For life, males are inferior with sex cells as they provide a paltry amount of food reserve in embryonic life. So, one may find females to be incognizant of the privilege they have in sexual selection, while males may be found to be incognizant of the privilege they have in safety and the evasion of sexual predation. Irrespectively, it is the DNA both types of individuals inhered.

167 That would be (a) the x-y axis of Western-Europe to East-Asia (b) the domestication of plants and animals across the Fertile Crescent (today's Middle east) (c) cultural diffusion of agrarian techniques, metallurgy and writing systems (d) diseases many were compelled to resist (e) the distance between both lateral oceans and the vast Saharan desert separating off sub-Saharan Africa from other civilisations (which doesn't help for any type of cultural diffusion) (f) the arduous tropical issues to endure, and so on...

168 Not only will we notice they are all of European ancestry, but we also notice they are all men. I keep this significantly distinct from sex which I do believe to be a *completely* separate issue. Intersectionality would be as if, one day, a biologist and a sociologist crossed each other in which they both happened to be conducting an experiment on the nature of identical twins. The biologist was conducting an experiment on 'heritability scores', and the sociologist, statistics on 'how many identical twins are married after the age of 30'. To save time, resources, and get home early, they both decide to conduct their experiments together, then write their conclusion together. Completely ludicrous.

169 See note 160.

[170] As of 2019 in the United States, 64% of black or African American children live in single-parent families, as opposed to American Indians of 52% coming in second place, and Hispanic or Latino children third with 42% (Kids Count Data Center, 2019).

[171] I only use provincial attributable to its utility. This could be country, city, town, village, etc.

Chapter 7: On Intolerance and Intention

[172] I do believe many deny having racial preferences in mating, are denying having racial preferences, and will deny it, attributable to the type of virtue-signalling that concerns uncomfortable truths (which ironically is only solved through the initial detection of such a truth in the first place).
I do believe the remedy of racial preferences of the psyche in a lifetime would be achieved through long-term phenomenological cohabitation with societies entailing of distinguished features within our species. It complicates itself further as linguistics plays a part, and all which is assimilated and accommodated throughout one's operational stages obtrudes. One rarely can afford or has sufficient time to achieve this conducive to our ephemeral lifespan, along with the truth that preferences will most certainly always exist.

[173] See more on *The Marshall Project* here: https://www.themarshallproject.org/next-to-die/al

[174] Before the United Kingdom is a country, Great Britain is an island entailing of other primates that we know of as our species, *Homo sapiens*. The biology hardwired in his mind will be wary of both survival and reproduction, henceforth, in such a calculative sense.

[175] This is a clear division of the idealist and realist philosophy of what should be. The political left at times seem to exhaust the idealist philosophy of how the world should be with the mistakable elision of how important the sciences are inculcated in such affairs. This sense is within biology, anthropology, as big as cosmology, and all the way down to the smallest within microbiology.

[176] Here is yet another instance of how responsibility cannot exist. It seems that everything is a 'mitigating circumstance' if it's a product of the environment or physiology.

[177] Cornell Law School. Visit: https://www.law.cornell.edu/supremecourt/text/106/583

[178] Once again, I ascribe this to the distribution of hierarchies. For the male, the white man did not wish to represent himself higher in the dominance hierarchy with an acquisition of a white female. Likewise a black female with the acquisition of a white male in the hierarchy.

[179] *Richard Perry LOVING et ux., Appellants, v. COMMONWEALTH OF VIRGINIA* (1967) was an infamous case of two residents of Virginia having married outside of the state. Richard Perry, a white man, and Mildred Jeter, a black womyn, were married in the District of Columbia akin to their own laws. On establishing their marital abode on the return to Virginia, Loving and Jeter were met by the Circuit Court of Caroline County and an indictment that charged them with the violation of the laws of interracial marriage. An appeal to the Supreme Court was soon imminent for the Lovings.

[180] Cornell Law School. Visit: https://www.law.cornell.edu/supremecourt/text/388/1

[181] As a compatibilist free will thinker, I do not believe it can be ascribed to historical chance. I do not see there being any other way of a tropical nation ruling a temperate region akin to all of the causes of why temperate nations rule today.

[182] My focus on the black ontic inscription as opposed to Asians, who fulfil a larger percentage of the population in England and Wales, or any others is attributable to two implicit reasons:

- The first is that such forms of racism must be respectively dissected as separately as possible, which would warrant another part of this book, as I find Islamophobia to be its own evil that shrouds the UK.
- The second, and quite importantly, is that even within countries such as Pakistan and India (who comprise a large part of minorities in the population), they come with their own malicious ontic inscriptions on black subjects in the UK, even as they may ironically suffer themselves.

[183] The feasibility to be anonymous in the online world also helps those exercising the id in all its forms.

[184] A menial task found across the west is the policing of small neighbourhoods with petty and insubordinate delinquents. There is a frank and manifest philosophy of *the snitch*, paramount in understanding how humans as a species that cooperate flexibly in large numbers can somehow be curtailed. As I have timely professed, the political drawing board is premised on philosophy, namely the handling of information and what is officially defined as epistemology. But where trust is lost, cooperation is adulterated. We have in the twenty-first century assiduously allowed thinkers to alleviate themselves from a cruel history people slowly no longer hold legal responsibility for, notwithstanding both the sufferers and the beneficiaries of where one is misfortunately or fortuitously placed in the hierarchy, which is a hard pill for egalitarians to swallow. What has emerged from suffering, inequality, misunderstanding, and political problems such as that of the Windrush scandal, and even irresponsibility, has been the philosophy of the *snitch*. The *snitch* is a philosophy most manifest in the most uncivilised of senses civilisation as a legislative practice of immediate death. In countries approximating higher forms of civilisation the *snitch philosophy* regresses to where the nation endeavours to progress. It's often that neighbourhoods predominated by characteristics akin to their in-group hold the *snitch* philosophy and in turn protect the criminals that not only keep the individual behind, but the neighbourhood, the city, the country, and the species from progressing concertedly. Conversely, those

individuals who allowed for such policies to pass in inciting the Windrush scandal to transpire, not only keep themselves behind, but first the neighbourhood, the city, the country, and the human species from its concerted progression. Steven Pinker points this out of Thomas Hobbes' work in *Leviathan*, of which he names 'the violence triangle', between the aggressor, the victim, and a bystander. Between an aggressor and a victim there can only be *predation* and *retaliation* (war), of which the third, the bystander not involved, is the law (Pinker, 2012). A *Leviathan* simply cannot be held if both aggressors and victims, in so far as they are at a state of war, do not trust bystanders.

[185] Northern Ireland has the highest fertility rate amongst UK nations.

[186] The form of epistemology between induction and inference is sociologically significant for what is taken away from individuals and such decisions akin to incompatibilist free will. This is of both the migrant and the naturalised citizen in the country they now both reside in.

[187] As stressed through research on neurology, as a compatibilist free will thinker, I do not believe responsibility can exist as I fail to see a will/intention without a cause. What I believe the political right to miss are all environmental influences that cause the subject to believe what he believes. Stressed in 19th century Darwinian biology in the *variability* of the species, given the prefrontal cortex, along with all other functions of the brain are dissimilar, one cannot be responsible for physiological as well as environmental causes.

[188] As long as this issue is reverberated, it can only compel the influence for more British people to involve themselves in the excavation of black British history.

[189] Be that as it may, other steps are being taken to protect black history, as London's 'Black Cultural Archives' (BCA) were to receive £200,000 stopgap funding to ensure its survival, with a letter signed by more than 100 cross-party MPs to save the BCA in Windrush Square, Brixton, in south London (Weale, 2018).

On a Conclusion to Knowledge

Chapter 8: Science, Sexual Selection, and an Homage to Korea

[190] I've often believed such *epistemological purchase* to be pertinent to the popular *Lifeboat* game played around the world, in which one must choose who to push off a boat akin to what they could contribute to the island they would be stranded on. Be that as it may, let's envisage two types of inquisitive intellects in the world we live in today in the west, instead of being reduced to solely survival on an island. The first, June, who reads 200 **scientifically** academic articles. Julie, however, reads 200 **sociologically** academic articles. Does June contribute more than July? Meanwhile, let's now envisage June reading 200 **scientifically** academic articles and July reading 100 **scientifically** academic articles while reading 100

sociologically academic articles? Is July more useful than June? It seems today that where political organisations may arrogate the epistemic authority to enforce a policy, science must now today be essential. I would also say science is so much older than one may suspect and is found to certainly coincide and be conducive to our ethological trivialities, ensconced in scripture so many are unfortunately willing to dispel of.

[191] For instance, Stephen Hawking's black hole theorem was recently confirmed for the first time at Massachusetts Institute of Technology. Using computer simulation, physicists were able to detect the performance of gravitational waves by observing the collision of two black holes. Hawking's law holds that the area of an event horizon should never shrink. Thus the signal was a product of the collision of two spiraling black holes that should not hold a new surface area smaller than their parent black hole, and this held true of the theorem. Even the largest of matter, such as black holes, are complying with the laws of the universe, notwithstanding their colossal size.
It marked the first observational evidence of Hawking's theorem since 1971 when he first proposed his idea, leading to epistemic milestones being achieved in scientific communities of today (Chu, 2021). Thus gravitational-wave data technology continues to be improved today.

[192] Looking at another instance of laudable prescience, have a look at this story of the *Star Orchid*:

> 'The story goes that Darwin was sent a sample of the flower in 1862. Upon seeing its long, narrow nectar tube, he predicted that there must be an insect with a very long proboscis (a tongue-like part) that could reach deep within the hollow space to "drink" the nectar at the bottom. In so doing the insect would bump into the flower's sticky pollen, enabling its transfer from one flower to another.
>
> But no such insect had ever been seen in Madagascar where the orchid came from, or anywhere else. And many scientists believed Darwin was wildly wrong, so he was ridiculed for his prediction.
>
> Nonetheless, Darwin firmly believed that the star orchid had developed its long nectar tube as an adaptation to help ensure pollination because orchid flowers have their pollen in a single mass and cannot disperse it as other flowers do. The orchids need their specific insect pollinators to survive.
>
> Sure enough, about four decades after Darwin's prediction, and insect with the exact physical characteristics that Darwin had predicted was discovered. Called the Hawk Moth, its scientific name is *Xanthopan morganii praedicta*, which is Latin for 'predicted moth' in honor of Darwin' (Newman, 2012).

[193] In fact, amine serotonin, also known as hydroxytryptamine creatine sulfate, or 5HT has been linked with increasing aggression in a wide range of species that also include humans, along with genetic alterations of amine neuron fusion enhancing

aggression as well. If you inject or infuse 5HT into the hemolymph (liquid that works like blood in humans) of crayfish or some species of lobsters, subordinate animals restore fighting levels significantly, exceeding the levels of established dominance relationships that were already set out prior to infusion or injection (Huber et al., 1997; Livingstone et al., 1980).

[194] A free will balance sheet would look similar to a financial balance sheet. It would be compartmentalised into *income*, *expenses*, *assets*, and *liabilities*. For instance, for many across the west, they may find themselves subjected to working eight hours a day out of the twenty-four hours there are in the day. That would mean an approximation of eight hours for sleeping, and another eight for you to do as you please. But when calculating time it would take to commute to and from work and so on, it's quite easy to get bitter. Meanwhile, the worker's contract will issue them a salary sitting in the *income* section of the free will balance sheet. With this income, they can choose to trade with others of the same species for services such as transportation, instead of being compelled to walk to their workplace. They can also trade for services to eat or buy uncooked meat instead of hunt their own food, in which such service would sit as *assets* that save them such time, not to forget that eating is essential.
If one chooses to secede from the capitalist make-up in the west, one would forego the use of planes, busses, cars, medical services and so on, which has never been as accessible to the lay person as ever before.

[195] Thus most of our canonical precepts, morals, and virtues of the past and anthropologically around the world is predicated fervently on some form of abstinence or sacrifice, the value being the long-term reward.

[196] Quite evidently, DNA tests most certainly did not exist.

[197] See part I for Dawkins's notion of both *philanderer* and *faithful* males, and *Coy* and *fast* females. Or head directly to his bestseller *The Selfish Gene.* You can also refer to John Maynard Smith's evolutionarily stable strategies in *Evolution and the Theory of Games* to get a more quantifiable understanding as well.

[198] Again one could attribute this to secularisation across the 20th and 21st century. More than secularisation, which has and should be pertained to whether God exists or not, is the eschewal of ethics, which is the eschewal of all ethics on a start-from-scratch basis. The arrogance of many newer philosophies prevails, deprecating ethics deeply ensconced in religious scripture and the like that many of the discerning continue to discern and distil. In so far as many scientific discoveries undermine ideas that are now obsolete, humans have been sexually reproducing for millions of years.

[199] I do not allude to or advocate for an outlawing of contraception. I will however refer to an analogy to better understand this notion. Contraception would be like a man paying a mortgage with no interest. One would reasonably argue it then ceases to be a mortgage! The down payment would be the amount of cash he pays before ultimately consummating the relationship. Once long-term contraception is habitually involved, no interest is paid. It may seem beneficial for both parties involved, however, in the long run, he may endure the 'best he can get' until something better comes along, and this may be manifested in infidelity. Likewise,

she may wait for the 'best she can get'. Novelty is the psychological ramification of contraception for young vicenarians. If she suffers from many ending relationships well into her tricenarian years, she may find her value in the dating market depreciated, in that the same type of men she has demanded seem to be harder to compel in committing to her.
If she chastely refrains, she must undertake a harsher and timely vetting procedure for the male, and upon marrying her (which is the only way for the relationship to be successfully consummated) his interest on the mortgage is being paid in full by the amount he must cooperate in sustaining the marriage. He would also intuit the fear of starting the whole process again with other implacable *stocks* on the environmental dating market, so may be more compelled to make his marriage work.

[200] I often say those in a few centuries may see pervasive philosophies in the world today as *philosophically barbaric*. As Ortiz-Ospina and Roser portray in their statistical analysis, gay marriage is being more and more condoned in the world today, but holistically isn't that impressive to future historians and sociologists. Same-sex marriage isn't legally recognised in the whole of the Asian continent save Taiwan, in which same-sex marriage is legal, and Israel, where some rights to same-sex couples do exist. That's applied to a combined population of approximately 34 million people of a continent that champions a population of approximately 4.3 billion. The same is said of Africa, in which only South Africa has successfully legalised same-sex marriage out of population of over 1.2 billion who do not legally recognise same-sex marriage. Those who still endure the dearth of understanding of homosexuality across the animal kingdom unfortunately pervade a sociology that reaches their politics. Maybe they ought to cooperate with scientists on such matters.

[201] As Brené Brown expressed as a ramification of such matters, it tends to lead to what she refers to as *disengagement*. It's quite often that when one embarks on an intimate relationship, intra or extramaritally, when enduring a lack of cooperation or understanding from their counterpart, they are advised to consider the termination of a relationship. With termination being too drastic of an approach, it's disengagement that ends up taking place, simply curtailing cooperation and the cost of activity their partner desired so much in the agreement (Brown, 2015; Brown, 2020).

[202] You can watch the conversation between Stephen Colbert and Dennis Rodman at https://youtube.com/watch?v=w7hf5wbggdA&feature=share

[203] Reading such an account of a person born a year after I was (1993) in the wrong part of the world only appears as something made up. Being brought up in England across the nineties and two-thousands, England, most especially our senior citizens, endured the task of compelling young children to really try and understand the calamities of the First and Second World War. But to read Yeonmi Park's story was difficult, to know that while we were being compelled to understand such history, children like Yeonmi Park, younger than me, had to endure what she unfortunately did. The same could be said for many today as stories remain untold of this politically detestable nation.

204 Much can be said about South Korea, and so it would ultimately be akin to one's phenomenological perspective as opposed to what one may experience vicariously, understand sociologically, quantifiably, and so on and so forth. In so far as the minutiae of South Korea may not have been or will be ventured by the reader, idiosyncrasies rendering the country's countenance hold dicta to be foundationally and mysteriously true.

205 The *Yushin* constitution, also referred to as the Fourth Republic of Korea, centralised Korea into Park's authoritarian ruling which ended for Park in 1979 after assassination, and for South Korea in 1981.

206 Most especially within a Joseon dynasty who regarded Confucius' analects in such high regard.

207 For a more moderate account on the east Asian problem of Korea, you can refer directly to Tim Marshall's *Prisoners of Geography*, 'Korea and Japan' (Chapter 8).

208 The notion of *danil minjok* is of a South Korean population being united as one nation through bloodline, which is fundamentally anathema to multi-ethnicity for all its worth. Though others declare a more east Asian approach putting twentieth century differences aside, moving towards a de-westernisation, as the author Chang-rae Lee wrote in his novel *A Gesture Life*, 'whether Chinese or Japanese or Korean we were rooted of a common culture and mind and that we should put aside our differences and work together' (Chang, 1999, pp. 248-249).

209 Films such as *Parasite*, the South Korean black-comedy-thriller that also won an Oscar for Best Picture in 2020, had both accentuated and conveyed the caveats of classism throughout South Korea, which the country still unfortunately endures, propelling itself onto even more serious matters. I often say, most certainly as an offered disclaimer in such matters, that a hitchhiking liberal hippy who ventures as a free spirit, in an effort to experience the world, would be merely reduced to homeless in South Korea. Even so, it would then also depend on the colour of his or her skin.

210 It's germane to the notion of 'story-topping' within social interactions. For instance, I may tell you I encountered a red squirrel on my way to work. Another 'tops' this by telling us she encountered five red dancing squirrels on her way home from work!

211 In a western nation such as the United Kingdom, I do believe value does not tend to subside so much down the financial/economic ladder but instead, transmutes. Thus we observed in the UK the Mods, Rockers, Goths, Hippies, Hipsters, Chavs, Punks, and so on, who tend to purport their own bohemian revolutionary philosophies. This is not so much the case in South Korea.

212 I also write this subjective disclaimer on South Korea as an epistemological pragmatist.

213 The last execution in Korea was in 1997. Kim Dae-Jung, president of South Korea from 1998 to 2003 signed death warrants commuting the death sentences of

18 inmates on death row to life imprisonment. Five in 1999 and thirteen at the end of 2002. Kim was also the recipient of the Nobel Peace prize in 2000 (Bae, 2009).

[214] That also goes for the number of unvetted conspiracy theories Daniela Mahl, Mike S. Shafer, and Jing Zeng map out when looking across 10 conspiracy theories and communication of online communities. Those theories were *Agenda 21, Anti-Vaccination, Chemtrails, Climate Change Denial, Directed Energy Weapons, Flat Earth, Illuminati, Pizzagate, Reptilians,* and *9-11* (Mahl, Schafer, and Zeng, 2021).

Acknowledgements

As a causal determinist, I hope by now you may take me for understanding all actions in the world to be made up of influences. So a special thanks goes to all that made this book happen. In no particular order of importance, first, I'd like to thank over the twelve-hundred students in Spain who I had the pleasure of working with, who, over the course of five years, allowed me to guineapig many of the ideas touched on in this book. I sincerely encountered some of the sharpest and most astute minds on the courses we shared.

As promised to my wonderful *MS3S* class in Korea (Angela, Hoy, Yoonoo, Joey, Bella, Lucy, Leo, Sara Ji Hyon, Yoona, Ann, Amy, and Junghoon), who will soon be old enough to parse the pages of this book, thank you for all the pleasant classes and personalities that made Monday my favourite day of the week!

I'd like to thank my closest friends, family, and even lost acquaintances who have at times miraculously borne my uncanny ways of being and yet motivated me in ways that I could never dream of compensating. Thank you to Chris DiBiase for the illustrations that went into this book and thank you to David and Gwen Morrison at PublishNation. Finally, a big thank you to you who has made it this far, I hope you've been influenced in some helpful way by this epistemological quest.